Numerical Simulation of Pulsed Plasma Thruster

Jianjun Wu · Jian Li · Yuanzheng Zhao · Yu Zhang

Numerical Simulation of Pulsed Plasma Thruster

 Springer

Jianjun Wu
National University of Defense Technology
(NUDT)
Changsha, China

Jian Li
National University of Defense Technology
(NUDT)
Changsha, China

Yuanzheng Zhao
National University of Defense Technology
(NUDT)
Changsha, China

Yu Zhang
National University of Defense Technology
(NUDT)
Changsha, China

ISBN 978-981-97-7957-4   ISBN 978-981-97-7958-1   (eBook)
https://doi.org/10.1007/978-981-97-7958-1

This work was supported by National Natural Science Foundation of China.

This Springer imprint is published by the registered company Springer Nature Singapore Pte Ltd.
The registered company address is: 152 Beach Road, #21-01/04 Gateway East, Singapore 189721, Singapore

If disposing of this product, please recycle the paper.

Foreword

Microsatellites have unique advantages such as their small size, lightweight, high functional density, short research, and application cycles, making them one of the most active research directions in the aerospace field today. Maintaining relative intersatellite positions, achieving high-precision attitude and orbit control impose high requirements on propulsion system performance metrics (e.g., specific impulse, lifespan, and thrust). Traditional cold gas and chemical propulsion systems fail to fully meet these requirements, limiting the development and application of microsatellite technology to some extent. Pulsed plasma thrusters (PPTs), with advantages such as high specific impulse, long lifespan, and precise impulse bit control, are one of the preferred propulsion devices for microsatellite formation flying, network operation, and attitude control.

Professor Jianjun Wu has been engaged in space electric propulsion technology research since the late 1990s. Under Prof. Wu's leadership, the research team has carried out systematic and in-depth research on the basic theory and engineering application of PPTs, established a comprehensive experimental research system of electric propulsion technology, and developed multiple numerical simulation systems for the operation of PPTs. Over the past two decades, the team has conducted research on fundamental theoretical issues related to thruster operation and solved many challenges in PPT plasma generation, acceleration, and measurement, achieving systematic and innovative progress. In addition, Prof. Wu and his team have conducted exploratory research on various space electric propulsion systems, including magnetoplasma thrusters, microcathode arc thrusters, ionic liquid thrusters, and pulsed inductive plasma thrusters, developing complete research and evaluation systems in electric thruster design, plasma diagnostics, and microthrust/microimpulse measurement, among others.

This book is a systematic summary of a numerical simulation study of PPTs by Prof. Wu's team and presents models, algorithms, and an analysis of results related to the whole PPT operation process, including propellant ablation, discharge acceleration, and plume expansion. The modeling and numerical methods for multiphysics coupling in PPTs are explained in detail. This book is divided into three parts: Ablation, Discharge, and Plume. With a focus on the operation mechanisms of PPTs, this book discusses the establishment of an integrated simulation model for the propellant ablation, discharge acceleration, and plasma plume motion processes, introduces an independently developed simulation design platform, proposes a new method for laser–electromagnetic isolation ablation, reveals the ablation mechanism and phase transition evolution of solid propellant, and elucidates the transport mechanisms of propellant in the ablation, evolution, acceleration, and erosion processes. In addition, this book reports breakthroughs in key technologies related to plasma generation and enhancement, discharge and acceleration, and confinement and control and provides effective solutions to address challenging bottlenecks in thrust performance, such as low propellant supply control accuracy, low ionization rate, and low electromagnetic energy conversion efficiency. This book also lays a theoretical and technical foundation for the further development of high-performance PPTs.

This book is helpful for understanding and mastering the operating mechanism of PPTs as well as for evaluating and assessing their operational characteristics and propulsion performance, providing an essential theoretical basis and reference for the design, development, and engineering applications of PPTs. I believe that the publication of this book can further promote the development of space electric propulsion technology in China. Considering this, I recommend this book to interested experts, scholars, graduate students, and engineering professionals.

August 2023

Dr. Qifeng Yu
Professor
Academician of Chinese
Academy of Sciences
Changsha, China

Preface

Pulsed plasma thrusters (PPTs) are electromagnetic thrusters that use pulse discharge to generate an electromagnetic field. The ablation and ionization products of a propellant are accelerated and ejected under the combined action of the Lorentz and aerodynamic forces to generate thrust. The advantages of PPTs, which were the first electric thrusters used by humans in space applications, include their high specific impulse, low-power consumption, simple structure, and lightweight, making them highly suitable for space propulsion tasks with long mission durations that require high control precision. To date, many countries, including the USA, Russia, Japan, France, Italy, Argentina, and China, have conducted extensive and in-depth research on PPTs and successfully applied different thruster models on various types of satellites.

The PPT operation process involves a variety of physical phenomena, including spark discharge in vacuum, working fluid ablation, charged particle and electromagnetic field interactions, plasma transport, and the plasma–wall "sheath" effect. These processes are interconnected, making the internal operating mechanism extremely complex. Moreover, since thrusters operate in a non-steady state, there is coupling between multiple fields such as electricity, magnetism, force, heat, and light. The short-pulse discharge time and the small-scale spatial variation of plasma make experimental measurements very difficult. Therefore, from both basic theory and engineering application perspectives, there are several urgent problems that need to be solved, including, but not limited to, addressing the low propellant utilization of the thruster, large energy loss within the system, low efficiency, and plume contamination caused by thruster operation. Despite decades of research, a comprehensive understanding and mastery of the operation process and related mechanisms of PPTs are lacking.

The National Natural Science Foundation of China (NSFC) has recognized the importance of PPT development as an advanced electric propulsion technology and strategically planned research tasks, including "Simulation and experimental study on the operation process and plume characteristics of PPTs", "Theoretical and experimental study of the discharge ablation process of solid ablative PPTs", and "Research on novel PPTs using energetic propellant", providing long-term continuous support for research conducted by the author's team. This book stems from the team's many years of theoretical research, experimental measurements, and engineering applications of PPTs and provides a systematic summary of numerical simulation research on the operation process of PPTs. Based on the operation characteristics of PPTs, the operation process of a thruster is divided into three stages: propellant ablation, discharge acceleration, and plume expansion. Physical models for different stages are established, and theoretical analyses, numerical models, and calculation methods for these physical models are presented. By analyzing numerical simulation results of the thruster operation process, the characteristics of the phase transition of the propellant from solid to plasma, the conversion from electric to plasma kinetic energy, and the transfer of materials from the inside to the outside during the operation process of the thruster are maximally reconstructed, providing a theoretical basis, model reference, and method support for elucidating the mechanisms and principles related to the operation process of PPTs.

Changsha, China   Jianjun Wu
August 2023   Jian Li
                                                                                              Yuanzheng Zhao
                                                                                                        Yu Zhang

Acknowledgements This book contains the results of research projects conducted by many graduate students from the author's team. We are grateful to Dr. Zhen He, Dr. Ziran Li, Dr. Le Yang, Dr. Rui Zhang, Dr. Daixian Zhang, Dr. Le Yin, Dr. Zehua Xie, Dr. Hua Zhang, and Dr. Sheng Tan for their contributions of the book.

We would like to thank the Innovative Research Group Project of the National Natural Science Foundation of China (Grant No. T2221002) and Hunan Provincial Natural Science Foundation (Grant No. 2024JJ6456) for their support. It is our hope that this book can provide a beneficial reference and assistance for practitioners in the electric propulsion field.

Contents

About the Authors

Jianjun Wu is a professor of PI of electrical propulsion research at National University of Defense Technology (NUDT). In 1995, he obtained his doctorate in Aerospace Engineering from NUDT, preceded by a master's degree and a bachelor's degree from Tianjin University in 1991 and 1988, respectively. Since 2002, he has held a tenured position at NUDT and concurrently leads the Innovation Research Group funded by the National Natural Science Foundation of China. His primary research interests lie in space electric propulsion technology and rocket engine technology. He has completed more than 30 research projects and won National Outstanding Scientific and Technological Book Award, Invention Award at the Geneva International Exhibition, and five Gold Awards at the International and National Invention Exhibition. He has authorized more than 100 patents and published more than 300 papers and 10 monographs.

Jian Li is currently a lecturer at National University of Defense Technology (NUDT) in China and holds membership in the Chinese Aerospace Society. He was awarded a Doctorate in Aerospace Science and Technology (2020) and a master's degree (2016) from NUDT. His primary research directions are space electric propulsion technology, low-temperature plasma application technology, and pulse plasma propulsion technology. He authored over 20 publications in journals and 5 monographs.

Yuanzheng Zhao, *Ph.D.* candidate, is studying at National University of Defense Technology (NUDT) in China. He was awarded a master's degree (2021) from NUDT. His primary research directions are micropropulsion technology, plasma diagnosis, and plasma magnetohydrodynamic simulation. He possesses advanced proficiency in plasma optical emission spectroscopy. He authored nine publications in journals.

Yu Zhang, male, is an associate professor at National University of Defense Technology. His main research direction is space electric propulsion technology. He has been selected into the Young Talent Promotion Project of the Chinese Association for Science and Technology and is the main member of the innovation research group of the National Natural Science Foundation of China, young editorial board members of journals such as *Journal of Rocket Propulsion* and *Journal of Propulsion Technology*. He has published 30 SCI papers and authorized 35 patents. He won the first prize of Provincial Science and Technology Award, the gold medal of National Invention Exhibition, the special award of Geneva Invention Exhibition, the gold medal of Beijing Invention Innovation Competition, and the third prize of Hunan Provincial Teaching Achievement.

Chapter 1
Introduction

The advantages of pulsed plasma thrusters (PPTs), the first type of electric propulsion system for space applications, include their low-power consumption, fast response, simple structure, easy integration, convenient control, and precise and controllable thrust, making them especially suitable for tasks such as microsatellite attitude control, position keeping, orbit raising, and formation flying. In recent years, with the increase in microsatellite applications, the demand for advanced on-orbit propulsion technologies for use in microsatellites has increased. Therefore, PPT research and applications have received widespread attention, becoming a hot topic and an important research direction in microsatellite propulsion technologies [1].

Despite more than half a century of PPT research and space flights, some problems still limit further PPT development and application. The PPT operation process involves multiple physical fields, such as force, heat, light, electricity, and magnetism, and encompasses knowledge from various disciplines such as gas discharge, electromagnetism, plasma physics, and fluid mechanics. As a result, the internal operating mechanism of PPTs is extremely complex, and our understanding of it is insufficient, thereby leading to problems such as low propulsion efficiency (generally less than 10%) and plume deposition and contamination. To date, these problems have not been solved effectively. In this book, theoretical analysis and numerical simulation methods are used to establish analytical and numerical models for PPT process stages, including propellant ablation, discharge acceleration, and plume expansion, study the plasma transport mechanism and energy conversion law during thruster operation, and understand the underlying operating mechanism of PPTs, thereby laying a theoretical foundation for the scientific research and engineering application of high-performance PPTs.

© The Author(s) 2025
J. Wu et al., *Numerical Simulation of Pulsed Plasma Thruster*,
https://doi.org/10.1007/978-981-97-7958-1_1

1.1   PPT Numerical Model Research Progress

With the development of computer technology, numerical simulation has become a feasible and effective method for studying plasma flow processes. It has been shown that the calculation results using numerical models for PPTs can effectively reflect the experimental results, providing strong support for the design, optimization, and performance evaluation of thrusters and playing a key role in the study of the PPT operating mechanism. At present, the common numerical models used for PPT research include zero-dimensional, electromechanical, propellant ablation, magnetohydrodynamic, and particle models.

1.1.1   Zero-Dimensional Models

Zero-dimensional models are simplified mathematical models that are usually used to describe the parametric characteristics of the PPT operation process. Based on the energy balance relationship between the pulse current and plasma [2], zero-dimensional models for PPTs treat plasma as a part of the discharge circuit. By comprehensively considering factors such as the power supply energy, propellant flux, and discharge configuration, these models can describe the pulse discharge characteristics of electrical parameters, such as the voltage and current, and predict propulsion performance parameters, such as the pulse impulse, average thrust, propulsion efficiency, and energy utilization.

Michels et al. [3] established a zero-dimensional model for capacitor-powered coaxial plasma guns, as shown in Fig. 1.1, and compared the efficiency predicted by the model with the calorimetric exhaust efficiency obtained from experiments. The upper limit of the efficiency of this model is approximately 40%. The experimentally measured efficiency when the plasma gun operates normally is approximately half of that predicted by the simplified zero-dimensional model. This difference is due to the discrepancy between the model-predicted and actual mass of the propellant.

In a solid ablative PPT (APPT) study, Brito et al. [4] proposed a zero-dimensional model for ablated mass estimation. This model considers the energy balance between

Fig. 1.1 Equivalent circuit diagram of a plasma gun

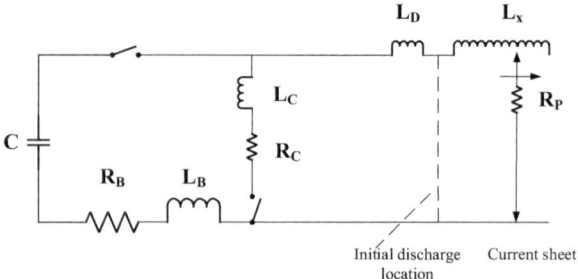

Fig. 1.2 Diagram of an
ideal APPT circuit

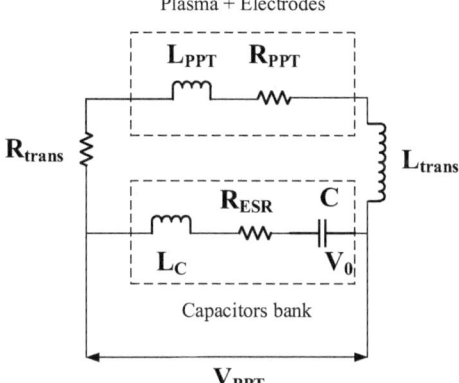

pulse current generation and transport, arc formation, heat loss caused by the Joule
effect, and plasma acceleration. As shown in Fig. 1.2, the ablated mass of polytetrafluoroethylene (PTFE) in the equilibrium equation is a function of electrical and geometrical parameters. The ideal arc thickness is introduced in the model, which is suitable
for propulsion performance estimation and preliminary thruster design. This model
can predict parameters such as the energy loss due to heat dissipation in the circuit,
average plasma temperature, and maximum current generated during the discharge
process. The simulation results indicate that this model can represent changes in the
PPT performance based on varying parameters to a certain extent, enabling a qual-
itative analysis of the PPT propulsion performance and providing reference for the
preliminary design of PPTs. In a comparison of the impulse bit calculated by using
the zero-dimensional model with that determined through experiments, the average
relative error was approximately 20%.

Zhu [2], Zhu et al. [5] established an inductor-resistor–capacitor (LRC) circuit-
based zero-dimensional model, which includes basic performance parameters such
as the PPT thrust, specific impulse, and efficiency. This model is used to analyze the
dynamic discharge process by assuming a change in conductance. The functional
relationships of the PPT efficiency and the kinetic energy of the propellant with
circuit parameters such as the discharge energy, current, and voltage are obtained,
thereby improving the accuracy of the zero-dimensional analysis model. A theoretical study of this variable-parameter zero-dimensional model showed that the total
efficiency η of the thruster is linearly related to the average exhaust velocity $\overline{u_e}$,
the unit conductance L', and the square root of the ratio of energy storage capacitance C to the initial conductance $L_{t=0}$, i.e., $\sqrt{C/L_{t=0}}$, and is inversely proportional to the resistivity ψ. . This zero-dimensional model was used to theoretically
analyze the operation process of a low-power water-propellant PPT in electromagnetic acceleration mode, and the influence of external circuit structural parameters
on the performance was investigated. A discharge parameter diagnostic experiment
was carried out to study the influence of the operating characteristics (e.g., discharge
type and operating energy threshold) and discharge parameters of water-propellant

PPTs on the propulsion performance and to verify the good operating characteristics of low-power water-propellant PPTs.

Zero-dimensional models provide a way for researchers to simplify and quickly analyze the operating principle and performance of PPTs. Despite certain limitations in the application of assumption-based zero-dimensional models, they remain an indispensable tool in the study of PPTs. With a better understanding of the operating principle of PPTs and the development of numerical simulation techniques, more accurate and reliable zero-dimensional models can be proposed and validated, providing a feasible and effective means for PPT design guidance, experimental validation, and performance optimization.

1.1.2   Electromechanical Models

Different from zero-dimensional models, electromechanical models for PPTs not only describe the energy balance relationship between the circuit and plasma but also use a plasma dynamics model to obtain information on the interaction and energy transfer relationship between the circuit pulse current and plasma generation and development.

Jahn [6] proposed a one-dimensional (1D) mathematical model, also known as the "slug" model, to describe the operation process of PPTs with parallel-plate electrodes. Most 1D models in subsequent studies were developed based on the "slug" model. In this model, the plasma acceleration process and PPT circuit components are simulated as an electromechanical system with interactions between the dynamic and electrical systems. Jahn used Kirchhoff's voltage law to describe the PPT discharge process and considered that all the ablated mass is accelerated in a "slug" form. When using this model, it is assumed that the ablated propellant in each pulse discharge is concentrated in the current sheet at the beginning of the discharge and accelerated and ejected under an electromagnetic force. The mass of the current sheet remains constant during the motion of the current sheet. This model can be used to simulate the operation process of a PPT and predict its macroscopic propulsion performance. However, a constant ablated propellant mass is assumed. This assumption is inconsistent with the actual variability of the ablated mass in the PPT operation. Therefore, this model fails to accurately reflect details such as the flow process of the plasma and the ablation process of the propellant during the PPT operation.

Waltz [7] replaced the Kirchhoff voltage equation in the "slug" model with the energy conservation equation to establish an improved electromechanical model. Then, he applied this model to research on the Lincoln Experimental Satellite 6 (LES-6) PPT at the Lincoln Laboratory. In a study on the influence of the electrode plate effect on a simulation, Leiweke [8] proposed an "edge" inductance model to improve the adverse impact on the simulation results caused by the assumption of semi-infinite electrode plates in the original electromechanical model. This model was then used in LES-8/9 simulations. A comparison of the experimental data showed that the simulation results of the original model were actually more consistent with the experimental results, that is, simply adding the edge inductance effect did not improve the original electromechanical model. Gatsonis and Demetriou [9] improved the "slug" model and developed a feedback controller to optimize the plasma jet velocity of PPTs. Damping and control terms were added to the momentum equation. An additional magnetic field was included as a control term. Similarly, Laperriere [10], Laperriere et al. [11] established an improved electromechanical model using a new plasma resistance model and a new inductance model within the discharge channel based on theoretical derivation and investigated the operation performance of PPTs under an additional magnetic field.

A 1D mathematical model for an axisymmetric gas-fed PPT (GFPPT), the earliest "snowplow" model, was proposed by Hart [12] in 1962. This model is similar to the "slug" model in that the entire system is regarded as an electromechanical system. The difference between these models is that the current sheet in the dynamic system is assumed to have a certain mass distribution in the acceleration channel. Michels and Ramins [13] used a variable mass distribution for the current sheet to further develop Hart's model. A noteworthy model after Michels's model is the improved model by Ziemer and Choueiri [14], in which the mass distribution equation is derived from the 1D dynamical theory of the expansion of gas into a vacuum.

Neither the "slug" model nor the "snowplow" model can make accurate predictions of propellant consumption during the operation process. Keidar and Boyd [15–18], Keidar et al. [19–21] developed an electromechanical model, which, after continuous improvements, can simulate the propellant ablation and plasma excitation processes during discharge. At a Teflon surface, particles have a low velocity and a high density, and local thermal equilibrium is established. Accordingly, Keidar et al. [22] developed a microscopic melting model and used it to calculate the melting discharge on the Teflon surface. The results showed that a low surface temperature led to a low melting rate, and a high plasma density could not be generated. Later, this model was continuously improved and developed into a 1D discharge model, providing information not only on the temperature change of the propellant but also on the composition and motion of plasma in the acceleration channel. This model can use unsteady inlet conditions to study PPT plumes [23–25].

While keeping the "slug" model unchanged, Vondra considered the aerodynamic effect on the acceleration process of the "slug" and added an aerodynamic term to the motion equations to establish an improved electromechanical model. This model was then used in the design of improved LES-6 PPTs [26]. Considering the actual motion of plasma in the discharge channel, Wei [27] treated the thickness of the current sheet in the discharge channel as varying with the discharge current, and established a "diffusion model" based on the "slug" model. Yang et al. [28] improved the electromechanical model by assuming that the ablated mass of the propellant varies with time. The ablated mass of the propellant in this model was calculated using empirical formulas. However, this approach reflects the specific propellant ablation process well. Shaw and Lappas [29] established an electromechanical analysis model of the discharge process for PPTs without propellant. This model considers plasma resistance variation over time during the discharge process and relates the electrode corrosion mass ejected by cathode spots to the discharge current. The predicted current waveforms matched the experimental measurements. Schönherr et al. [30] also improved the electromechanical model and established a computational model suitable for the configuration of a flared tongue-shaped electrode, based on which the propellant utilization efficiency of PPTs was studied.

In summary, compared to zero-dimensional models, electromechanical models fully consider the dynamics process of PPT plasma and, hence, can model and simulate the interaction between the circuit and plasma during the thruster discharge process and more objectively reflect the multiphysical field coupling characteristics of the PPT operation, providing accurate and reliable numerical simulation analysis tools for the design optimization and performance evaluation of PPTs. However, a direct connection between the specific solid propellant ablation process and the PPT discharge process has not yet been established using electromechanical models; this is one important direction worthy of further research.

1.1.3   Propellant Ablation Models

Compared to the use of gases or liquids, the use of solids as PPT propellants has many advantages. For example, the thruster does not require supply system-related components such as valves, tanks, or heaters, greatly simplifying the structure of the thruster and thus improving the operating reliability. At present, PTFE is used as a propellant for most PPTs in space missions. Because it is difficult to control the solid PTFE ablation process and the ablated propellant mass is related to the discharge energy, it is not possible to increase the discharge energy without changing the ablated mass. In addition, there is an ablation lag phenomenon. As a result, the utilization rate of PPT propellant is very low, which is a key factor restricting the improvement of the thruster performance. In comparison, solid propellant ablation models can provide boundary conditions for numerical simulations of the whole plasma flow field of PPTs and can also be used to study the ablation characteristics of PPT propellant, analyze the propellant utilization rate, and evaluate the influence of the propellant on the thruster propulsion performance.

At the beginning of a PTFE ablation model study, based on experimental data analysis, the PTFE ablation process was simplified as a sublimation process to establish a vapor pressure ablation model, and the ablated mass flux was described as a function of the ablated surface saturation vapor pressure, the plasma pressure difference, and the ablated surface temperature [31], that is,

$$\dot{m} = \frac{p_f - p_i}{\sqrt{2\pi m_f k_B T_s}} \tag{1.1}$$

This model can be used to estimate the ablated mass flux of the propellant based on limited experimental data. However, the simplicity of this model leads to large errors in the calculation results. Turchi and Mikellides [32] applied this ablation model to simulate the PPT operation process. Boundary conditions were subsequently provided for a computational model of the PPT operation process.

The above model does not reflect the mechanism of the PTFE ablation process, which severely limits research on high-precision numerical simulations of the PPT operation process. With the gradual deepening of the understanding of the PTFE ablation process, the modeling and simulation of this process are constantly improving [33, 34]. To reflect the dissociation of PTFE, Bespalov and Zalogin [35], based on a vapor pressure ablation model, considered the chemical reactions and heat conduction processes in the ablation process and proposed a new improved model as follows:

$$\dot{m} = \sqrt{\frac{A_p k \rho T_s^2 \exp(-E/RT_s)}{(h_s - h_\infty)E}} \tag{1.2}$$

Considering that the ablated end surface of the propellant gradually recedes as the ablation process progresses, Kemp [36] assumed that the ablated mass of the propellant is exponentially related to its temperature and linearly related to its surface temperature. The following improved model was proposed:

$$\dot{m} = \sqrt{\frac{A_p k_s \rho_s T_s^2 \exp(-B_p/T_s)}{(h_s - h_\infty)B_p}} \tag{1.3}$$

Due to the consideration of the recession velocity of the ablated surface, the Kemp model provides a framework for establishing a more detailed ablation model. However, this model does not consider the transformation of the material state of the propellant that occurs before dissociation. In addition, the model assumes a constant internal temperature gradient of the propellant, which is a rough assumption.

Clark [37] proposed a relatively comprehensive mathematical model that can be solved through numerical calculations. Different from previous models, this model considers changes in the state of PTFE at approximately 600 K and the linear variation of material properties with temperature. Clark divided PTFE into two different temperature zones, where the propellant is in a crystalline solid state or a molten state, respectively. The boundary conditions of the ablated surface were established according to the law of energy conservation, and FORTRAN was used to carry out a simulation study of the ablation process for the first time. In subsequent studies, the PTFE ablation model was continuously improved based on Clark's model. Stechmann [33] used Clark's propellant ablation model to provide boundary conditions for the simulation of the PPT operation process, and the simulation results were in good agreement with the experimental data. Although Clark's model can comprehensively describe the ablation characteristics of the PPT propellant, it still has the following problems. First, the model fails to associate PPT discharge characteristics with propellant ablation characteristics and thus cannot reflect the influence of the former on the latter. Second, the model does not consider the thermal permeability distribution of the plasma radiation source inside the propellant or the influence of the propellant surface properties (e.g., surface reflectance and absorptance) on the propellant ablation characteristics, and it cannot reflect the influence of carbonization and metal sputtering of the propellant surface on the ablation morphology.

By establishing a propellant ablation model, we can evaluate the lifespan and performance of PPTs and provide guidance for propellant selection and design optimization. In addition, ablation models can be used to predict the influence of thermal ablation and ion bombardment/ablation on the composition and structure of the propellant. Through the comprehensive use of this information, more effective protective measures can be developed to extend the service life of PPTs.

1.1.4 Magnetohydrodynamic Models

Ablation models reveal characteristics (e.g., mass flux and distribution) of the PPT propellant that influence the acceleration process in the thruster discharge channel. However, the complexity of propellant ablation increases the complexity of modeling the PPT operation process. Zero-dimensional models and electromechanical models have low computational costs and can predict performance parameters such as the impulse bit, specific impulse, and propulsion efficiency. However, zero-dimensional models rely on empirical parameters, and electromechanical models require a large number of assumptions, neither of which can provide a detailed description and analysis of the PPT operation process. As theoretical research and computer technology continue to develop, establishing magnetohydrodynamic (MHD) models that can accurately describe the plasma flow process and conducting numerical simulations of the operation process based on magnetohydrodynamics have become important directions for PPT research.

To study plasma characteristics during the operation of mN-class PPTs, Palumbo and Guman [38] developed MHD equations. The basic MHD equations consist of the basic electromagnetic field, fluid mechanics, and plasma state equations. The electromagnetic field is described by the Maxwell equations and Ohm's law, and the flow field is described by the Navier–Stokes (NS) equation, that is,

$$\frac{\partial}{\partial t}\begin{bmatrix} \rho \\ \rho U \\ B \\ \rho e \end{bmatrix} + \nabla \cdot \begin{bmatrix} \rho U \\ \rho UU + \left(p + \frac{B^2}{2\mu_{e0}}\right)\overline{\overline{I}} - \frac{BB}{\mu_{e0}} \\ UB - BU \\ \left(\rho e + p + \frac{B^2}{2\mu_{e0}}\right)U - \frac{B}{\mu_{e0}}(U \cdot B) \end{bmatrix} = \begin{bmatrix} 0 \\ \nabla\overline{\overline{\tau}} \\ -\nabla \times (\nu_e \nabla \times B) \\ \nabla\left(\overline{\overline{\tau}} \cdot U\right) + \nabla \cdot q + \nu_e \frac{(\nabla \times B)^2}{\mu_{e0}} \end{bmatrix}$$

$$(1.4)$$

It is generally difficult to obtain analytical field quantity solutions using magnetohydrodynamics equations. Therefore, the equations need to be simplified. For example, the common assumption of plasma quasi-neutrality states that fluid is a conducting medium and satisfies the condition of electrical neutrality, i.e., the positive and negative charges per unit volume of neutral plasma are equal in number. In reality, quasi-neutrality means that the material can be regarded as essentially charge neutral. Strictly speaking, not all the electromagnetic forces between particles disappear; instead, they are approximately considered to diminish within the characteristic length scale of the plasma system under investigation. For specific research problems, when some approximation conditions are satisfied within the range of typical parameters applied in magnetohydrodynamics, some corresponding simplifications can be made. Spanjers et al. [39, 40], Spanjers [41] showed that a simplified 1D MHD model can provide simple information on the operation process of a thruster. However, to obtain more accurate information on the energy transfer between the plasma, propellant, and electrodes, as well as the development and changes of the plasma in the acceleration channel, it is necessary to conduct a higher-dimensional numerical simulation study to gain a deeper understanding of the physical processes of the PPT operation. With the gradual development of theoretical modeling and numerical simulations, high-dimensional numerical simulations of PPTs have become possible. One representative example is the Multiblock Arbitrary Coordinate Hydromagnetic (MACH2) numerical simulation program based on MHD equations.

MACH2 is a 2.5-dimensional unsteady MHD program. This program is the primary theoretical tool used in the U.S. for PPT operation process analysis [42, 43]. MACH2 was first developed by the Plasma Theory and Computation Center of the U.S. Air Force Research Laboratory (AFRL). This program has been applied to the theoretical analysis of the operation process of PPTs by Turchi et al. from Ohio State University [32]. Based on the MHD equations, MACH2 can solve the problems of low-temperature heat conduction, and thermal radiation in plasmas. This program employs an arbitrary Lagrangian–Eulerian method for variable mesh processing and performs flow field calculations in the Lagrangian step and mesh modification in the Eulerian step. MACH2 can be used to calculate the distribution of the plasma discharge states and yields more detailed and effective results than those obtained from experiments. The main reason for inaccurate calculation results is the non-equilibrium state caused by the uneven distribution of the particle density in the plasma within the PPT discharge channel.

The MACH2 program assumes that heavy particles and electrons have the same temperature to obtain the plasma equation of state, which is far from the actual operation process. Thomas used the Chemical Equilibrium with Applications (CEA) program combined with the Shah equation to solve the plasmoid state in PPTs and improved the MACH2 program [42]. In the MACH2 program, only the transport characteristics of electrons are considered in the calculation model of plasma thermal conductivity, and the energy transport characteristics of heavy particles near the propellant surface through chemical actions are not considered. Schmahl proposed a dual-temperature model based on the assumption of local thermodynamic equilibrium to study the motion state of multicomponent PTFE plasma [44]. The results showed that when the temperature exceeded 5000 K, the calculation results of the dual-temperature model were significantly better than those of the MACH2 program [45].

For solid ablative PPTs, the ablation process of the solid propellant is extremely complex. The ablation model provides the boundary conditions for the simulation of the plasma plume in the channel. The quality of the ablation model has a strong impact on the simulation results of the whole APPT operation process. Researchers at Ohio State University studied an ablation model for APPT solid propellant. It was assumed that there exists a saturated vapor layer near the ablated surface of PTFE and that the temperatures of the vapor layer and ablated surface of the solid propellant are the same. The pressure was calculated from the saturation vapor pressure curve of PTFE. The temperature distribution of PTFE was calculated by specifying the net heat flux, and the velocity of the ablation boundary of the propellant was calculated by the pressure gradient. Then, the ablated mass of the propellant was determined [46–49].

In modeling the PPT operation process, the MACH2 program does not consider the influence of the non-neutral region and thus is unable to reflect the non-local thermodynamic equilibrium effect in the plasma plume. Due to the assumption of continuity of the fluid medium, this program fails to simulate the motion process of the plume outside the discharge channel. Furthermore, the MACH2 2D calculation does not allow a simulation study of the electrode edge effect of the parallel-plate thruster. Lin and Kamiadakis [50, 51] of Brown University proposed a high-order simulation method and used the Air Force Office of Scientific Research (AFOSR) computational software to study the plasma flow, external circuits, and Teflon ablation of PPTs. The continuum-based method was used to address the viscous effect, and modified NS equations as well as appropriate velocity slip and temperature jump boundary conditions were used at the wall. A method was introduced to ensure that the pressure remains positive during density jumps. A spectral element spatial discretization method was used for structured and unstructured meshes to simulate 1D, 2D, and 3D fully coupled PPT flows.

The simulation of the plume development process of the PPT using MHD models can clarify the composition and motion state of the plasma plume and thus enable a deeper understanding of the operating mechanism of PPTs. Gatsonis and Hastings [52] established a hydrodynamic model to describe a plasma plume. Then, they conducted a numerical simulation study of a large-scale plasma plume to predict its motion state. This information was used to analyze the interaction between the PPT plasma plume and an associated spacecraft. Brukhty et al. [53], based on the experimental data and the PLASIM model developed by Robinson et al. [54], provided expressions for describing the electric potential and electron density, which have been widely used in numerical simulations of electric thruster plumes.

Surzhikov and Gatsonis [55] constructed a 3D MHD model for PPT plume simulation and solved a numerical model using a time-splitting method to calculate the variation patterns of the plasma plume pressure, plasma density, and magnetic pressure over time. Because the model assumes that a plume is a single fluid, the motion process of different components in the plume cannot be calculated. Thus, this model fails to reflect the influence of different particle motion states on the PPT performance. Based on the basic assumptions of the local thermodynamic equilibrium and the macroscopic electrical neutrality of plasma, Yang [56] established a 1D unsteady MHD mathematical model for the simulation of the PPT operation process and conducted a simulation study of this process. While this model can identify the macroscopic characteristics of the PPT operation, it cannot reflect the specific motion and ablation processes of the plasma plume.

1.1.5 Particle Models

To study the motion characteristics of rarefied gases, Bird proposed the direct simulation Monte Carlo (DSMC) method in 1963 [57]. This model can be used to simulate the vacuum plume field. In a vacuum environment, the plasma plume exhibits a tendency to freely expand and has a large density gradient. The governing equations describing its motion state are the full Boltzmann equations. The DSMC method is compatible with the Boltzmann equations and has been widely applied in numerical studies of plasma plume in vacuum.

To simulate plasma driven by an electromagnetic field, the particle-in-cell (PIC) method is the most widely used method. Birdsall and Langdon [58], Hockney and Eastwood [59], applied the PIC method to the numerical simulation of collisionless plasma motion. However, in the PIC method, very small calculation time steps are required to capture the electron trajectories, and the Debye length of the plasma plume affects the solution process of the Poisson equation, severely constraining the application of this method.

Samanta Roy [60], Samanta Roy et al. [61, 62] improved the PIC model by assuming that the plasma plume comprises charged ions, neutral particles, and particles stripped from the ablated surface, maintaining charge neutrality. Additionally, it was assumed that electrons conform to a Boltzmann distribution. The electric field was obtained from the Poisson equation, and the neutral particles were obtained from an approximate analytical model. A DSMC-PIC hybrid algorithm was developed by exploiting the advantages of the DSMC method. This algorithm was applied to the simulation of ion engine plumes.

In the simulation of plumes using the quasi-neutral PIC-DSMC method, to avoid the limitations of the time and space step sizes of the PIC method, Oh [63] solved for the electromagnetic field using the Boltzmann relationship, in which the collision of neutral particles, including charge exchange collision, was simulated using the DSMC method with a time counter (TC), and other forms of collision were simulated using a hard sphere model. This method can simulate a steady plasma plume and has been continuously used and improved. Keidar and Boyd [64] of the University of Michigan combined a discharge model with the PIC-DSMC hybrid method to develop a numerical model that can be used for the full-field simulation of the PPT operation process. This model can be used to simulate the PPT discharge process as well as the entire process of the generation and development of the plasma plume. The calculation results of the discharge model provide boundary conditions for the plasma plume simulation.

The main limitations of the PIC method arise from the limitation of the Debye length on solving the Poisson equation and the time step limitation on the electron motion velocity. Moreover, DSMC cannot effectively address the problem of self-consistency in the plasma electromagnetic field. To address the shortcomings of the DSMC and PIC methods while exploiting their advantages, Yin [65], Gatsonis and Yin [66, 67] proposed a PIC-DSMC hybrid fluid algorithm. This algorithm uses the DSMC method to calculate the collision process of particles in the plasma plume and the PIC method to calculate the motion process of charged particles in the electromagnetic field. The electric field distribution is solved using the Boltzmann equation. This model comprehensively considers the collisions and Coulomb forces between particles in the plasma plume and can calculate the velocity and density of each component in the plume as well as the electric field distribution. Gatsonis and Gagne [68] further introduced the electron energy equation to this model so that the temperature of the plasma plume can be solved.

In summary, many theoretical models have been developed for the PPT discharge process. These models can address certain issues, but further development is needed. Zero-dimensional models can only meet the requirements of performance estimation and preliminary thruster design. Neither the "slug" nor the "snowplow" models can predict propellant consumption. Electromechanical models have a simple structure and a short computational cycle. On simple experimental data, these models can predict macroscopic performance parameters of PPTs that are difficult to measure. However, these models are too simple to reflect microscopic processes such as the ablation process of the propellant and the formation and development processes of the plasma plume. The MACH2 program is based on magnetohydrodynamics and accounts for the three temperatures, low-temperature heat conduction, and thermal radiation of the plasma. In addition, MACH2 can simulate the distribution of the plasma plume and obtain more detailed information than that obtained from experimental data. However, this program cannot reflect the specific processes of propellant ablation and the formation and development of the plasma plume. The 3D MHD equations can be used to simulate the plasma region where the continuum fluid assumption can be applied. However, these equations cannot be used to simulate the real motion processes in the transition region and the molecular free-flow region, and the computational cycle is long. The DSMC and PIC methods can be combined with the discharge model to simulate the entire operation process of the PPT and the formation and development process of the plasma plume. However, the calculation process is complex and the calculation cycle is long. Therefore, to thoroughly explore the intrinsic mechanism of PPTs, there is a need to use different methods to carry out more appropriate model and algorithm research according to the characteristics of the different stages of the thruster operation process.

1.2   Organization of the Book

Based on the operating characteristics of PPTs, this book discusses the establishment of physical models for the PPT propellant ablation, discharge acceleration, and plume expansion stages and proposes assumptions and solution methods for these models. The validity of these methods is verified, and numerical simulations and theoretical analyses of the operation process of the thruster are carried out. This book is divided into three parts: ablation, discharge, and plume. The main contents of this book are outlined as follows.

This chapter describes both domestic and international research progress on numerical PPT models and provides an overview of the main contents of this book.

Part 1 Ablation:

In Chap. 2, the physical properties of the PTFE propellant of the thruster are analyzed, a physical model for PPT arc ablation is established, and numerical simulations are carried out on the temperature distribution evolution pattern as well as the propellant ablation lag and particle emission phenomena during the PTFE ablation process. In Chap. 3, numerical simulations are performed on the laser ablation process of PTFE in the laser-supported PPT ignition process, and a propellant laser ablation model considering the non-Fourier effect is established with a double-layer dynamic structure consisting of the liquid and solid phases of the propellant. The PTFE ablation process is divided into two stages. A solution is obtained by a finite volume method using a fully implicit scheme on a non-uniform mesh. The reliability of the model is verified by comparing analytical and numerical solutions, providing theoretical and model support for an in-depth understanding of the laser ablation mechanism of polymer materials. In Chap. 4, based on the study in Chap. 3, the ablation process of aluminum (Al) propellant in the PPT is modeled and simulated. Under intense laser radiation with a nanosecond pulse width, the non-Fourier heat conduction and phase change ablation processes of Al are investigated, and the thermal evaporation and phase explosion mechanisms during laser ablation are considered. A non-Fourier heat conduction equation based on an enthalpy method is established to investigate the influences of the shielding effect of the Al plasma, laser parameters, non-Fourier effect, and various factors (e.g., laser fluence, laser wavelength, and background gas pressure) on the ablation process.

Part II Discharge:

In Chap. 5, the traditional electromechanical model, an electromechanical model considering the mass accumulation of the current sheet, and an electromechanical model considering an additional magnetic field are constructed to conduct a performance analysis of the PPT discharge process. The characteristics and application conditions of the three models are described in detail. Subsequently, numerical simulations are conducted on the operation process of the PPT under different conditions to investigate the influence of the PPT operating parameters on the operation process and the overall performance. In Chap. 6, theoretical and numerical studies of the discharge process of the PPT is carried out based on the MHD model. An MHD model of the PPT discharge process is established, including a discharge circuit model, a two-phase ablation model, a magnetofluid flow model, and a thermochemical and transport model. Numerical calculation methods for the model are proposed, and numerical simulations and theoretical analyses are carried out on the PPT discharge process. The flow acceleration characteristics of the plasma in the discharge channel are investigated, and the discharge characteristics and thruster performance under different current waveforms are evaluated.

Part III Plume:

In Chap. 7, a particle simulation method is used to develop a hybrid DSMC/PIC fluid algorithm for the PPT plume expansion process. The plume field of the PPT under different initial voltages is simulated to obtain the distribution of the plume field over time as well as information on the particle velocity, axial line, and mass reflux rate, and the influence of charge exchange collision (CEX) on the flow field is analyzed. In Chap. 8, the hybrid DSMC/PIC fluid algorithm is used to carry out a numerical simulation of the plume field of the PPT based on the inlet conditions of the MHD model. A computational study of the PPT is conducted using the PPT 1D dual-temperature MHD discharge model and the 3D dual-temperature MHD model as the inlet models and the hybrid particle plume model to obtain information on the axial line and mass reflux rate of the plume field under different initial voltages and different capacitances, providing a theoretical and methodological basis for the study of PPT plume expansion and reflux phenomena.

References

1. Yang L, Li Z, Yin L, et al. Review of pulsed plasma thruster. J Rocket Propul. 2006;32(2):32–6.
2. Zhu P. Design and study of low-power water-fed pulsed plasma thrusters. Nanjing University of Science and Technology; 2011.
3. Michels CJ, Heighway JE, Johanse AE. Analytical and experimental performance of capacitor powered coaxialplasma guns. AIAA J. 1966;4(5):823–30.
4. Brito CM, Elaskar SA, Brito HH, et al. Zero-dimensional model for preliminary design of ablative pulsed plasma Teflon thrusters. J Propul Power. 2004;20(6):970–7.
5. Zhu P, Hou L, Zhang W. On the energy balance and efficiency of water-fed pulsed plasma thruster. J Propul Technol. 2011;2:292–300.
6. Jahn RG. Physics of electric propulsion. New York: McGraw-Hill; 1968.
7. Waltz PM. Analysis of a pulsed electromagnetic plasma thruster. USA: Massachusetts Institute of Technology (Doctor); 1969.
8. Leiweke RJ. An advanced pulsed plasma thruster design study using one-dimensional slug modelling. New Mexico; 1996.
9. Gatsonis NA, Demetriou MA. Prospects of plasma flow modelling and control for micro pulsed plasma thrusters. In: AIAA 2004-3464, 40th AIAA/ASME/SAE/ASEE joint propulsion conference and exhibit. Fort Lauderdale, FL; 2004.
10. Laperriere DD. Electromechanical modelling and open-loop control of parallel-plate pulsed plasma microthrusters with applied magnetic fields. USA: Worcester Polytechnic Institute (Master); 2005.

11. Laperriere DD, Gatsonis NA, Demetriou MA. Electromechanical modelling of applied field micro pulsed plasma thrusters. In: AIAA 2005–4077; 2005.

12. Hart PJ. Plasma acceleration with coaxial electrode. Phys Fluids. 1962;5(1):38–47.

13. Michels CJ, Ramins P. Performance of coaxial plasma gun with various propellants. Phys Fluids. 1964;7(11):S71–4.

14. Ziemer JK, Choueiri EY. Dimensionless performance model for gas-fed pulsed plasma thrusters. In: AIAA 98–3661, 34th AIAA/ASME/SAE/ASEE joint propulsion conference and exhibit. Cleveland, OH; 1998.

15. Keidar M, Boyed ID. Electrical discharge in the teflon cavity of a coaxial pulsed plasma thruster. Trans Plasma Sci. 2000;28(2):376–85.

16. Keidar M, Boyd ID. Analyses of Teflon surface charring and near field plume of a micro-pulsed plasma thruster. In: AFRL-PR-ED-TP-2002-185; 2001.

17. Keidar M, Boyd ID. Ionization non-equilibrium and ablation phenomena in a micro-pulsed plasma thruster. In: AIAA 2002-4275, 38th AIAA/ASME/SAE/ASEE joint propulsion conference and exhibit. Indianapolis, IN; 2002.

18. Keidar M, Boyd ID. Ablation study in the capillary discharge of an electrothermal gun. J Appl Phys. 2006;99(5):053301–8.

19. Keidar M, Boyd ID, Beilis II. Model of an electrothermal pulsed plasma thruster. J Propul Power. 2003;19(3):424–30.

20. Keidar M, Boyd ID, Lepsetz N. Performance study of the ablative Z-pinch pulsed plasma thuster. In: AIAA 2001-3898, 37th AIAA/ASME/SAE/ASEE joint propulsion conference and exhibit. Salt Lake City, Utah; 2001.

21. Keidar M, Boyed D, Beilis II. On the model of Teflon ablation in an ablation-controlled discharge. J Phys D Appl Phys. 2001;34:1675–7.

22. Keidar M, Fan J, Boyd ID. Vaporization of heated materials into discharge plasmas. J Appl Phys. 2001;89(6):3095–8.

23. Keidar M, Boyd DI, Antonsen E, et al. Electromagnetic effects in the near-field plume exhaust of a micro-pulsed-plasma thruster. J Propul Power. 2004;20(6):961–9.

24. Keidar M, Boyd ID. Device and plume model of an electrothermal pulsed plasam thruster. In: AIAA 2000-3430, 36th AIAA/ASME/SAE/ASEE joint propulsion conference and exhibit. Huntsville, AL; 2000.

25. Keidar M, Boyd ID. Electronagnetic effects in the near field plume exhaust of a pulsed plasma thruster. In: AIAA 2001-3638, 37th AIAA/ASME/SAE/ASEE joint propulsion conference and exhibit. Salt Lake City, UT; 2001.

26. Vondra R, Thomassen K, Solbes A. Analysis of solid Teflon pulsed plasma thruster. AIAA 70-179; 1970.

27. Wei R. Diffusion model for PPT and solution to the simplified system of MHD equations. Chin J Space Sci. 1982;2(4):319–26.

28. Yang L, Liu X-Y, Wu Z-W, et al. Analysis of Teflon pulsed plasma thrusters using a modified slug parallel plate model. In: AIAA 2011-6077; 2011.

29. Shaw P, Lappas V. Modelling of a pulsed plasma thruster; simple design, complex matter. In: Space propulsion conference, San Sebastian Spain; 2010.

30. Schönherr T, Komurasaki K, Herdrich G. Propellant utilization efficiency in a pulsed plasma thruster. J Propul Power. 2013;29(6):1478–87.

31. Loh K, Kushari A. Effect of pressure level on the performance of an auto-initiated pulsed plasma thruster. Plasma Sci Technol. 2010;12(4):466–72.

32. Turchi PJ, Mikellides PG. Modelling of ablation-fed pulsed plasma thrusters. In: AIAA 95-2915; 1995.

33. Stechmann DP. Numerical analysis of transient Teflon ablation in pulsed plasma thrusters. Worcester Polytechnic Insitute; 2007.

34. Gatsonis NA, Juric D, Stechmann DP. Numerical analysis of Teflon ablation in pulsed plasma thrusters. In: AIAA 2007-5227; 2007.

35. Bespalov VL, Zalogin GN. Ablation of Teflon in a stream of dissociated air. AIAA J. 1979;17(2):35682.

36. Kemp NH. Surface recession velocity of an ablating polymer. AIAA J. 1968;6(9):1790–1.
37. Clark BL. An experimental and analytical investigation of Teflon ablation heat conduction parameters by the method of non-linerar estimation. Cornell University; 1971.
38. Palumbo DJ, Guman WJ. Continuing development of the short-pulsed ablative space propulsion system. In: AIAA 72-1154; 1972.
39. Spanjers GG, Malak JB, Leiweke RJ, et al. Effect of propellant temperature on efficiency in the pulsed plasma thruster. J Propul Power. 1998;14(4):11.
40. Spanjers GG, Antonsen EL, Burton RL, et al. Advanced diagnostics for millimetre-scale micro pulsed plasma thrusters; 2002.
41. Spanjers GG. Investigation of propellant in efficiency in a pulsed plasma thruster. In: Proceedings of the 32nd AIAA/ASME/SAE/ASEE joint propulsion conference; 1996.
42. Thomas HD. Numerical simulation of pulsed plasma thrusters. USA: The University of Tennessee (Doctor); 2000.
43. Mikellides YG. Theoretical modelling and optimization of ablation-fed pulsed plasma thrusters. USA: The Ohio State University (Doctor); 1999.
44. Schmahl CS. Thermochemical and transport processes in pulsed plasma microthrusters: a two-temperature analysis. USA: The Ohio State University (Doctor); 2002.
45. Schmahl CS, Turchi PJ, Mikellides PG, et al. Development of equation-of-state and temperature properties for molecular plasma in pulsed plasma thruster part 2: two-temperature equation-of-state for Teflon. In: AIAA paper 98-3662, 34th AIAA joint propulsion conference. Cleveland, OH; 1998.
46. Mikellides IG, Turchi PJ. Optimizaiton of pusled plasma thrusters in rectangular and coaxial geometries. In: IEPC 99-211, 26th international electric propulsion conference. Kitakyushu; 1999.
47. Mikellides PG, Neilly C. Pulsed inductive thruster, part 1: modelling, validation and performance analysis. In: AIAA 2002-4091, 38th AIAA/ASME/SAE/ASEE joint propulsion conference and exhibit. Indianapolis, IN; 2002.
48. Mikellides PG, Turchi PJ. Modelling of late-time ablation in Teflon pulsed plasma thrusers. In: AIAA 96-2733, 32nd AIAA/ASME/SAE/ASEE joint propulsion conference and exihibit. Lake Buena Vista, FL; 1996.
49. Mikellides PG, Turchi PJ. Theoretical investigation of pulsed plasma thrusters. In: AIAA 98-3807, 34th AIAA/ASME/SAE/ASEE joint propulsion conference and exhibit. Cleveland, OH; 1998.
50. Lin G, Karniadakis GE. High-order modelling of micro-pulsed plasma thrusters. In: AIAA 2002-2872, 3th theoretical fluid mechanics meeting. St. Louis, Missouri; 2002.
51. Lin G, Kamiadakis GE. A discontinuous galerkin method for two-temperature plasmas. Comput Methods Appl Mech Eng. 2006;195:3504–27.
52. Gatsonis NA, Hastings DE. Evolution of the plasma environment induced around spacecraft by gas releases: three-dimensional modelling. J Geophys Res. 1992;97(10):14989–5005.
53. Brukhty VI, Kirdyashev KP, Svetlitskaya OE. Electromagnetic interference measurements within the T-100 endurance test. IEPC 95-73; 1995.
54. Robinson RS, Kaufman HR, Winder DR. Plasma propagation simulation near an electrically propelled spacecraft. J Spacecraft. 1982;19(5):445–50.
55. Surzhikov ST, Gatsonis NA. Plasma flow through a localized heat release region; 1999.
56. Yang L. Theoretical and experimental study of the operation process of pulsed plasma thrusters. Graduate School of National University of Defence Technology; 2007.
57. Bird GA. Molecular gas dynamics and the direct simulation of gas flows. Oxford: Clarendon Press; 1994.
58. Birdsall CK, Langdon AB. Plasma physics via computer simulation. New York: McGraw-Hill; 1985.
59. Hockney RW, Eastwood JW. Computer simulation using particles. New York: Adam Hilger; 1988.
60. Samanta Roy RI. Numerical simulation of ion thruster plume backflow for spacecraft contamination assessment. USA: MIT (Doctor); 1995.

61. Samanta Roy RI, Hastings DE, Gatsonis GA. Ion-thruster plume modelling for backflow contamination. J Spacecraft Rocket. 1996;33(4):525–34.
62. Samanta Roy RI, Hastings DE, Gatsonis GA. Numerical simulation of spacecraft contamination and interactions by ion-thruster effluents. J Spacecraft Rocket. 1996;33(4):535–42.
63. Oh DY. Computational modelling of expanding plasma plumes in space using a PIC-DSMC algorithm. USA: MIT (Doctor); 1997.
64. Keidar M, Boyd ID. Plasma generation and plume expansion for a transmission-mode micro-laser ablation plasma thruster. In: AIAA 2003-4567, 39th AIAA/ASME/SAE/ASEE joint propulsion conference and exhibit. Huntsville, Alabama; 2003.
65. Yin X. Axisymmetric hybrid numerical modelling of pulsed plasma thruster plumes. USA: Worcester Polytechnic Institute (Doctor); 1999.
66. Gatsonis NA, Yin X. Hybrid (particle-fluid) modelling of pulsed plasma thruster plumes. J Propul Power. 2001;17(5):945–58.
67. Gatsonis NA, Yin XM. Axisymmetric DSMC-PIC simulation of quasineutral partislly ionized jets. In: AIAA 97-2535, 32nd AIAA thermophysics conference. Atlanta, GA; 1997.
68. Gatsonis NA, Gagne M. Electron temperature effects on pulsed plasma thruster plume expansion. In: AIAA 2000-3428, 36th AIAA/ASME/SAE/ASEE joint propulsion conference and exhibit. Huntsville, Alabama; 2000.

Part I
Ablation

Chapter 2
Numerical Simulation of the Arc Ablation Process of PTFE Propellant

Polytetrafluoroethylene (PTFE) is commonly employed as a propellant in PPTs due to its favorable vacuum physical properties. These properties include its non-adhesiveness and non-brittleness at low temperature, low outgassing rate in vacuum, and inherent self-lubrication. PTFE is ablated and ionized under the action of the discharge arc, and the products are accelerated and ejected from the thruster under the combined action of the Lorentz force and aerodynamic force, thus generating thrust. Due to factors such as propellant ablation lag, the utilization efficiency of PTFE is very low, resulting in a low level of propulsion efficiency (about 10%), which is one of the crucial factors limiting the widespread application of PPTs in microsatellites. Therefore, establishing a simulation model that can accurately reflect the ablation process of the PPT propellant and conducting a theoretical analysis of the propellant ablation process are necessary for understanding the intrinsic mechanism of propellant ablation in the thruster and improving the propulsion performance of PPTs.

With a focus on the arc ablation phenomenon of a solid propellant in the PPT operation process, this chapter begins with an analysis of the physical properties of the PTFE propellant. Based on changes in the phase characteristics with temperature, the PPT ablation process is divided into three stages: discharge initiation without propellant melting, continuous discharge with propellant melting, and post-discharge. A PPT propellant ablation model is established. Based on this model, numerical simulation and analysis are conducted on the evolution pattern of the PTFE ablation temperature distribution, the propellant ablation lag, and particle emission phenomena during the PPT discharge process, providing a theoretical basis for an in-depth understanding of the physical laws related to arc ablation of the propellant.

2.1 PTFE Propellant Ablation Model

PTFE is a fully symmetric, non-polar polymer material [1]. A PTFE molecule is composed of covalently bonded carbon (C) and fluorine (F) atoms, and the chemical formula is C_2F_4. The molecular structure is shown in Fig. 2.1. The C–C chain backbone of PTFE is surrounded by a protective layer of F atoms. The unique structural characteristics of PTFE and the bond energy of the C–F bond of up to 466 kJ/mol mean that PTFE has chemical stability, thermal stability, and chemical inertness unmatched by other materials. In addition, PTFE can be used in the temperature range of -190 to 260 °C, making it suitable for low-temperature environments in outer space [2]. The very low water absorption, excellent dielectric properties, and aging resistance of PTFE make it a very suitable propellant for PPT.

PTFE has poor thermal conductivity, with a thermal conductivity coefficient of only 0.24 W/(m K). During the discharge process of the PPT, the low thermal conductivity of PTFE results in a large temperature gradient over its micron-scale thickness, which in turn yields large particles in the molten propellant. These large neutral particles, which are not effectively ionized, result in low propellant utilization efficiency, affecting the performance of the thruster. In addition, it is difficult to form PTFE. It is also difficult for PTFE to undergo secondary processes. PTFE has a large coefficient of linear expansion and is prone to deformation and cracking when combined with other materials. These characteristics pose challenges for the doping modification of PTFE [3].

The melting point of PTFE is approximately 600 K, which is much greater than that of other thermoplastic materials. When heated to a temperature below 600 K, PTFE exists as a long chain in a crystalline solid state, while when heated to a temperature higher than 600 K, PTFE transitions from a solid state to a molten state. In the molten state, PTFE has a high viscosity that reaches 10^{11} to 10^{12} Pa s, which is much higher than that of typical plastics at molding temperatures (10^3 to 10^4 Pa s) [3]. In this state, PTFE is not sufficiently fluid, can still retain its original shape, and is sensitive to external stress. After PTFE undergoes a phase change, due to the relatively low energy of the C–C bonds in the polymer, these bonds will dissociate at a temperature of 720 K to generate small-molecule fluorocarbon gas, and most of the dissociation

Fig. 2.1 Molecular structure of PTFE

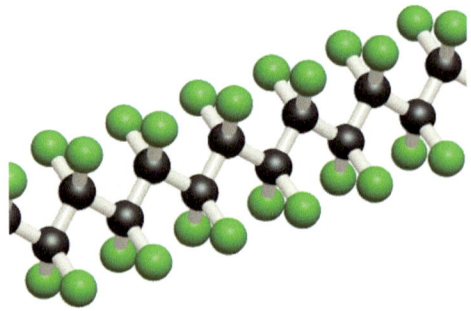

products are C_2F_4 monomers. C_2F_4 monomers have a very high vapor pressure. Once separated from the polymer carbon chain, C_2F_4 will immediately detach from the PTFE surface to form a high-density gas layer near the PTFE surface. This process does not have a specific temperature occurrence point or energy threshold, so this phenomenon persists after the phase change [4, 5]. The presence of this gas layer provides a favorable gas environment for PPT discharge.

2.1.1 Endothermic Heating Stage

Figure 2.2 shows a schematic diagram of propellant ablation and heating in the unmelted stage of PTFE. The plasma arc formed by the PPT discharge interacts with the surface of the propellant. The heat flux between the ablated end surface of the PTFE propellant and the plasma arc is assumed to be I_0. PTFE is heated, but since the melting point temperature T_m has not been reached, it remains in a solid state without undergoing a change of state.

Let R_r be the surface reflection coefficient of the propellant. Then, the heat flux at the ablated end surface s of the propellant is $(1 - R_r)I_0$. Let the absorption coefficient of the solid propellant for radiation energy be α_c and the heat flux at depth x, i.e., where the plasma arc penetrates into the propellant, be $I(x)$. Then,

$$dI(x)/dx = -\alpha_c I(x) \tag{2.1}$$

When $x = s$, $I(s) = (1 - R_r)I_0$. Thus, we have

$$I(x) = (1 - R_r)I_0 e^{-\alpha_c(s-x)} \tag{2.2}$$

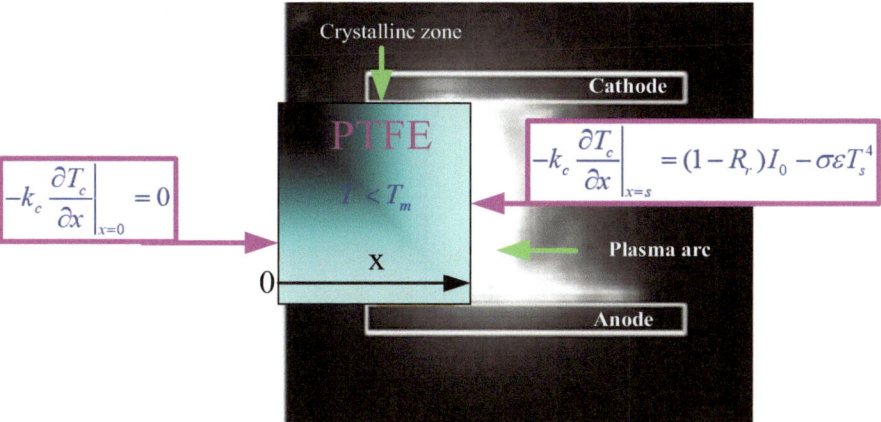

Fig. 2.2 Schematic diagram of propellant ablation in the unmelted stage

The temperature at any point x inside the propellant at time t, $T_c(x, t)$, is given by

$$\rho_c(T)C_c(T)\frac{\partial T_c(x, t)}{\partial t} = \frac{\partial}{\partial x}\left(k_c(T)\frac{\partial T_c(x, t)}{\partial x}\right)$$

$$+ (1 - R_r)\alpha_c I_0 e^{-\alpha_c(s-x)} \tag{2.3}$$

where $\rho_c(T)$ is the density of solid PTFE, $C_c(T)$ is the isobaric heat capacity per unit mass of solid PTFE, and $k_c(T)$ is the thermal conductivity of solid PTFE.

The boundary conditions of the ablated end surface and tail of the propellant are

$$-k_c\frac{\partial T_c}{\partial x}\bigg|_{x=s} = (1 - R_r)I_0 - \sigma\varepsilon T_s^4 \tag{2.4}$$

$$-k_c\frac{\partial T_c}{\partial x}\bigg|_{x=0} = 0 \tag{2.5}$$

where σ is the Stefan–Boltzmann constant ($\sigma = 5.67 \times 10^{-8}$ W/(m^2 K^4)), ε is the PTFE surface emissivity, T_s is the ablated end surface temperature, and s is the propellant length.

2.1.2   Phase Change Stage

With increasing discharge energy, the temperature at the ablated end surface of PTFE rapidly rises above the melting point T_m ($T_m = 600$ K). The solid and molten PTFE coexists at this stage, as shown in Fig. 2.3. For a solid PTFE propellant, Eq. (2.3) remains applicable for describing the internal temperature of the propellant.

At the solid–molten propellant interface s_m,

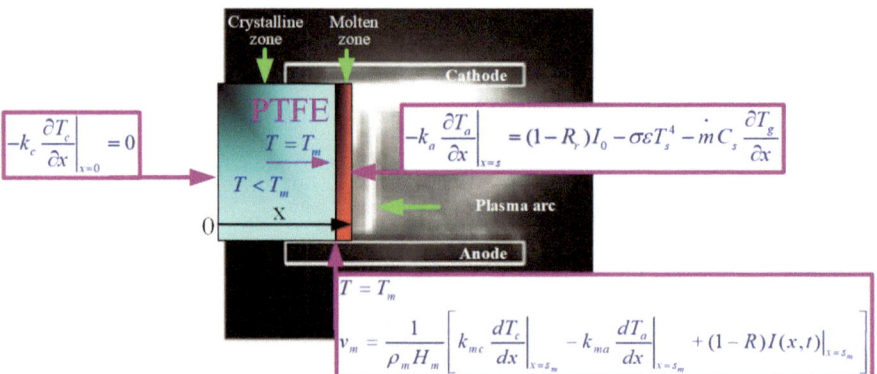

Fig. 2.3 Schematic diagram of propellant ablation in the melting stage

$$T_c(x = s_m) = T_a(x = s_m)T_m \tag{2.6}$$

The temperature at any point x in the molten propellant at time t, $T_a(x, t)$, can be determined by the heat conduction Eq. (2.7), that is,

$$\rho_a(T_a)C_a(T_a)\frac{\partial T_a}{\partial t} = \frac{\partial}{\partial x}\left(k_a(T_a)\frac{\partial T_a}{\partial x}\right) + Q_p$$

$$+ (1 - R_r)\alpha_a I_0 e^{-\alpha_a(s-x)} \tag{2.7}$$

where $\rho_a(T_a)$ is the density of the molten PTFE, $C_a(T_a)$ is the isobaric heat capacity per unit mass of molten PTFE, T_a is the temperature of molten PTFE, $k_a(T_a)$ is the thermal conductivity coefficient of the molten PTFE, and α_a is the absorption coefficient of the molten PTFE for radiation energy.

Q_p is described by the Arrhenius equation [6], that is,

$$Q_p = -A_p\rho_a(T_a)H_p(T_a)\exp(-B_P/T_a) \tag{2.8}$$

where A_p is the frequency factor, $H_p(T_a)$ is the depolymerization energy per unit mass of PTFE, B_p is the activation temperature, and

$$B_P = E_T/R \tag{2.9}$$

where ET is the activation energy, R is the gas constant, and $R = 83.14$ J/(kg K).

The boundary condition at the front end face of the molten PTFE propellant is

$$-k_a\frac{\partial T_a}{\partial x}\bigg|_{x=s} = (1 - R_r)I_0 - \sigma\varepsilon T_s^4 - \dot{m}C_sT_s \tag{2.10}$$

In this stage, the initial value of the temperature at any point in the propellant is obtained from the final temperature distribution in the first stage.

The solid–molten propellant interface s_m is determined by Eq. (2.11):

$$\rho_m H_m v_m = k_{mc}\frac{dT_c}{dx}\bigg|_{s_m} - k_{ma}\frac{dT_a}{dx}\bigg|_{s_m} + (1 - R_r)I(x, t)|_{x=s_m} \tag{2.11}$$

where H_m is the latent heat of the phase change, ρ_m is the average density of the solid and molten PTFE at T_m, and v_m is the velocity of the interface motion given below

$$v_m = \frac{1}{\rho_m H_m}\left[k_{mc}\frac{dT_c}{dx}\bigg|_{x=s_m} - k_{ma}\frac{dT_a}{dx}\bigg|_{x=s_m} + (1 - R_r)I(x, t)|_{x=s_m}\right] \tag{2.12}$$

s_m is

$$s_m = s - \int v_m \mathrm{d}t \tag{2.13}$$

The recession velocity v of the ablated end surface is calculated as

$$v = -\frac{\dot{m}}{\rho_{\mathrm{ref}}} \tag{2.14}$$

where ρ_{ref} is the reference density of PTFE ($T_c = 298$ K).

The mass flux $\dot{m}$ of the ablated propellant is

$$\dot{m} = A_p \int_{s_m}^{s} \rho_a \exp(-B_p / T_a) \mathrm{d}x \tag{2.15}$$

The recession Δs of the ablated end surface is

$$\Delta s = \int v \mathrm{d}t$$

2.1.3 Natural Cooling Stage

After the PPT discharge ends, there is no more energy transferred to the surface of the propellant, and the heat flux I_0 is zero. When the propellant temperature is higher than 600 K, the molten propellant still maintains a high temperature, and the propellant is in the ablation lag stage. The propellant ablation process is shown in Fig. 2.4. The heat conduction equation in this stage is obtained by ignoring the terms of the external and internal heat sources in the heat conduction equation and various boundary conditions for the second stage. When the propellant temperature is lower than 600 K, the heat conduction equation for the first stage, with external and internal heat source terms ignored, starts to apply.

When the propellant temperature is higher than the melting temperature, the internal temperature can be described by Eq. (2.16).

$$\rho_m(T)C_m(T)\frac{\partial T_m(x, t)}{\partial t} = \frac{\partial}{\partial x}\left(k_m(T)\frac{\partial T_m(x, t)}{\partial x}\right) + Q_m(x, t) \tag{2.16}$$

When the propellant temperature is lower than the melting temperature, the internal temperature is described by Eq. (2.17).

$$\rho_s(T)C_s(T)\frac{\partial T_s(x, t)}{\partial t} = \frac{\partial}{\partial x}\left(k_s(T)\frac{\partial T_s(x, t)}{\partial x}\right) \tag{2.17}$$

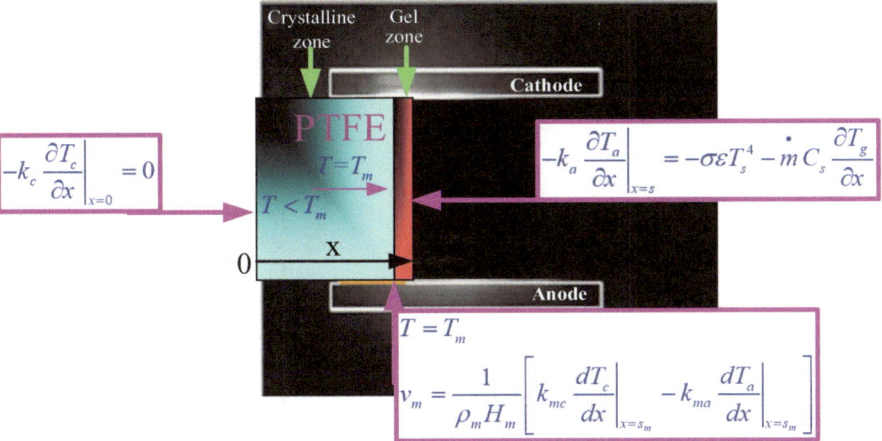

Fig. 2.4 Schematic diagram of the ablation of the molten propellant after the end of discharge

The corresponding boundary conditions of the ablated surface are

$$-k_m(T)\frac{\partial T_m(x,t)}{\partial x}\bigg|_{x=s} = -\sigma\varepsilon\left(T^4 - T_\infty^4\right) - \dot{m}C_m(T)T \tag{2.18}$$

$$-k_s(T)\frac{\partial T_s(x,t)}{\partial x}\bigg|_{x=s} = -\sigma\varepsilon\left(T^4 - T_\infty^4\right) \tag{2.19}$$

The boundary condition at the tail of the propellant remains unchanged, and the adiabatic boundary condition still applies.

$$-k_s(T)\frac{\partial T_s(x,t)}{\partial x}\bigg|_{x=0} = 0 \tag{2.20}$$

2.2 Numerical Calculation Method

2.2.1 Coordinate Transformation

During the operation process of a PPT, as the solid propellant is continuously ablated, both the ablation interface and the phase interface are constantly changing. To adapt to the situation of moving interfaces, coordinate transformation is needed in the solution for PTFE ablation. For solid and molten PTFE, the coordinate system (x, t) is transformed into the coordinate systems (ξ, τ) and (η, τ), respectively. The transformation process is as follows:

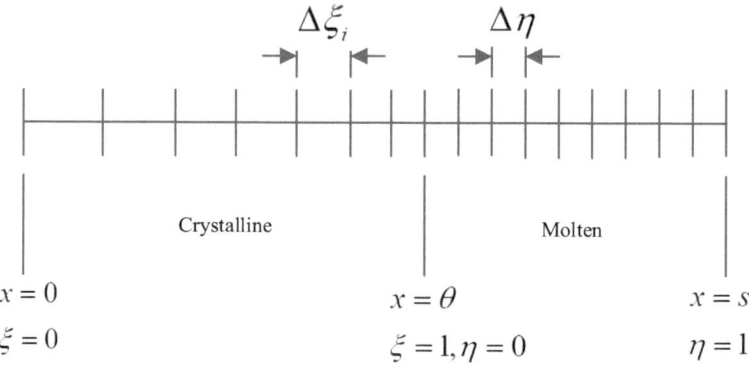

Fig. 2.5 Schematic diagram of meshing

$$\begin{cases} \tau = t \\ \xi = x/\theta(t) \\ \eta = [x - \theta(t)]/[s(t) - \theta(t)] \end{cases} \tag{2.21}$$

$\xi = 0$ or $x = 0$ represents the unablated surface of Teflon, $x = \theta$ is the interface between the molten and solid PTFE, where $\xi = 0$ and $\eta = 0$, and $x = s$ is the ablated surface of PTFE, which is also the end surface in the molten state, where $\eta = 1$. During the ablation process, the temperature distribution of the propellant block is extremely non-uniform. There is a large temperature gradient near the ablated surface of the propellant, and the temperature gradient gradually decreases as the distance from the ablated surface increases. Therefore, when generating a mesh, it is necessary to refine the mesh in regions with large temperature gradients to ensure the accuracy of the calculation and to decrease the number of cells in regions with small temperature gradients to reduce the computational burden (Fig. 2.5).

In this book, a geometric progression technique is used to mesh the solid PTFE region, and a uniform step size is used to mesh the molten region near the ablated surface where there is a large temperature gradient. For the molten and solid propellant, the numbers of cells are set to n_1 and n_2, respectively. The geometric ratio for the solid region is set to $q = 1.05$. For the dimensionless mesh, we have

$$\begin{cases} n_1 \Delta\eta = 1 \\ \Delta\xi_m + q\Delta\xi_m + q^2\Delta\xi_m + \cdots + q^{n_2-1}\Delta\xi_m = 1 \end{cases} \tag{2.22}$$

where $\Delta\eta$ is the mesh step size for the molten region and $\Delta\xi_m = (1 - q)/(1 - q^{n_2})$ is the minimum mesh step size for the solid region.

For the solid and molten regions, the chain rule yields

$$\begin{cases} \frac{\partial}{\partial x} = \frac{1}{\theta}\frac{\partial}{\partial \xi} \\ \frac{\partial}{\partial t} = \frac{\partial}{\partial \tau} - \frac{\xi\dot{\theta}}{\theta}\frac{\partial}{\partial \xi} \end{cases} \tag{2.23}$$

and

$$
\begin{cases}
\frac{\partial}{\partial x} = \frac{1}{s-\theta} \frac{\partial}{\partial \eta} \\
\frac{\partial}{\partial t} = \frac{\partial}{\partial \tau} - \frac{(1-\eta)\dot{\theta}+\eta\dot{s}}{s-\theta} \frac{\partial}{\partial \eta}
\end{cases}
\tag{2.24}
$$

Then, the heat conduction equations for the solid and molten regions can be respectively transformed into

$$
\rho_s(T)C_s(T)\frac{\partial T_s(\xi,\tau)}{\partial \tau} = \frac{1}{\theta^2}\frac{\partial}{\partial \xi}\left(k_s\frac{\partial T_s(\xi,\tau)}{\partial \xi}\right)
$$
$$
+ \rho_s(T)C_s(T)\frac{\xi\dot{\theta}}{\theta}\frac{\partial T_s(\xi,\tau)}{\partial \xi}
\tag{2.25}
$$

and

$$
\rho_m(T)C_m(T)\frac{\partial T_m(\eta,\tau)}{\partial \tau} = \left(\frac{1}{s-\theta}\right)\frac{\partial}{\partial \eta}\left(k_m\frac{\partial T_m(\eta,\tau)}{\partial \eta}\right)
$$
$$
+ \rho_m(T)C_m(T)\frac{(1-\eta)\dot{\theta}+\eta\dot{s}}{s-\theta}\frac{\partial T_m(\eta,\tau)}{\partial \eta}
$$
$$
+ Q(\eta,\tau)
\tag{2.26}
$$

2.2.2 Model Validation

It is assumed that a rectangular heat source with a power density $I_0 = 1.0 \times 10^5$ W/m^2 and a pulse width $t_p = 1$ s acts uniformly on the end surface of the propellant. The constant physical parameters of PTFE are adopted, including a density $\rho_c = 1914.0$ kg/m^3, a specific heat capacity $c_c = 707.9$ J/(kg K), a thermal conductivity $k_c = 0.2477$ W/(m K), and an initial temperature of 300 K. Under the action of an external pulsed heat source, the analytical solution for the propellant temperature distribution is as follows [7]:

$$
T(x,t) = T_0 + \frac{2I_0}{k_c}\sqrt{\frac{at}{\pi}}\left[\exp\left(-\frac{x^2}{4at}\right) - \frac{x}{2}\sqrt{\frac{\pi}{at}}\mathrm{erfc}\left(\frac{x}{2\sqrt{at}}\right)\right]
$$
$$
- [t - t_p]\frac{2I_0}{k_c}\sqrt{\frac{a(t-t_p)}{\pi}}
$$
$$
\left[\exp\left(-\frac{x^2}{4a(t-t_p)}\right) - \frac{x}{2}\sqrt{\frac{\pi}{a(t-t_p)}}\mathrm{erfc}\left(\frac{x}{2\sqrt{a(t-t_p)}}\right)\right]
\tag{2.27}
$$

where the thermal diffusivity is $\alpha = k_c/\rho_c C_c$ and the thermal diffusion length is $\delta = 2\sqrt{\alpha t}$. $[t - t_p] = 1$ when $t > t_p$, and $[t - t_p] = 0$ when $t \leq t_p$.

As shown in Fig. 2.6, the numerical solution of the temperature change at different depths inside the propellant is highly consistent with the analytical solution. Under the condition of an external steady-state heat flow, Kemp derived the calculation formulas for the ablated mass flux and the ablated end surface recession velocity of PTFE as follows [8]

$$\dot{m} = \sqrt{\frac{A_p \rho_c k T_s^2}{B_p(h_s - h_{-\infty})}} \exp\left(-\frac{B_p}{T_s}\right) \tag{2.28}$$

$$v = \sqrt{\frac{A_p k T_s^2}{B_p \rho_s(h_s - h_{-\infty})}} \exp\left(-\frac{B_p}{T_s}\right) \tag{2.29}$$

where $h_s - h_{-\infty}$ is the enthalpy difference before and after ablation of the propellant.

Using our developed computer program, the ablated mass flux and ablated surface recession velocity of PTFE when stable ablation of PTFE is achieved are calculated under constant heat flux. As shown in Fig. 2.7, the numerical and analytical solutions agree well, indicating that our program can accurately calculate the ablation characteristics of PTFE.

The discharge process of a PPT can be equivalent to that of an RLC circuit. According to the study results of Alexeev et al. [9], 5% of the input PPT discharge channel energy E_{tr} is taken as the interaction energy between the PPT discharge arc and the propellant surface, and the ablated end surface of the propellant has an area $A_g = hw$. Then, $I_0 = 0.05$ V $(t)I(t)/A_g$. Typical operating conditions of the PPT experimental prototype designed in our laboratory are used as an example. The

Fig. 2.6 Variations in the temperature of the propellant at different locations

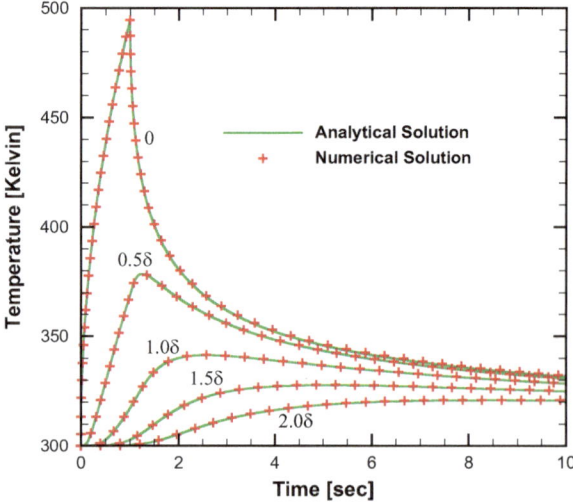

Fig. 2.7 Ablated mass flux of the propellant and ablated surface recession velocity

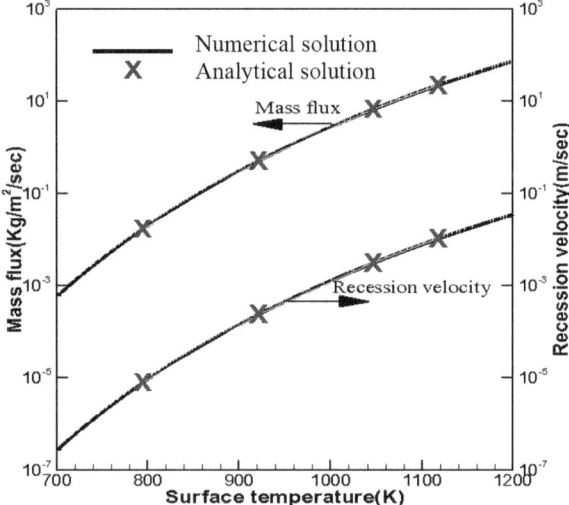

structural parameters, electrical parameters, and PTFE physical parameters of the PPT used in the calculation are shown in Table 2.1 [10, 11].

2.3 Numerical Simulation Results and Analysis

2.3.1 Basic Physical Process of the PPT Propellant Ablation

Figure 2.8 shows the variations in heat flux and propellant ablated mass flux as a function of time, and Fig. 2.9 presents the variations in propellant ablated end surface temperature as a function of time.

As observed in Fig. 2.8, after the main discharge of the PPT begins, as the amplitude of the main discharge current increases, the heat flux increases, and the temperature of the ablated end surface of the propellant rises rapidly. Since the propellant temperature is less than 600 K, the ablated mass flux of the propellant is zero. As the heat flux continues to increase, after a relaxation time of approximately 0.5 μs, the ablated end surface of the propellant reaches a temperature of 600 K. The ablated mass flux of the propellant increases rapidly with increasing heat flux, reaching a maximum at 0.73 μs. The ablated end surface of the propellant reaches its maximum temperature of 1439 K at 0.84 μs. Slightly lagging behind the maximum heat flux, the ablated propellant reaches a maximum mass flux at 0.86 μs. During the first heat flow oscillation cycle, the ablated mass of the propellant increases rapidly, during which the ablated mass of the propellant accounts for more than 95% of the total ablated mass.

Table 2.1 Calculation parameters

Parameter	Numerical value	Unit
Electrode width w	15	mm
Electrode spacing h	45	mm
Capacitance C	12	μF
Charging voltage V_0	1500	V
Equivalent resistance of the circuit R_{eq}	25.28	mΩ
Equivalent inductance of the circuit L_{eq}	51.33	nH
Thermal conductivity of solid PTFE λ_c	$(5.023 + 6.11 \times 10^{-2}T) \times 10^{-2}$	W/ (m K)
Thermal conductivity of molten PTFE λ_a	$(87.53–0.14T + 5.82 \times 10^{-5}T^2) \times 10^{-2}$	W/ (m K)
Density of solid PTFE ρ_c	$(2.119 + 7.92 \times 10^{-4}T\text{-}2.105 \times 10^{-6}T^2) \times 10^3$	kg/m^3
Density of molten PTFE ρ_a	$(2.07–7 \times 10^{-4}T) \times 10^3$	kg/m^3
Reference density of PTFE ρ_{ref}	1933	kg/m^3
Specific heat of solid PTFE C_c	$514.9 + 1.563T$	J/ (kg K)
Specific heat of molten PTFE C_a	$904.2 + 0.653T$	J/ (kg K)
Absorption coefficient of solid PTFE α_c	0.056	cm^{-1}
Absorption system of molten PTFE α_a	0.22	cm^{-1}
Surface emission coefficient of PTFE ε	~0.92	
Phase transition latent heat of PTFE H_m	5.86×10^4	J/kg
Depolymerization energy per unit mass of PTFE H_p	$1.774 \times 10^6–279.2T$	J/kg
Activation energy E_T	3.473	MJ/kg
Frequency factor A_p	3.1×10^{19}	s^{-1}
Activation temperature B_p	41,769	K

Figure 2.9 shows that as the oscillation of the heat flux decreases, the oscillation of the propellant surface temperature decreases. Although heat is continuously transferred to the PTFE surface, the heat flux has an insignificant influence on the ablated mass flux of the propellant, the ablation rate decreases rapidly, and the position of the ablated end surface remains essentially unchanged. As shown in Fig. 2.10, more heat is continuously transferred to the interior of the PTFE, the thickness of the molten PTFE continues to increase, and the subsequent increase in energy is not effectively used for the ablation of the propellant. At approximately 4 μs, as the heat flux decreases, the thickness of the molten PTFE gradually decreases. At 15 μs, the heat flux essentially decreases to zero, at which point the PPT discharge basically

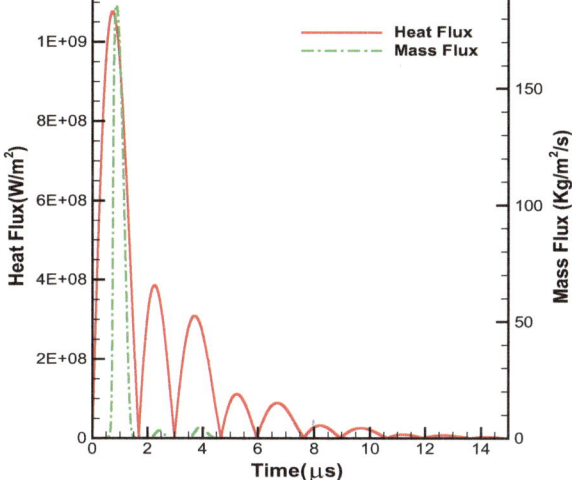

Fig. 2.8 Variations in heat flux and ablated mass flux of the propellant as a function of time

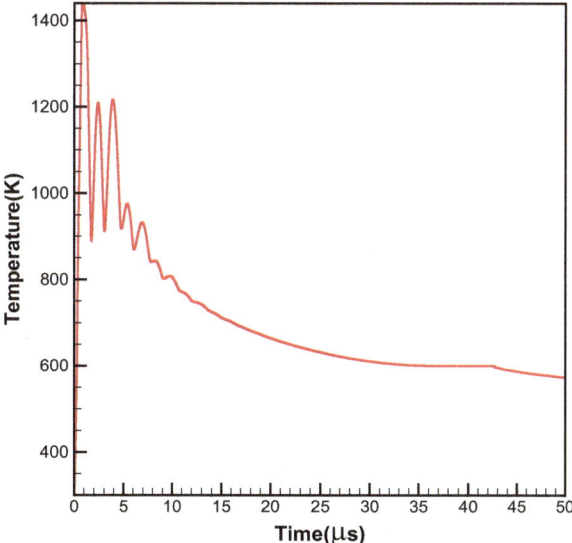

Fig. 2.9 Variation in temperature at the ablated end surface of the propellant as a function of time

ends. However, at this time, the temperature of the ablated end surface of the propellant remains at a high temperature of 700 K, and the PTFE propellant at a depth of approximately 1.6 μm is still in the molten state.

Under the action of external stresses such as thermal stress, molten PTFE will splash in the form of large particles, resulting in the emission of propellant particles. Although the emission of propellant particles consumes a large amount of propellant, the thrust that is generated can be ignored. This is obviously an unfavorable factor that causes the low propellant utilization and system efficiency of the PPT thruster.

Fig. 2.10 Variations in the positions of the ablated end surface and phase interface as a function of time

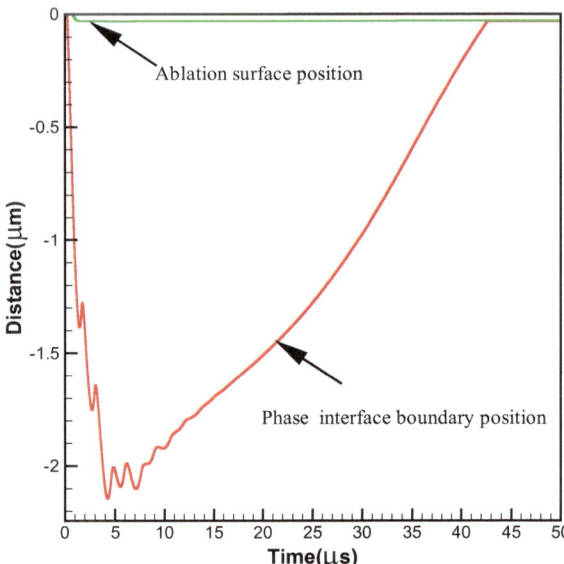

Clearly, the damped oscillation of the discharge current not only wastes system energy and increases the residual energy on the surface of the propellant but also causes ablation lag and increases propellant consumption.

In addition, the reverse discharge of the PPT discharge current not only reduces the acceleration effect of the Lorentz force, causing the products of the ablated and ionized propellant in the early stage to be retained on the end surface of the propellant but also results in low energy utilization efficiency of the PPT. Therefore, reducing current oscillation, changing the heat flow supply method, and decreasing the thickness of the molten propellant after discharge are possible effective ways to reduce propellant loss and improve thruster system efficiency.

2.3.2 Influence of Discharge Characteristics on the Ablation Characteristics of the Propellant

Keeping other calculation parameters unchanged, the equivalent inductance values of the external circuit are set to 4, 24, and 44 nH. Due to the change in the equivalent inductance of the discharge circuit, the PPT discharge exhibits different damping characteristics. As shown in Fig. 2.11, under the same initial energy, as the circuit inductance decreases, the damping of the discharge circuit increases, the peak heat flux increases, the oscillation decreases, and the energy is released rapidly.

As shown in Figs. 2.12 and 2.13, the temperature of the ablated end surface of PTFE rises rapidly under the action of heat flux. The circuit inductance decreases, the ablation relaxation time of PTFE gradually shortens, and the peak temperature of

Fig. 2.11 Variations in heat flux with time under different circuit inductances

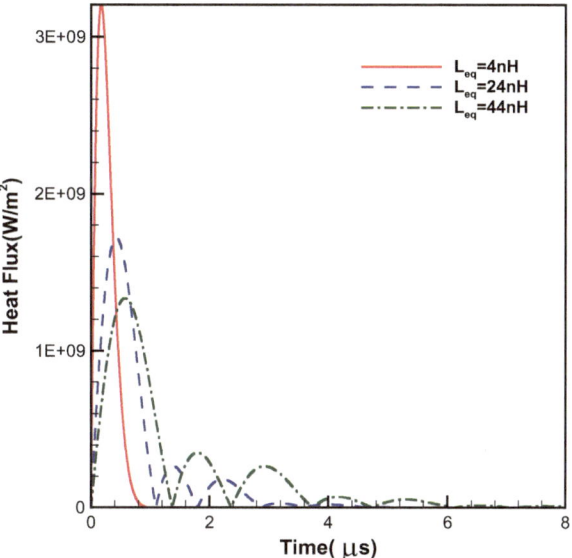

the ablated end surface gradually increases, indicating a larger ablated mass flux of the propellant under lower inductance conditions. Due to the larger peak discharge current under lower inductance conditions, a larger discharge current implies a higher current density and a stronger induced magnetic field, which are favorable factors for promoting the ionization and acceleration of the ablated propellant.

Fig. 2.12 Variations in temperature at the ablated end surface of the propellant with time under different circuit inductances

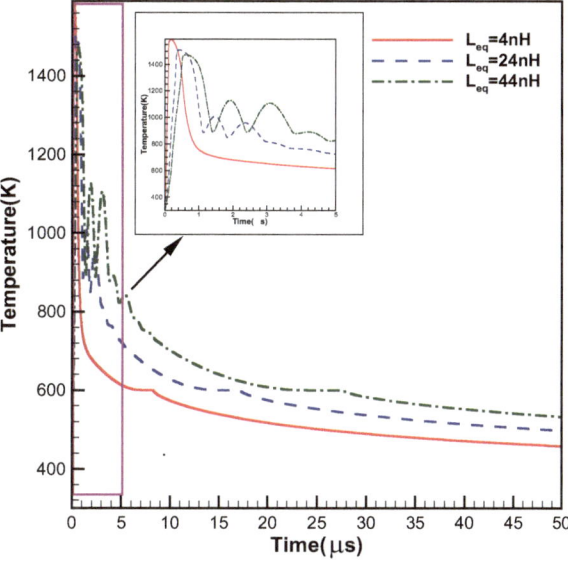

Fig. 2.13 Variations in the
ablated mass flux of the
propellant with time under
different circuit inductances

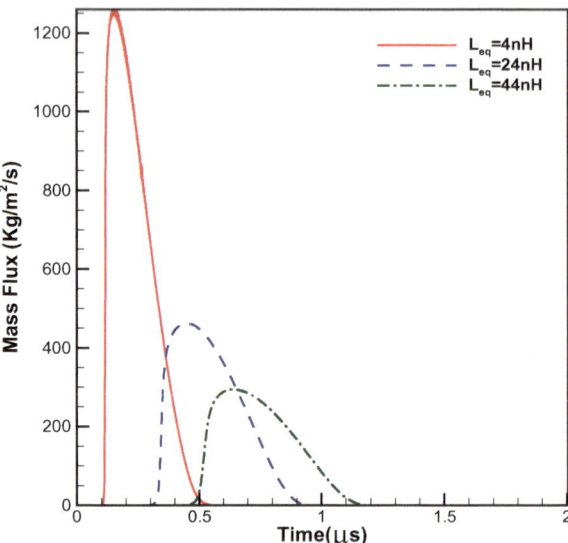

It is observed in Fig. 2.12 that with decreasing circuit inductance, the surface
temperature of the propellant decreases rapidly over time, and the time during which
the temperature at the end surface of the propellant is above 600 K is greatly
reduced. Compared with that under low circuit inductance, the oscillation of heat
flux under high circuit inductance intensifies, and the discharge energy cannot be
quickly released. Therefore, not all the energy is used for the ablation of the propellant. Instead, some of this energy is continuously transferred to the interior of the
propellant, resulting in a continued increase in its internal temperature. On the other
hand, low inductance reduces the oscillation of the discharge waveform and promotes
the concentrated release of the discharge energy. As shown in Fig. 2.14, there is a
large amount of ablated propellant, while the thickness of the molten propellant is
small, helping to reduce the loss of propellant caused by particle emission and thus
achieving the goal of improving the efficiency of the PPT system.

Fig. 2.14 Variations in the ablated end surface and the phase interface position of the propellant with time under different circuit inductances

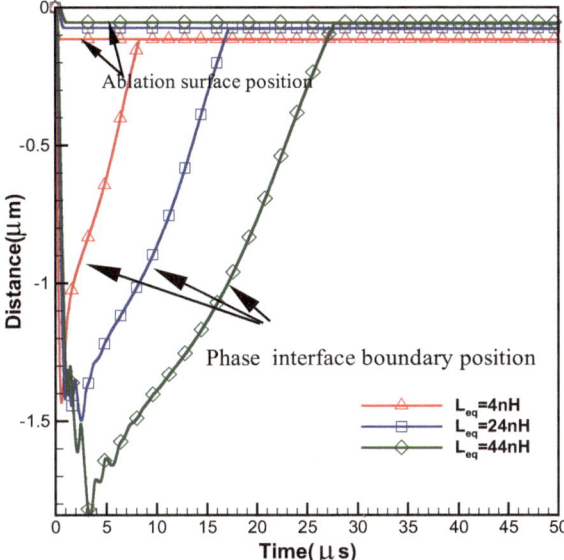

References

1. Song W. Study on the properties of polytetrafluoroethylene composites modified and reinforced with aramid fiber filler. Zhenjiang: Jiangsu University; 2009.
2. Zhang Y, Li H, Zhang H. Fluorine-containing functional materials. Beijing: Chemical Industry Press; 2008.
3. Xu S. Mechanical properties of PTFE/Al energetic reactive materials. Changsha: National University of Defense Technology; 2010.
4. Wentink TJ. High temperature behaviour of Teflon. Avco Everett Research Laboratory; 1959.
5. Pope RB. Simplified computer model for predicting the ablation of Teflon. J Spacecraft. 1975;12(2):83–8.
6. Stechmann DP. Numerical analysis of transient Teflon ablation in pulsed plasma thrusters. Worcester: Worcester Polytechnic Institute; 2007.
7. Zhang DX, Zhang R, He Z, Wu JJ, Zhang F. Numerical investigation on laser ablation characteristics of PTFE in advanced propulsion systems. Appl Mech Mater. 2012;229–231:727–31.
8. Kemp NH. Surface recession velocity of an ablating polymer. AIAA J. 1968;6(9):1790–1.
9. Alexeev YA, Kazeev MN, Kozlov VF. Propellant energy flux measurements in pulsed plasma thruster. In: IEPC-2003-0039; 2003.
10. Aral N. Transient ablation of Teflon in intense radiative and convective environments. AIAA J. 1979;17(6):634–40.
11. Arai N. An analytical investigation of the trasient ablation of Teflon in convective and radiative environments. Institute of Space and Aeronautical Science, University of Tokyo; 1979.

Chapter 3
Numerical Simulation of the Laser Ablation Process of PTFE Propellant

To avoid the ablation lag problem of pulsed plasma thrusters (PPTs), researchers have used laser ablation as a replacement for spark plug ignition and proposed a new laser-sustained PPT (LS-PPT) [1, 2]. Arai [3], Aral et al. [4], Arai and Karashimat [5] studied the transient ablation problem of the PTFE ablative thermal protection layer on a blunt-body spacecraft during reentry under intense radiative and convective environments. They also elucidated the surface recession of the PTFE layer and the internal temperature and time evolution pattern of the molten layer. Finally, they constructed a phase change ablation model of a one-dimensional (1D) bilayer PTFE [6], considering the transmittance of the crystalline and molten layers as well as the bulk absorption of radiant energy by the bilayer structure. Stechmann [7] slightly modified the ablation model proposed by Arai et al. and used a volume fraction method to capture the phase change interface between the crystalline and molten layers and obtained the ablation pattern along the surface of the PTFE propellant in the PPT. Galfetti [8] used microthermocouples to measure the temperature variation pattern of PTFE samples under continuous laser radiation and compared the results with the results calculated using a heat conduction model. In the above studies, the energy applied to the PTFE target was quasi-continuous, with a long time scale and a space scale on the order of centimeters or meters, and the heat flux generally ranged from 10^2 to 10^5 W/cm^2. However, in laser ablation treatment (LAT) [9] and laser-ablated PPT studies, an intense pulsed laser, with a higher laser intensity and a time scale of several nanoseconds or less, is generally used for propellant ablation,. Under extreme ablation conditions such as a very high temperature gradient, large heat flux, and very short time scale, thermal waves propagate at a finite velocity [10], the mechanism of heat conduction cannot be described by classical heat conduction equations [11, 12], and non-Fourier effects in heat conduction and phase transition processes become significant.

The non-Fourier heat conduction equation can be solved analytically only under a few geometric and boundary conditions [13–17]. Furthermore, the intense laser ablation process is very complex, making it difficult to accurately measure the rapid

© The Author(s) 2025
J. Wu et al., *Numerical Simulation of Pulsed Plasma Thruster*,
https://doi.org/10.1007/978-981-97-7958-1_3

evolution process and internal temperature pattern of the target through experimental methods. Therefore, a numerical simulation method is used in this chapter to establish a laser ablation model for a propellant, study heat conduction and phase transition characteristics of the propellant, analyze variations in the temperature field, molten layer thickness, and phase interface recession velocity of the target, and reveal the patterns of influence of factors such as the non-Fourier effect, laser parameters, and target absorption properties on heat conduction and phase transition.

3.1 Propellant Ablation Model Considering the Non-Fourier Effect

Consider a laser beam irradiating the surface of a PTFE target with a laser intensity I_0, as shown in Fig. 3.1. The temperature of the PTFE target starts to increase under the action of the laser beam. The heat conduction and phase transition ablation processes are divided into two distinct stages according to the temperature of the target. In the first stage, the surface temperature is lower than $T_m = 600$ K, and a phase transition has not yet occurred. To obtain the temperature field of the PTFE propellant with a single-crystalline layer structure, a monolayer ablation model is established for numerical simulation. When the surface temperature of the target reaches T_m, the target begins to undergo a phase transition. Thus, the second stage begins. In the second stage, the target is divided into a crystalline layer deep inside and a molten layer near the surface, and a bilayer ablation model similar to Arai's model can be used for simulation. The initial length of the PTFE target is δ, and the locations of the phase interface and the ablation interface are θ and s, respectively. The recession velocities of the phase interface and the ablation interface are v_m and v, respectively. The temperature distributions of the two layers can be obtained by solving the heat conduction equation with the corresponding boundary conditions. The non-Fourier effect of heat conduction is considered in the calculations using the monolayer and bilayer models. In addition, both the reflection of the laser by the target and the bulk absorption of laser energy by the target need to be considered in the model.

In 1867, Maxwell [10] established the thermal wave model. In the 1960s, Cattaneo [11] and Vernotte [12] introduced the thermal relaxation time τ_0 and established the non-Fourier heat conduction law, i.e., the generalized Fourier law, which can be expressed as

$$\tau_0 \frac{\partial q}{\partial t} + q = -\lambda \nabla T \tag{3.1}$$

where $\mathbf{q}$ is the heat flux vector and λ is the thermal conductivity of the target.

From the non-Fourier heat conduction law, we know that

$$q + \tau_0 \frac{\partial q}{\partial t} = -\lambda \nabla T \tag{3.2}$$

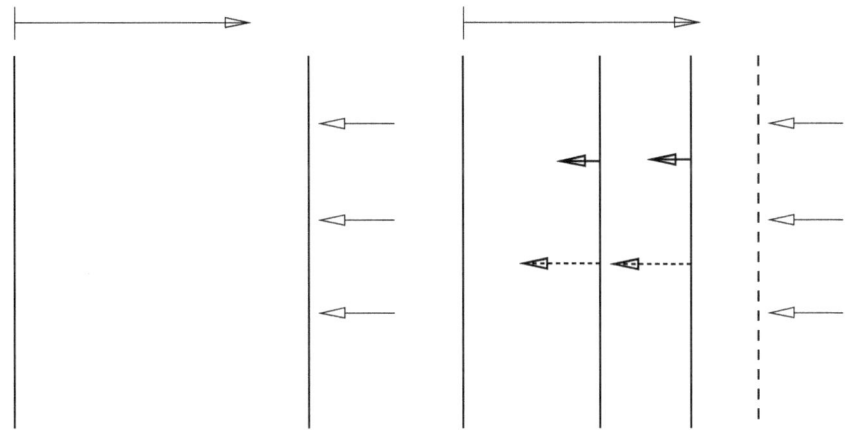

Fig. 3.1 Physical model and coordinate system **a** first ablation stage and **b** second ablation stage

where τ_0 is the thermal relaxation time and T is the temperature.

Considering the conservation of energy in the heat conduction process, the heat conduction equation can be expressed as

$$\frac{\partial(\rho cT)}{\partial t} = -\nabla \cdot q + S \tag{3.3}$$

where S is the energy source term.

Equation (3.3) can be transformed into

$$\frac{\partial^2(\rho cT)}{\partial t^2} = -\frac{\partial}{\partial t}(\nabla \cdot q) + \frac{\partial S}{\partial t} \tag{3.4}$$

Equation (3.2) can be transformed into

$$\nabla \cdot q + \tau_0 \nabla \cdot \left(\frac{\partial q}{\partial t}\right) = -\nabla \cdot (\lambda \nabla T) \tag{3.5}$$

Equation (3.5) can be transformed into

$$\nabla \cdot q = -\nabla \cdot (\lambda \nabla T) + \tau_0 \frac{\partial^2(\rho cT)}{\partial t^2} - \tau_0 \frac{\partial S}{\partial t} \tag{3.6}$$

Substituting Eq. (3.6) into Eq. (3.3) gives the hyperbolic heat conduction equation as

$$\frac{\partial(\rho c T)}{\partial t} + \tau_0 \frac{\partial^2(\rho c T)}{\partial t^2} - \tau_0 \frac{\partial S}{\partial t} = \nabla \cdot (\lambda \nabla T) + S \tag{3.7}$$

For 1D heat conduction problems, Eq. (3.7) can be simplified to

$$\frac{\partial(\rho c T)}{\partial t} + \tau_0 \frac{\partial^2(\rho c T)}{\partial t^2} - \tau_0 \frac{\partial S}{\partial t} = \frac{\partial}{\partial x}\left(\lambda \frac{\partial T}{\partial x}\right) + S \tag{3.8}$$

When the thermal relaxation time is set to 0, that is, $\tau_0 = 0$, Eq. (3.8) reduces to the classical heat conduction equation as

$$\frac{\partial(\rho c T)}{\partial t} = \frac{\partial}{\partial x}\left(\lambda \frac{\partial T}{\partial x}\right) + S \tag{3.9}$$

The thermal diffusivity is

$$a = \frac{\lambda}{\rho c} \tag{3.10}$$

In addition, the thermal wave propagation velocity can be defined as

$$C_h = \sqrt{\frac{a}{\tau_0}} \tag{3.11}$$

The thermal relaxation time τ_0 can be used to characterize the average time effect of microscopic relaxation processes (such as electron–electron, electron–photon, and photon–photon scattering) and is a macroscopic parameter of a series of microscopic processes. How to relate the macroscopic relaxation time to microscopic physical processes is still a problem that requires in-depth study in heat conduction processes [17]. Chester's work demonstrated that the thermal wave propagation velocity C_h is approximately 55.7% of the sound speed C_s [18], that is,

$$C_h = 55.7\% C_s \tag{3.12}$$

By combining Eqs. (3.11) and (3.12), the thermal relaxation time τ_0 can be estimated as

$$\tau_0 = \frac{3a}{C_s^2} \tag{3.13}$$

Assuming that the characteristic length of the target is δ_0, the characteristic times of thermal wave propagation and diffusion are, respectively, defined as

$$\begin{cases} t_W = \frac{\delta_0}{C_h} \\ t_D = \frac{\delta_0^2}{a} \end{cases} \tag{3.14}$$

To characterize the non-Fourier effect, a dimensionless number can be defined as

$$N_{DW} = \frac{t_D}{t_W} = \frac{\delta_0}{\sqrt{a\tau_0}} \tag{3.15}$$

Obviously, for the Fourier heat conduction problem, the dimensionless number N_{DW} is infinite. However, for a finite thermal wave propagation velocity, the dimensionless number N_{DW} is a positive finite value. When $N_{DW} \to 1$, the characteristic times t_W and t_D are comparable, and the non-Fourier effect becomes significant. For a polymer propellant, the thermal diffusivity a is approximately 10^{-7} m²/s, C_s is approximately 10^3 m/s, and C_h is approximately 10^2–10^3 m/s; hence, τ_0 is between 10^{-11} and 10^{-13} s. Therefore, if the characteristic length δ_0 is on the order of nm and the characteristic time t_W is on the order of ps, the non-Fourier effect is significant and should be considered.

3.1.1 The First Ablation Stage (Non-Fourier Heat Conduction Without Considering the Phase Transition)

In the first ablation stage, the temperature of the target is lower than the phase transition temperature T_m, indicating that the target has a crystalline monolayer structure, as shown in Fig. 3.1a. The temperature of the target can be described by the heat conduction equation (Eq. 3.8). The heat source term can be expressed as

$$S = Q_\tau = (1 - R)\alpha I_0 \exp[-\alpha(\delta - x)] \tag{3.16}$$

where Q_τ is the laser deposition power density, I_0 is the initial laser intensity on the surface of the target, and R and α represent the reflection and absorption coefficients, respectively.

The heat flux boundary condition is applied to the surface of the target, i.e.,

$$-\lambda \frac{\partial T}{\partial x}\bigg|_{x=\delta} = -(1 - R)I_0 \tag{3.17}$$

The adiabatic boundary condition is used on the back surface of the target, that is,

$$-\lambda \frac{\partial T}{\partial x}\bigg|_{x=0} = 0 \tag{3.18}$$

The initial conditions for the first ablation stage are

$$\begin{cases} T(0 \leq x \leq \delta)|_{t=0} = T_0 \\ \frac{\partial T}{\partial t}\big|_{t=0} = 0 \end{cases} \tag{3.19}$$

3.1.2 The Second Ablation Stage (Non-Fourier Heat Conduction Considering Phase Transition)

When the surface temperature is higher than the phase transition temperature T_m, the target changes from a crystalline state to a molten state. The phase interface begins to recess from the surface toward the interior of the target, and an ablated surface starts to form on the surface of the target. As a result, a bilayer structure in the molten and crystalline states forms near the surface of the target, as shown in Fig. 3.1b. The temperature distributions of both the crystalline layer and the molten layer can be described by the heat conduction equation (Eq. 3.8), but with different boundary conditions.

The heat source term is given by

$$S = \begin{cases} Q_\tau & 0 \leq x \leq \theta^- \\ Q_\tau - Q_p & \theta^+ \leq x \leq s \end{cases} \tag{3.20}$$

where the heat source generated by the laser transmitting through the target is

$$Q_\tau = (1 - R)\alpha I_0 \exp[-\alpha(s - x)] \tag{3.21}$$

In addition, the energy released by a unit volume of polymer during the depolymerization process is

$$Q_p = A_P H_p \rho \exp(-B_p/T) \tag{3.22}$$

where A_p is a preexponential factor, the activation temperature of the depolymerization reaction is $B_P = E_A/\hat{R}$, in which E_A is the activation energy, and $\hat{R}$ is the universal gas constant.

In the calculation process, the initial position of the ablated surface of the target is set to $s = \delta$.

The heat flux boundary condition is applied to the surface of the target, i.e.,

$$-\lambda \frac{\partial T}{\partial x}\bigg|_{x=s} = -\beta I_0 \tag{3.23}$$

The adiabatic boundary condition is used on the back surface of the target, that is,

$$-\lambda \frac{\partial T}{\partial x}\bigg|_{x=0} = 0 \tag{3.24}$$

According to the conservation of energy at the phase interface, we obtain

$$q_{\theta^-} = q_{\theta^+} + \beta I(\theta) + \rho_m H_m \frac{d\theta}{dt} \tag{3.25}$$

That is,

$$\lambda_s \frac{\partial T}{\partial x}\bigg|_{\theta-} - \lambda_g \frac{\partial T}{\partial x}\bigg|_{\theta+} = \Phi - \tau_0 \frac{\partial \Phi}{\partial t} \tag{3.26}$$

where

$$\Phi = q_{\theta-} - q_{\theta+} = \beta I(\theta) + \rho_m H_m \frac{d\theta}{dt} \tag{3.27}$$

The motion velocity of the phase interface can be expressed as

$$v_m(t > t_m) = \frac{d\theta}{dt} = \frac{1}{\rho_m H_m}[\Phi - \beta I(\theta)] \tag{3.28}$$

where t_m is the time required for the surface temperature of the target to increase from the initial temperature T_0 to the phase transition temperature T_m. In Eqs. (3.27) and (3.28), the laser intensity can be expressed as $I(\theta) = I_0 \exp[-\alpha(s - \theta)]$.

For Fourier heat conduction, by applying the conservation of energy at the phase interface, we obtain

$$\lambda_s \frac{\partial T}{\partial x}\bigg|_{\theta-} = \lambda_g \frac{\partial T}{\partial x}\bigg|_{\theta+} + \beta I(\theta) + \rho_m H_m \frac{d\theta}{dt} \tag{3.29}$$

The motion velocity of the phase interface is

$$v_m(t > t_m) = \frac{d\theta}{dt} = \frac{1}{\rho_m H_m}\left[\lambda_s \frac{\partial T}{\partial x}\bigg|_{\theta-} - \lambda_g \frac{\partial T}{\partial x}\bigg|_{\theta+} - \beta I(\theta)\right] \tag{3.30}$$

The temperature of the phase interface satisfies

$$T(x = \theta^+, t > t_m) = T(x = \theta^-, t > t_m) = T_m \tag{3.31}$$

Considering that the temperature is continuous at the end of the first ablation stage and the start of the second ablation stage, the initial condition for the temperature in the second ablation stage is

$$\begin{cases} T(0 \leq x \leq s, t = t_m^+) = T(0 \leq x \leq s, t = t_m^-) \\ \frac{\partial T}{\partial t}(t = t_m^+) = \frac{\partial T}{\partial t}(t = t_m^-) \end{cases} \tag{3.32}$$

The initial velocities at the phase interface and the ablation interface are

$$\begin{aligned} v_m(t = t_m) &= \frac{d\theta}{dt} = -\frac{1}{\rho_m H_m}\left(\lambda_s \frac{\partial T}{\partial x}\bigg|_{x=s} + \beta I_0\right) \\ v(t = t_m) &= \frac{ds}{dt} = 0 \end{aligned} \tag{3.33}$$

The instantaneous position of the phase interface is

$$\theta(t \geq t_m) = \delta + \int_{t_m}^{t} v_m dt \tag{3.34}$$

As the temperature of the target continues to rise, the ablated mass rate at the surface of the target is

$$\dot{m}(t \geq t_m) = -\rho_0 \frac{ds}{dt} = \int_{\theta}^{s} A_p \rho \exp\left(-\frac{B_p}{T}\right) dx \tag{3.35}$$

where ρ_0 is the reference temperature of the target.

The recession velocity and position of the ablated surface of the target are

$$v(t \geq t_m) = \frac{ds}{dt} = -\frac{\dot{m}}{\rho_0} \tag{3.36}$$

$$s(t \geq t_m) = \delta + \int_{t_m}^{t} v dt \tag{3.37}$$

3.2 Numerical Calculation Method

3.2.1 Coordinate Transformation

Since the positions of the ablation interface and the phase interface both change over time, the coordinate system needs to be transformed to facilitate the calculation of this type of motion interface problem. The original coordinate system (x, t) of the crystalline and molten layers is transformed into (ξ, τ) and (η, τ), respectively. $\xi = 0$ or $x = 0$ represents the back surface of the crystalline layer; $\xi = 1$, $\eta = 0$, or $x = \theta$ represents the phase interface between the crystalline and molten layers; and $\eta = 1$ or $x = s$ represents the outer surface of the molten layer, also known as the ablated surface. The crystalline and molten layers are each meshed using a geometric progression technique, with the geometric ratios denotes as q_1 and q_2, respectively. These ratios are set in the calculation to 1.005 and 1.0, respectively.

Applying the chain rule for each layer of the target yields

$$\begin{cases} \frac{\partial}{\partial x} = \frac{1}{\theta} \frac{\partial}{\partial \xi} \\ \frac{\partial}{\partial t} = \frac{\partial}{\partial \tau} - \frac{\xi \dot{\theta}}{\theta} \frac{\partial}{\partial \xi} \end{cases} \tag{3.38}$$

and

$$
\begin{cases}
\dfrac{\partial}{\partial x} = \dfrac{1}{s-\theta}\dfrac{\partial}{\partial \eta} \\[2mm]
\dfrac{\partial}{\partial t} = \dfrac{\partial}{\partial \tau} - \dfrac{(1-\eta)\dot\theta + \eta\dot s}{s-\theta}\dfrac{\partial}{\partial \eta}
\end{cases}
\tag{3.39}
$$

The heat conduction equations for the crystalline and molten layers can be transformed to the following forms:

$$
\frac{\partial(\rho c T)}{\partial \tau} + \tau_0 \frac{\partial^2(\rho c T)}{\partial \tau^2} - \tau_0 \frac{\partial S}{\partial \tau} = \frac{1}{\theta^2}\frac{\partial}{\partial \xi}\left(\lambda \frac{\partial T}{\partial \xi}\right)
$$
$$
- \tau_0 \left(\frac{\xi\dot\theta}{\theta}\right)^2 \frac{\partial^2(\rho c T)}{\partial \xi^2} + 2\tau_0 \frac{\xi\dot\theta}{\theta}\frac{\partial^2(\rho c T)}{\partial \tau \partial \xi}
$$
$$
+ \left[\frac{\xi\dot\theta}{\theta} - \tau_0\xi\left(\frac{\dot\theta}{\theta}\right)^2\right]\frac{\partial(\rho c T)}{\partial \xi} - \tau_0 \frac{\xi\dot\theta}{\theta}\frac{\partial S}{\partial \xi} + S
\tag{3.40}
$$

and

$$
\frac{\partial(\rho c T)}{\partial \tau} + \tau_0 \frac{\partial^2(\rho c T)}{\partial \tau^2} - \tau_0 \frac{\partial S}{\partial \tau} = \left(\frac{1}{s-\theta}\right)^2 \frac{\partial}{\partial \eta}\left(\lambda \frac{\partial T}{\partial \eta}\right)
$$
$$
- \tau_0 \left[\frac{(1-\eta)\dot\theta + \eta\dot s}{s-\theta}\right]^2 \frac{\partial^2(\rho c T)}{\partial \eta^2}
$$
$$
+ 2\tau_0 \frac{(1-\eta)\dot\theta + \eta\dot s}{s-\theta}\frac{\partial^2(\rho c T)}{\partial \tau \partial \eta}
$$
$$
+ \frac{(1-\eta)\dot\theta + \eta\dot s}{s-\theta}\left(1 - \tau_0\frac{\dot s - \dot\theta}{s-\theta}\right)\frac{\partial(\rho c T)}{\partial \eta}
$$
$$
- \tau_0 \frac{(1-\eta)\dot\theta + \eta\dot s}{s-\theta}\frac{\partial S}{\partial \eta} + S
\tag{3.41}
$$

Setting the relaxation time $\tau_0 = 0$, Eqs. (3.40) and (3.41) can be expressed as

$$
\frac{\partial(\rho c T)}{\partial \tau} = \frac{1}{\theta^2}\frac{\partial}{\partial \xi}\left(\lambda \frac{\partial T}{\partial \xi}\right) + \frac{\xi\dot\theta}{\theta}\frac{\partial(\rho c T)}{\partial \xi} + S
\tag{3.42}
$$

and

$$
\frac{\partial(\rho c T)}{\partial \tau} = \left(\frac{1}{s-\theta}\right)^2 \frac{\partial}{\partial \eta}\left(\lambda \frac{\partial T}{\partial \eta}\right) + \frac{(1-\eta)\dot\theta + \eta\dot s}{s-\theta}\frac{\partial(\rho c T)}{\partial \eta} + S
\tag{3.43}
$$

3.2.2   Equation Discretization

The governing equation and the corresponding boundary conditions are discretized by the finite volume method. The resulting fully implicit discrete equation can be expressed as

$$a_P T_P^{n+1} - a_E T_E^{n+1} - a_W T_W^{n+1} = a_P^0 T_P^n + S_u \tag{3.44}$$

where $a_p = a + a_E + a_W - S_p$. Other coefficients, such as a_W and a_E, in the equation can be obtained in a similar manner.

The coefficients of the discrete equation corresponding to the interior points ($0 < \xi < 1$ or $i = 2, 3, \ldots, n_1 - 1$) of the crystalline layer in the second stage are listed as follows:

$$a_E = \rho c \tau_0 \frac{C_h^2 - (\xi \dot{\theta})^2}{\theta^2 \delta \xi_{PE}}$$

$$\qquad + \rho c \omega \left[\left(1 + \frac{2\tau_0}{\Delta\tau} \right) \frac{\xi \dot{\theta}}{\theta} - \tau_0 \xi \left(\frac{\dot{\theta}}{\theta} \right)^2 \right]$$

$$a_W = \rho c \tau_0 \frac{C_h^2 - (\xi \dot{\theta})^2}{\theta^2 \delta \xi_{WP}}$$

$$\qquad - \rho c (1 - \omega) \left[\left(1 + \frac{2\tau_0}{\Delta\tau} \right) \frac{\xi \dot{\theta}}{\theta} - \tau_0 \xi \left(\frac{\dot{\theta}}{\theta} \right)^2 \right]$$

$$a_p^0 = \rho c \left(1 + \frac{\tau_0}{\Delta\tau} \right) \frac{\Delta \xi_P}{\Delta\tau}$$

$$S_P = 0$$

$$S_u = \rho c \frac{\tau_0}{\Delta\tau} \Delta\xi_P \left(\frac{\partial T}{\partial \tau} \right)^n$$

$$\qquad - 2\rho c \frac{\tau_0}{\Delta\tau} \frac{\xi \dot{\theta}}{\theta} \left[\omega T_E^n + (1 - 2\omega) T_P^n - (1 - \omega) T_W^n \right]$$

$$\qquad + \left[\left(1 + \frac{\tau_0}{\Delta\tau} \right) \Delta\xi_P \Delta\tau - (1 - 2\omega) \tau_0 \Delta\tau \frac{\xi \dot{\theta}}{\theta} \right] \left(\frac{\partial S_P}{\partial \tau} \right)^n$$

$$\qquad + (1 - \omega) \tau_0 \Delta\tau \frac{\xi \dot{\theta}}{\theta} \left(\frac{\partial S_W}{\partial \tau} \right)^n - \omega \tau_0 \Delta\tau \frac{\xi \dot{\theta}}{\theta} \left(\frac{\partial S_E}{\partial \tau} \right)^n$$

$$\qquad + (1 - \omega) \tau_0 \frac{\xi \dot{\theta}}{\theta} S_W^n$$

$$\qquad + \left[\Delta\xi_P - (1 - 2\omega) \tau_0 \frac{\xi \dot{\theta}}{\theta} \right] S_P^n - \omega \tau_0 \frac{\xi \dot{\theta}}{\theta} S_E^n \tag{3.45}$$

where $\Delta\tau$ is the time step, $\delta\xi_{Wp}$, $\delta\xi_{pE}$, and $\Delta\xi_p$ are space steps, S_E, S_W, and S_p denote the average source terms in the right, local, and left cells, respectively, and the geometrical parameter is $\omega = 1/\left(1 + \frac{1}{q_1}\right)$.

3.2.3 Stability Analysis

The fully implicit discrete equation (Eq. 3.44), in the first stage is unconditionally stable. However, due to the existence of phase transition and nonlinear boundary conditions, the discrete equation in the second stage is not unconditionally stable. It can be proven that to ensure the stability of Eq. (3.44) in the second stage of ablation, the following condition should be satisfied:

$$a_P > 0, \quad a_E > 0, \quad a_W > 0 \tag{3.46}$$

For the second stage of ablation, the discrete equation corresponding to the crystalline layer should satisfy

$$\frac{\Delta\tau\left|C_h^2 - \dot{\theta}^2\right|}{\min(\Delta\xi_i)} < \left|\dot{\theta}\theta\right| \tag{3.47}$$

For the second stage of ablation, the discrete equation corresponding to the molten layer should satisfy

$$\frac{\Delta\tau\left|C_h^2 - \dot{\theta}^2\right|}{\min(\Delta\eta_j)} < \left|(s - \theta)\dot{\theta}\right| \tag{3.48}$$

Based on the boundary conditions of the crystalline layer and the molten layer in the two ablation stages, the discrete equation (Eq. 3.44), can be solved using the tridiagonal matrix algorithm (TDMA).

3.2.4 Validation with Numerical Examples

To validate the reliability of the calculation method, numerical calculations were conducted for two classical examples of non-Fourier heat conduction, and the results were compared with known analytical solutions. The custom-made PTFE target was used in the calculations. Most of its thermophysical parameters were obtained from measurement in the present book and some are from the work of Arai [3], Aral et al. [4], Arai and Karashimat [5], as shown in Table 3.1. In the table, the subscript 1 represents the crystalline propellant, and the subscript 2 represents the molten propellant.

Table 3.1 Thermochemical and optical properties of the custom-made PTFE target

Description	Property	Value
Phase transition temperature	r_m/K	600
Thermal conductivity	1/(W/(m K))	1.0
Thermal conductivity	2/(W/(m K))	0.8
Thermal conductivity of solid phase at T_m	s/(W/(m K))	0.36
Thermal conductivity of gel phase at T_m	g/(W/(m K))	0.24
Mass density	y_g/(kg/m^3)	1700
Mass density	y_g/(kg/m^3)	500
Reference density	y_r/(kg/m^3)	1933
Mean density at T_m	y_r/(kg/m^3)	1957
Thermal capacity	c_r/(J/(kg K))	500
Thermal capacity	c_g/(J/(kg K))	800
Latent heat of solid to gel phase transition	H_m/(J/kg)	5.86×10^4
Depolymerization energy per unit mass	H_p/(J/kg)	1.77×10^6–$279 \times r$
Preexponential frequency factor	A_p/s^{-1}	3.1×10^{19}
Depolymerization activation temperature	B_p/K	3.7×10^4
Reflectivity on exposed surface	R/%	0.1
Thermal relaxation time	d_r/ps	10–100
Absorption coefficient	d_q/cm^{-1}	0.22

1. Non-Fourier heat conduction under isothermal boundary conditions

Consider non-Fourier heat conduction for a target with a semi-infinite flat plate. The surface temperature rises instantaneously from the initial temperature T_0 to T_m and then remains at this temperature, i.e.,

$$T(x, t)|_{x=\delta} = \begin{cases} T_m & t > 0 \\ T_0 & t \leq 0 \end{cases} \tag{3.49}$$

After a temperature gradient is applied on the surface of the target, there is a certain delay in the establishment of heat flow within the target, resulting in a non-Fourier effect or thermal relaxation behavior in heat conduction. Under the conditions that the thermophysical parameters of the target are constant and the internal heat sources are not considered, the analytical solution for the temperature field of the target can be expressed as [19]

$$T(x, t) = \begin{cases} T_0 & t \leq (\delta - x)/C_h \\ T_0 + T_m e^{-C_h(\delta-x)/(2a)} + \Delta T & t > (\delta - x)/C_h \end{cases} \tag{3.50}$$

where

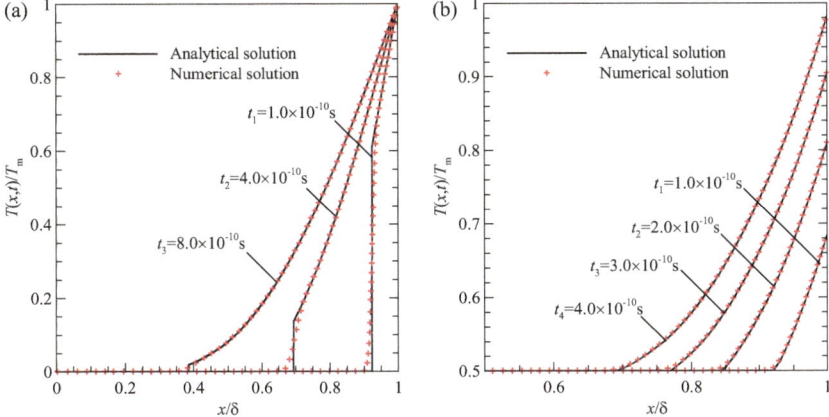

(a) Under the isothermal boundary condition and (b) under the constant flux boundary condition

Fig. 3.2 Analytical and numerical solutions for the target temperature. **a** Under the isothermal boundary condition and **b** under the constant flux boundary condition

$$\Delta T = \frac{C_h(\delta - x)T_m}{2a} \int_{\frac{x}{C_h}}^{t} \frac{J_1\left(\frac{C_h^2}{2a}\sqrt{\tau^2 - (\delta - x)^2/C_h^2}\right)}{\sqrt{\tau^2 - (\delta - x)^2/C_h^2}} e^{-\frac{C_h^2\tau}{2a}} d\tau \tag{3.51}$$

Here, the first-order Bessel function of the first kind is $J_1(z) = \frac{1}{\pi}\int_0^{\pi} \exp(z\cos\theta)\cos\theta d\theta$.

Figure 3.2a shows the numerical solutions and the corresponding analytical solutions, and good agreement is observed between the two.

2. Non-Fourier heat conduction under constant flux boundary conditions

Suppose that the phase transition is not considered and the heat flux boundary condition is applied on the surface of the target, i.e.,

$$-\lambda \frac{\partial T}{\partial x}\bigg|_{x=\delta} = q|_{x=\delta} + \tau_0 \frac{\partial q}{\partial \tau}\bigg|_{x=\delta} = -\beta I_0 \tag{3.52}$$

When a heat flux is applied to the target surface, heat flow inside the target does not occur immediately; instead, this change occurs gradually within a certain thermal relaxation time. The analytical solution for the temperature of the target can be obtained as [16]

$$T(x, t) = \frac{\beta I_0 C_h}{\lambda} \int_{\frac{x}{C_h}}^{t} J_0\left(\frac{C_h^2}{2a}\sqrt{\tau^2 - (\delta - x)^2/C_h^2}\right) \exp\left(-\frac{\tau}{2\tau_0}\right) d\tau$$

$$+ \frac{2\beta I_0 \alpha a}{\lambda\sqrt{1 + 4a\alpha^2}} \exp[-\alpha(\delta - x)] \int_{\frac{x}{C_h}}^{t} \exp\left(-\frac{t}{2\tau_0}\right)$$

$$\sinh\left(\sqrt{1 + 4a\alpha^2 \tau_0}\frac{t - \tau}{2\tau_0}\right)$$

$$\times J_0\left(\frac{C_h^2}{2a}\sqrt{\tau^2 - (\delta - x)^2/C_h^2}\right) d\tau$$

$$+ \frac{2\beta I_0 \alpha a}{\lambda\sqrt{1 + 4a\alpha^2}} \exp[-\alpha(\delta - x)]$$

$$\times \int_0^t \exp\left(-\frac{\tau}{2\tau_0}\right)\sinh\left(\sqrt{1 + 4a\alpha^2\tau_0}\frac{\tau}{2\tau_0}\right)d\tau, \qquad (3.53)$$

where J_0 (z) is the zero-order Bessel function of the first kind, that is, $J_0(z) = \int_0^\pi \cosh(z\cos\theta)d\theta/\pi$.

As shown in Fig. 3.2b, the numerical solutions and analytical solutions match well. The two numerical examples presented in Fig. 3.2 demonstrate that the non-Fourier heat conduction model for PTFE ablation by intense laser is reliable, and the relevant theoretical and calculation results are credible.

3.3 Numerical Simulation Results and Analysis

3.3.1 Temperature Evolution of the Target Under Non-Fourier and Fourier Heat Conduction Conditions

Figure 3.3 shows the temperature evolution at different target depths ($\delta_1 = 0$, $\delta_2 = 20$ nm, and $\delta_3 = 40$ nm). The calculations are conducted with a time step of 2.5 ns, an incident laser intensity of $I_0 = 1.0 \times 10^{10}$ W/m^2, and thermophysical parameters shown in Table 3.1. The surface temperature of the target ($\delta_1 = 0$) increases rapidly with time, and a phase transition occurs at 0.7 ns. The temperature inside the target exhibits different patterns. As shown in Fig. 3.3b, the thermal relaxation time has a great impact on the evolution of the internal temperature. The longer the thermal relaxation time is, the later the onset of temperature rise inside the target. This is because a larger τ_0 leads to a delayed arrival of the thermal wave. For Fourier heat conduction ($\tau_0 = 0$), both internal and surface temperatures start to rise at time 0 without delay, as the thermal wave velocity is infinite when $\tau_0 = C$. In addition, the onset of phase transition calculated by the non-Fourier heat conduction model is slightly later than that calculated by the Fourier heat conduction model. Figure 3.3

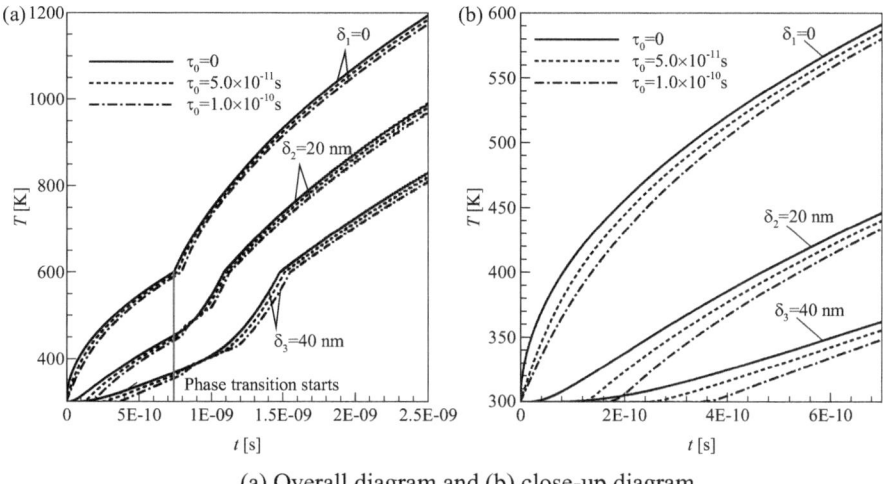

(a) Overall diagram and (b) close-up diagram

Fig. 3.3 Target temperature evolution **a** overall diagram and **b** close-up diagram

indicates that as the thermal relaxation time increases, the influence of the non-Fourier effect on the temperature evolution becomes more significant.

Figure 3.4 shows the temperature distribution of the target at 1.5 and 2.5 ns. Compared to the second calculation example using the constant flux boundary condition, the internal heat source is considered in the calculation. Therefore, the internal temperature and the surface temperature of the target can both be considered to increase at the same time. As shown in Fig. 3.4, the temperature distribution of the target varies slightly under different thermal relaxation time conditions. Both the surface temperature and internal temperature decrease as the thermal relaxation time increases.

Figure 3.5 shows the velocity variations at the phase interface and ablation interface over time. When a phase transition starts to occur on the target surface, the velocity v_m at the phase interface rises rapidly. The maximum velocity is in the range of 60–80 m/s. The larger τ_0 is, the larger the maximum value of v_m. Subsequently, the velocity at the phase interface gradually decreases and then stabilizes. At approximately 1.8 ns, as the temperature of the ablated surface rise above 1000 K, the recession velocity of the ablated surface begins to increase. As τ_0 increases, the delay in the onset time of the phase transition and the recession time of the ablated surface increase.

Figure 3.6 presents the variation in the ablated mass flux. When the surface temperature is not too high, the ablation of the target is not noticeable. The ablated mass flux begins to increase at approximately 1.8 ns. A larger τ_0 results in a smaller ablated mass flux.

Fig. 3.4 Temperature distributions at $t = 1.5$ and 2.5 ns

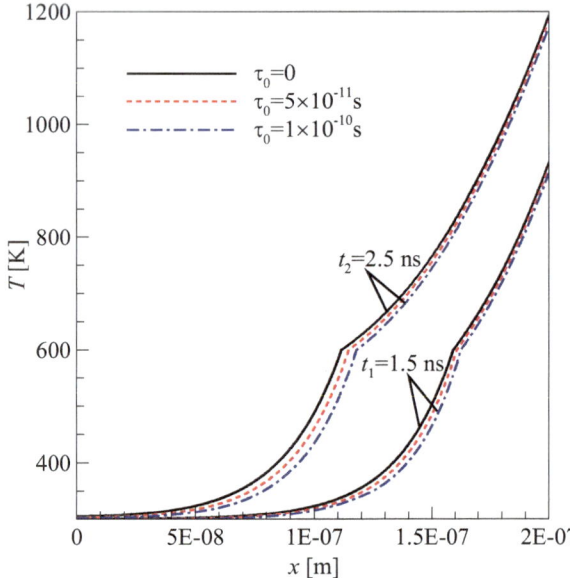

Fig. 3.5 Recession velocities of the phase and ablation interfaces

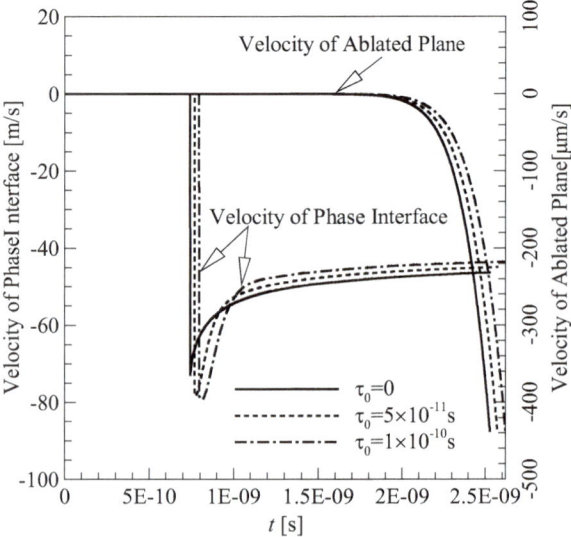

3.3.2 *Influence of the Laser Intensity on Temperature Evolution*

As shown in Fig. 3.7, laser intensities of 1.0×10^{10}, 8.0×10^{9}, and 6.0×10^{9} W/m^2 are considered. It is evident that the laser intensity has a great impact on the target

Fig. 3.6 Ablated mass flux

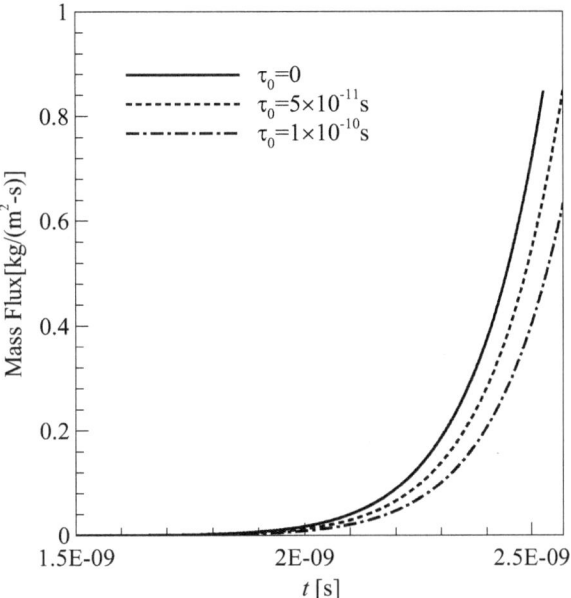

temperature, manifested by a faster increase in both surface and internal temperatures as the laser intensity increases. At the same time and depth, higher laser intensities result in higher target temperatures. As shown in Fig. 3.7b, under different laser intensities, the target temperature at a depth of $\delta_2 = 20$ nm consistently starts to increase at time $\frac{\delta_2}{C_h} = 1.8 \times 10^{-10}$ s. This is because the thermal wave propagation velocity of the target is determined by the thermal diffusivity and thermal relaxation time and is hence not affected by the laser intensity.

3.3.3 Influence of the Absorption Coefficient on Temperature Evolution

A Fig. 3.8 shows the variation in surface temperature over time for different absorption coefficients. Since the absorption coefficient α characterizes the laser absorption ability of the target, a larger absorption coefficient indicates a stronger absorption of laser energy by the target. The laser energy absorption by the target can be considered as surface absorption when the absorption coefficient is large and bulk absorption otherwise. The amount of laser energy deposited at a certain target depth depends on both the local absorption coefficient and the energy transmitted from the surface direction of the target.

If the absorption coefficient is sufficiently large, more absorption occurs on the surface of the target; as a result, less energy is deposited in the depth of the target with

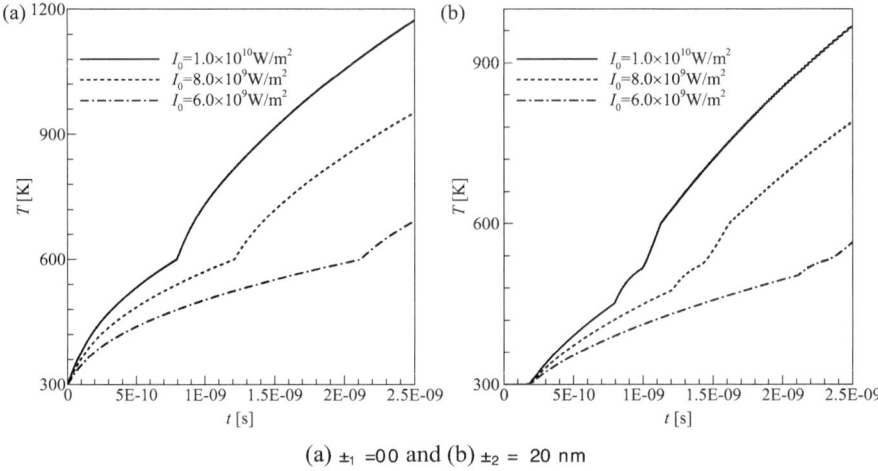

(a) $\pm_1 = 00$ and (b) $\pm_2 = 20$ nm

Fig. 3.7 Temperature evolution under different laser intensities

Fig. 3.8 Evolution of temperature under different absorption coefficients

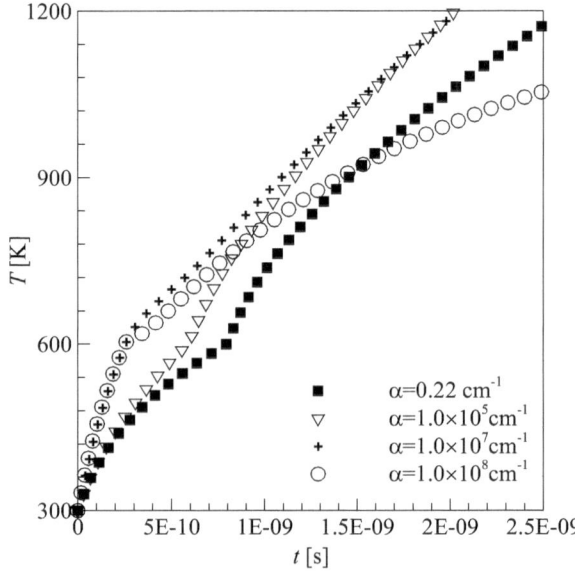

increasing absorption coefficient. When the absorption coefficient is not very large, more absorption occurs inside the target; then, more energy is deposited deep in the target as the absorption coefficient increases. As shown in Fig. 3.8, before the phase transition occurs, the larger the absorption coefficient is, the faster the increase in the surface temperature of the target. The rate of increase in the surface temperature of the target does not change much for relatively large and small absorption coefficients. As shown in Fig. 3.9, the time required for the target to achieve the phase transition

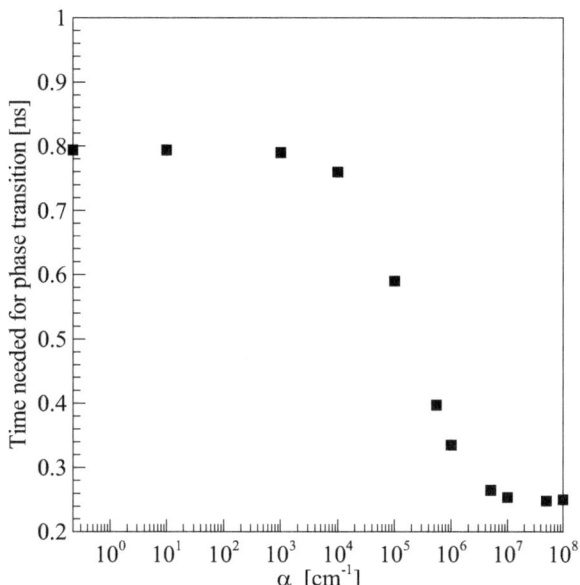

Fig. 3.9 Time required to achieve phase transition under different absorption coefficients

decreases with increasing absorption coefficient, and minimal change is observed for very small and very large absorption coefficients. However, when the absorption coefficient is between 1.0×10^4 and 1.0×10^6 cm^{-1}, the time required for the target to reach the phase transition changes rapidly. After the phase transition, the influence of the absorption coefficient on the temperature of the target becomes complex. In the case of surface absorption, a larger the absorption coefficient results in less energy reaching the depth of the target. In the case of bulk absorption, a larger absorption coefficient results in more energy being deposited locally, leading to higher local temperatures. Therefore, the temperature of the target increases faster when $\alpha = 1.0 \times 10^5$ cm^{-1} than when $\alpha = 0.22$ cm^{-1}.

3.3.4 Influence of the Mesh and Time Steps on the Calculation Results

Figure 3.10 shows the influence of the mesh and time steps on the calculation results. Let n_1 and n_2 represent the numbers of cells in the solid and liquid layers, respectively. In the calculations, the numbers of cells for the three cases are as follows: case 1, $n_1 = 50$, $n_2 = 10$; case 2, $n_1 = 200$, $n_2 = 50$; and case 3, $n_1 = 1000$, $n_2 = 200$.

A Fig. 3.10a presents the variation in temperature over time at different depths. The temperature distribution varies slightly among the three mesh cases, and the calculation results converge to stable values with an increase in the number of cells.

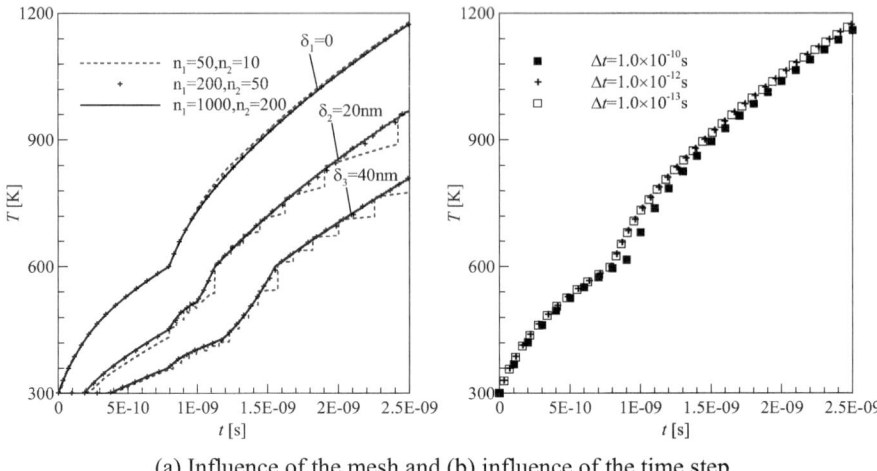

(a) Influence of the mesh and (b) influence of the time step

Fig. 3.10 Temperature under different mesh and time step settings **a** Influence of the mesh and **b** influence of the time step

A Fig. 3.10b illustrates the influence of different time steps ($dt = 0.1$, 1100 ps) on temperature. It is observed that the calculation results of the surface temperature are basically independent of the time step.

References

1. Arai N. Transient ablation of Teflon in intense radiative and convective environments. AIAA J. 1979;17(6):634–40.
2. Zhang D. Theoretical, experimental, and simulation study of laser-sustained pulsed plasma thrusters. Changsha: Graduate School of National University of Defense Technology; 2014.
3. Arai N. A Study of Transient Thermal Response of Ablation Materials. ISAS Rept. 1976;32:544.
4. Aral N, Karashima K-I, Sato K. Transient Ablation of Teflon Hemispheres. AIAA J. 1977;15(11):1656–7.
5. Arai N, Karashimat K. Transient thermal response of ablating bodies. AIAA J. 1979;2(2):572–9.
6. Clark BL. A Parametric Study of the Transient Ablation of Teflon. J Heat Trans T ASME. 1972;94:347–54.
7. Stechmann DP. Numerical analysis of transient Teflon ablation in pulsed plasma thrusters. Worcester Polytechnic Institute; 2007.
8. Galfetti L. Experimental measurements and numerical modelling of conductive and radiative heat transfer in polytetrafluoroethylene. In: RTO applied vehicle technology panel (AVT) symposium on advanced flow management: part B-heat transfer and cooling in propulsion and power systems; 2001.
9. Phipps CR, Birkan M, Bohn W, Eckel HA, Horisawa H, Lippert T, Michaelis M, Rezunkov Y, Sasoh A, Schall W, Schaming S, Sinko J. Review: laser-ablation propulsion. J Propul Power. 2010;26(4):609–37.

10. Maxwell JC. On the dynamical theory of gases. Philos Trans Roy Soc Lond. 1867;157:49–88.
11. Cattaneo C. Sur une forme de l'équation de la chaleur e´liminant le paradoxe d'une propagation instantanée. C R Acad Sci. 1958;247(4):431–3.
12. Vernotte P. Les paradoxes de la théorie continue de l'équation de la chaleur. C R Acad Sci. 1958;246(22):3154–5.
13. Kundu B, Lee KS. Fourier and non-Fourier heat conduction analysis in the absorber plates of a flat-plate solar collector. Sol Energy. 2012;86:3030–9.
14. Abdel-Hamid B. Modelling non-Fourier heat conduction with periodic thermal oscillation using the finite integral. Appl Math Model. 1999;23:899–914.
15. Gembarovic J, Gembarovic J. Non-Fourier heat conduction modelling in a finite medium. Int J Thermophys. 2004;25(4):1261–7.
16. Zhang DM, Li L, Li ZH, Guan L, Tan X, Liu D. Non-Fourier heat conduction studying on high-power short-pulse laser ablation considering heat source effect. Eur Phys J Appl Phys. 2006;33:91–6.
17. Tang DW, Araki N. On non-Fourier temperature wave and thermal relaxation time. Int J Thermophys. 1997;18(2):493–504.
18. Chester M. Second sound in solids. Phys Rev. 1963;131(5):2013–5.
19. Jiang R. Transient shock effects in heat conduction, mass diffusion, and momentum transfer. Beijing: Science Press; 1997. p. 55–62.

Chapter 4
Numerical Simulation of the Nanosecond Laser Ablation of Al Propellant

Pulsed plasma thrusters (PPTs) using gaseous propellants [1–4] typically far outperform those using solid propellants in terms of parameters such as specific impulse and propulsion efficiency [5]. Therefore, during the PPT operation, first, the propellant is transformed from a solid state to a gaseous or plasma state to ensure that what is actually ionized in the discharge channel or discharge chamber of the thruster is not the solid propellant but rather the gaseous or plasma propellant. In general, the solid propellant cannot be completely converted to a gaseous or plasma state during the laser ablation process. To increase the gas and plasma components in the discharge channel, an intense laser with a nanosecond pulse width is used as the energy source for propellant ablation. This approach is important for improving the PPT propulsion performance.

Research on the mechanism of propellant ablation under intense laser radiation is of crucial academic and engineering value in many fields, such as laser machining, laser surgery, laser coating, laser nanomaterial preparation, and laser propulsion. As shown in Fig. 4.1, laser ablation refers to the process in which the surface material of a propellant is stripped off or particles are emitted under the laser action. Depending on the different value ranges of parameters such as the fluence, pulse width, and wavelength of the laser as well as the absorption coefficient and surface reflectivity of the target, the laser ablation process involves physicochemical processes such as cavitation damage in the liquid phase and mechanical fragmentation in the solid phase caused by thermal evaporation, phase explosion, photophysical ablation, thermal depolymerization, and thermal shock, as well as processes such as plasma generation, dynamic shielding, and plasma absorption. As the temperature of the target gradually increases from room temperature, evaporation and gasification are considered. Evaporation, a physical process of particle emission that may occur in the propellant at any temperature, is the diffusion of particles on the surface of a propellant due to the presence of concentration and temperature gradients. As shown in Fig. 4.1, when the temperature rises to the melting point, in addition to evaporation, the solid–liquid phase transition, i.e., melting, should also be considered. Melting

© The Author(s) 2025
J. Wu et al., *Numerical Simulation of Pulsed Plasma Thruster*,
https://doi.org/10.1007/978-981-97-7958-1_4

is a solid–liquid phase transition that occurs when the temperature of the propellant rises to the melting temperature and is accompanied by the recession of the phase interface. When the temperature increases to the atmospheric boiling point, the liquid phase of the target may not boil. Instead, the liquid phase of the target may be in a superheated state, which may be caused by several factors, e.g., (1) the evaporation pressure near the surface or an increase in the boiling point of the liquid phase region of the propellant and (2) propellant with an interior so pure that it lacks nuclei for gasification. However, the superheated state is metastable. When perturbed, uniform bubble nucleation may occur in the liquid phase region of the propellant, and bubbles are rapidly generated and diffuse to the outer surface of the target, where boiling occurs. When the liquid phase temperature reaches $0.8T_{cr}$ (critical temperature of Al $T_{cr} = 6063$ K), the superheated liquid phase undergoes the so-called dielectric transition, and the absorption coefficient of the laser decreases sharply. Therefore, the target in the dielectric transition region becomes semi-transparent. As the dielectric transition region extends into the liquid phase, a dielectric transition layer is formed near the liquid phase surface, referred to as the D–T layer in this book, as shown in Fig. 4.2b. When the temperature rises to near the critical temperature ($0.9T_{cr}$), the superheated state may transform into a more violent boiling state, resulting in explosive sputtering and particle ejection involving bubbles as well as liquid and solid particles, i.e., the so-called phase explosion. At this time, the ablation rate of the propellant includes the joint contributions of gasification and phase explosion. Relevant studies have shown that, under the action of a laser with a low fluence, the ablation rate is mainly determined by normal evaporation and gasification, while in the case of laser with a high fluence, the ablation rate is mainly determined by particle sputtering and ejection caused by phase explosion. Relevant experimental studies have revealed that there is a certain laser fluence threshold F_{th} such that the ablation rate increases abruptly when the laser fluence F is greater than F_{th}. For the Al target, $F_{th} = 5.2$ J/cm^2 was experimentally measured [6, 7]. In addition, during the laser ablation process, the plasmaization of the target and the shielding and absorption of the plasma dynamically change the intensity of the laser reaching the ablated surface of the target.

This chapter focuses on numerical simulation of the physical processes of nanosecond laser ablation of the propellant and in-depth study of the underlying mechanism. By establishing a model for the nanosecond laser ablation of Al, we investigate the non-Fourier heat conduction and phase transition ablation processes of Al under intense laser radiation with a nanosecond pulse width. The non-Fourier effect often cannot be ignored for ultra-intense laser ablation. Therefore, we need to consider the influence of the non-Fourier effect on laser ablation. To solve problems such as phase transition and ablation, a non-Fourier heat conduction equation based on the enthalpy method is established to study the influence of factors such as the Al plasma shielding effect, laser parameters, and non-Fourier effect on the ablation process.

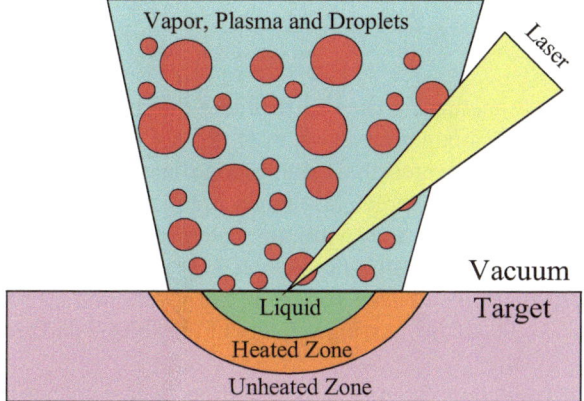

Fig. 4.1 Schematic diagram of laser ablation of a target

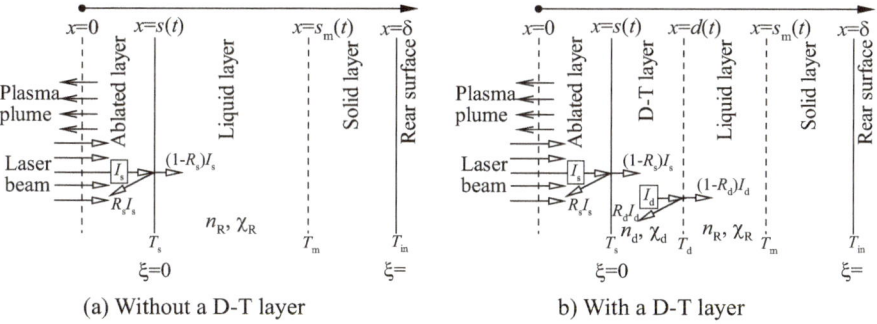

(a) Without a D-T layer   b) With a D-T layer

Fig. 4.2 Schematic diagram of a model for laser ablation of a target

4.1 Non-Fourier Heat Conduction and Phase Explosion Model Based on the Enthalpy Method

4.1.1 Normal Evaporation and Phase Explosion

1. Non-Fourier heat conduction based on the enthalpy method

As shown in Figs. 4.1 and 4.2, the energy deposition process of the laser in the target satisfies conservation of energy, that is,

$$\frac{\partial H}{\partial t} = -\nabla \cdot \mathbf{q} + S \tag{4.1}$$

where H is the enthalpy per unit volume (J/m^3) and q is the heat flux (W/m^2).

The one-dimensional (1D) non-Fourier heat conduction differential equation is

$$\frac{\partial H}{\partial t} + \tau_0 \frac{\partial^2 H}{\partial t^2} - \tau_0 \frac{\partial S}{\partial t} = \frac{\partial}{\partial x}\left(\lambda \frac{\partial T}{\partial x}\right) + S \tag{4.2}$$

Here, the enthalpy method is used to facilitate the analysis and calculation of physical problems, with a coordinate system of (x, t). Considering the recession of the ablated surface of the target, the moving coordinate system (ξ, τ) is fixed at the liquid–gas interface of the target. As shown in Fig. 4.2, the target has an initial thickness of δ in the x direction and an infinite scale perpendicular to the x direction. Considering the recession of the gas–liquid interface (i.e., the ablated surface) of the target, let the position of the ablated surface in the x direction be $x = s(t)$, with an initial position of $x = 0$, and the moving velocity of the ablated surface be $v_s(t)$; then, $v_s(t) = ds(t)/dt$.

The moving coordinate system (ξ, τ) is fixed on the ablated surface $(x = s\ (t))$ and undergoes the following coordinate transformation:

$$\begin{cases} \tau = t \\ \xi = \frac{x-s(t)}{\delta-s(t)} \end{cases} \tag{4.3}$$

From Eq. (4.3), we have

$$\begin{cases} \frac{\partial}{\partial x} = \frac{1}{\delta-s(t)} \frac{\partial}{\partial \xi} \\ \frac{\partial}{\partial t} = \frac{\partial}{\partial \tau} - \frac{v_s(t)(1-\xi)}{\delta-s(t)} \frac{\partial}{\partial \xi} \end{cases} \tag{4.4}$$

From Eq. (4.4) transformed from Eq. (4.2), the 1D hyperbolic heat conduction equation in terms of enthalpy in the moving coordinate system can be obtained as follows:

$$\begin{aligned}
\frac{\partial H}{\partial \tau} + \tau_0 \frac{\partial^2 H}{\partial \tau^2} - \tau_0 \frac{\partial S}{\partial \tau} &= \frac{1}{(\delta - s)^2} \frac{\partial}{\partial \xi}\left(\lambda \frac{\partial T}{\partial \xi}\right) \\
&+ \frac{v_s(1-\xi)}{\delta - s}\left[2\tau_0 \frac{\partial^2 H}{\partial \xi \partial \tau} - \frac{\tau_0 v_s(1-\xi)}{\delta - s} \frac{\partial^2 H}{\partial \xi^2}\right. \\
&\left. + \left(\frac{\tau_0 v_s}{\delta - s} + 1\right)\frac{\partial H}{\partial \xi} - \tau_0 \frac{\partial S}{\partial \xi}\right] + S
\end{aligned} \tag{4.5}$$

The initial conditions for Eq. (4.5) are

$$\begin{cases} T(\xi, \tau)|_{\tau=0} = T_{\text{ini}} \\ \frac{\partial T}{\partial \tau}|_{\tau=0} = 0 \end{cases} \tag{4.6}$$

The boundary conditions at the back face $(x = \delta$ or $\xi = 1)$ are

$$-\lambda \frac{\partial T}{\partial x}\bigg|_{x=\delta} = 0; \quad \frac{\lambda}{s(t)} \frac{\partial T}{\partial \xi}\bigg|_{\xi=1} = 0 \tag{4.7}$$

The temperature of the target rises rapidly under laser radiation. For the Al target, only the isothermal phase transition process is considered. When the temperature is lower than the melting or freezing point ($T_m = 933.47$ K), that is, $T < T_m$, the enthalpy per unit volume is

$$H = \int_{T_{\text{ref}}}^{T} \rho C_p dT + \Delta H_0 \tag{4.8}$$

where ΔH_0 is the zero-point enthalpy and $T_{\text{ref}} = 298.15$ K is the reference temperature.

When the temperature rises to the melting or freezing point, the isothermal phase transition process begins. Let f be the volume fraction of the liquid phase. At the onset of the solid–liquid transition, $f = 0$, and the enthalpy per unit volume is

$$H_{m,0} = \int_{T_{\text{ref}}}^{T_m} \rho C_p dT + \Delta H_0 \tag{4.9}$$

During the solid–liquid transition process, the enthalpy per unit volume in the control volume is

$$H = H_{m,0} + f \rho_m L_m \tag{4.10}$$

where the latent heat of fusion $L_m = 399.9$ kJ/kg and the phase transition density $\rho_m = f_\rho(T_m^+) + (1 - f_\rho(T_m^-))$.

When the solid–liquid transition ends, $f = 1$, and the enthalpy per unit volume is

$$H_{m,1} = H_{m,0} + \rho_m L_m \tag{4.11}$$

After the solid–liquid transition process is complete, the temperature of the target continues to rise, and the enthalpy per unit volume is

$$H = H_{m,1} + \int_{T_m}^{T} \rho C_p dT \tag{4.12}$$

As shown from the numerical calculation process below, it is necessary to back calculate the temperature and liquid volume fraction from the known enthalpy. In this process, calculation steps such as integration and iteration are needed in combination with Eqs. (4.8) to (4.12) to obtain the temperature T from the enthalpy H, and then the phase state of the target is determined.

The liquid volume fraction f can also be expressed in terms of enthalpy as

$$f = \begin{cases} 0 & H < H_{m,0} \\ \frac{H - H_{m,0}}{\rho_m L_m} & H_{m,0} \le H \le H_{m,1} \\ 1 & H > H_{m,1} \end{cases} \qquad (4.13)$$

2. Normal evaporation mechanism

As the temperature increases, the evaporation on the surface of the target intensifies. The surface ablated mass rate caused by evaporation and gasification can be obtained from the Herz–Knudsen equation and the Clausius–Clapeyron equation as follows:

$$\dot{m}_{vap}(\tau) = \beta \left(\frac{m_a}{2\pi k_B T_s(\tau)} \right)^{1/2} p_b \exp\left[\frac{m_a L_v}{k_B} \left(\frac{1}{T_b} - \frac{1}{T_s(\tau)} \right) \right] \qquad (4.14)$$

where T_s is the ablated surface temperature and β is the evaporation viscosity coefficient, characterizing the influence of reflux, with a value of 0.82 in this book. The boiling point T_b at a reference pressure p_b of 1.01325×10^5 Pa is 2792.15 K. The enthalpy of evaporation L_v is 10.897 MJ/kg, and the mass of the gas particles, m_a, is 4.48×10^{-26} kg.

3. Phase explosion mechanism

When the temperature of the target rises to near the critical temperature T_{cr} ($0.9T_{cr}$), the superheated state may transform to a more violent boiling state, resulting in explosive sputtering and ejection of particles containing bubbles as well as liquid and solid particles, known as phase explosion. At this time, the portion of the target with a temperature greater than $0.9T_{cr}$ is stripped off due to phase explosion, and the ablation rate of the propellant should include the contributions from both gasification and phase explosion. These variables should be included in the ablation depth calculation. Relevant research shows that there are two main physical conditions for the formation of phase explosions: first, the scale of the bubble nucleus exceeds the critical radius R_c, causing the bubble size to increase steadily; second, there are a sufficient number of bubble nuclei.

    In the initial stage of nucleation, consider a few gasification nuclei as isolate islands in the sea of the liquid phase. Under certain conditions, these nuclei undergo processes such as growth, fragmentation, reduction, and disappearance. Let the bubble formed at a nucleus have a temperature of T and an internal pressure of p_g. T should be approximately equal to the surrounding liquid temperature T_1; otherwise, intense heat conduction and convection will occur. The dynamic growth of the bubble radius over time can be expressed as

$$R(\tau) = \left[\frac{2L_v \rho_v}{3\rho_l} \frac{T_l - T_{sat}(p_l)}{T_{sat}(p_l)} \right]^{0.5} \tau \qquad (4.15)$$

where ρ_l and T_l are the density and temperature of the superheated liquid, respectively. The pressure of the superheated liquid is $p_l \approx 0.54 p_{sat}(T_l)$, the saturation pressure is $p_{sat}(T_l) = p_b \exp\left(\frac{\beta_s m_a L_v}{k_B T_l}\right)$, and $T_{sat}(p_l)$ is the saturation temperature of the superheated liquid under pressure.

The radius corresponding to the stable existence of bubbles is referred to as the critical radius and can be expressed as

$$R_c = \frac{2\sigma_s}{p_g - p_l} \tag{4.16}$$

where p_g is the pressure inside the bubble and σ_s is the surface tension [N/m]. The critical radius of the bubble can be expressed as

$$R_c = \frac{2\sigma_s}{p_{sat}(T_l) \exp\{[p_l - p_{sat}(T_l)]m_a/(\rho_l k_B T_l)\} - p_l} \tag{4.17}$$

When the current radius of the nucleated bubble is less than the critical radius R_c, the difference between the internal and external pressures of the bubble will not be sufficient to overcome the surface tension confinement, causing gas phase condensation inside the bubble and a decrease in the size of the bubble until it disappears. A bubble can grow only when the current bubble radius is greater than the critical radius. Therefore, the criterion for the growth of the nucleated bubble is

$$R > R_c \tag{4.18}$$

As the difference between the internal and external pressures of the nucleated bubble increases, the critical radius of the bubble decreases continuously, making it easier for the bubble to grow and causing more nuclei for gasification to form. Therefore, the number of gasification nuclei will increase as the difference between internal and external pressures of the bubble or the superheat ΔT increases.

It has been shown that the bubble nucleus generation rate can be expressed as [8]

$$\frac{dN_n(\tau)}{d\tau} = 1.5 \times 10^{38} \exp\left[-\frac{\Delta G(T)}{k_B T}\right] \exp\left(-\frac{\tau_{hn}}{\tau}\right) \text{nuclei}/(\text{m}^3 \text{ s}) \tag{4.19}$$

where T is the temperature of the target at the gasification nucleus. ΔG characterizes the energy required to generate a stable gasification nucleus and can be expressed as $\Delta G = \frac{16\pi}{3}\sigma_s^3/(\rho_v L_v \beta_s)^2$ [J]. σs is the surface tension. For liquid Al, the surface tension is expressed as $\sigma_s = a_s - b_s T$, where $a_s = 1.135$ N/m and $b_s = 1.34 \times 10^{-4}$ N/(m·K). The gas density inside the bubble is $\rho_v = \frac{m_a}{k_B T} p_b \exp\left(\frac{\beta_s m_a L_v}{k_B T}\right)$. The degree of superheat at the bubble is defined as $\beta_s = \frac{T}{T_b} - 1$. The time constant τ_{hn} characterizes the relaxation time of a large number of bubble nuclei required for the formation of phase explosion and can be estimated as $\tau_{hn} = 1 \sim 100$ ns.

Equation (4.19) shows that a higher temperature leads to the generation of a larger number of bubbles in the liquid region and a higher probability of phase explosion. Since the temperature is strongly affected by the power density, pulse width, and wavelength of the laser, these parameters are important factors affecting phase explosion.

Based on the above analysis of the phase explosion mechanism, the ablated mass flux caused by phase explosion can be estimated as

$$\dot{m}_{\exp}(\tau) = 2 \times 10^{38} \pi [\delta - s(\tau)] \int_0^1 r_c^3 \rho_v \exp\left(-\frac{\Delta G_n}{k_B T}\right) d\xi \qquad (4.20)$$

The ablation rate of the target includes the joint contribution of normal gasification/evaporation and phase explosion, i.e.,

$$\dot{m}_t(\tau) = \dot{m}_{\mathrm{vap}}(\tau) + \dot{m}_{\exp}(\tau) \qquad (4.21)$$

The recession velocity and position of the ablated surface are

$$\begin{cases} v_s(\tau) = \dfrac{\dot{m}_t(\tau)}{\rho_c} \\ s(\tau) = \int_0^\tau v_s(\tau) d\tau \end{cases} \qquad (4.22)$$

$x = s\,(t)$ at the boundary of the ablated surface, and the center point of the mesh is arranged on the boundary line. Applying the law of conservation of energy to this half mesh gives

$$\frac{\partial H}{\partial t} \frac{\Delta x}{2} = S \frac{\Delta x}{2} - q_e - (H + \rho L_v) v_s \qquad (4.23)$$

where the left side of the equation represents the energy change per unit time, the first term on the right side of the equation is the laser deposition energy per unit time, the second term is the heat flow on the right boundary of the half mesh, and the third term is the energy taken away by the ablated target per unit time, including the energy of the target itself and the latent heat of the liquid–gas transition.

For laser ablation with a nanosecond pulse width, a phase explosion may occur, greatly increasing the ablation rate. When the temperature near the ablated surface reaches approximately $0.9T_{\mathrm{cr}}$, phase explosion occurs, causing the particles near the ablated surface to be stripped out.

Equation (4.5) can be solved based on the initial conditions (4.6) and boundary conditions (4.23). However, before doing so, the thermophysical parameters of the target and the laser heat source need to be determined.

4.1.2 Thermophysical Parameters of Al

The thermophysical parameters of the Al target can be obtained by the following methods.

1. Density

The density of the target can be obtained from the literature [9, 10] and the Guggenheim formula as follows:

$$\begin{cases} \rho(T) = 2852.5 - 0.5116T, & T \leq T_m \\ \rho(T) = \rho_{cr}[1 + 0.75(1 - T/T_{cr}) + 6.5(1 - T/T_{cr})^{1/3}], & T > T_m \end{cases} \quad (4.24)$$

where the critical density ρ_{cr} is 430 kg/m^3, which is obtained from the literature [11].

2. Electrical conductivity and thermal conductivity

When the temperature is lower than $0.8T_{cr}$, the electrical conductivity is

$$\sigma(T) = \frac{1}{\eta(T)} \quad (4.25)$$

where $\eta(T)$ is the resistivity (Ω m).

By fitting the experimental data from the literature [10, 11], we obtain

$$\eta(T) = \begin{cases} (-0.3937 + 1.1035 \times 10^{-2}T) \times 10^{-8}, & T < T_m \\ (12.4729 + 1.3605 \times 10^{-2}T) \times 10^{-8}, & T_m \leq T < 0.8T_{cr} \end{cases} \quad (4.26)$$

When the temperature is higher than $0.8T_{cr}$, the electrical conductivity

$$\sigma(T) = 2.5 \times 10^4 \text{ S/m} \quad (4.27)$$

Based on the literature [11] and the Wiedeman–Franz law, the thermal conductivity of the metal target is

$$\lambda(T) = \begin{cases} 226.67 + 0.033T, & T \leq 400K \\ 226.6 - 0.055T, & 400K < T < T_m \\ 2.45 \times 10^{-8}\sigma(T)T, & T \geq T_m \end{cases} \quad (4.28)$$

3. Specific heat capacity

The specific heat capacity of the target is

$$C_p(T) = \begin{cases} 762 + 0.467T, & T < T_m \\ 921, & T \geq T_m \end{cases} \quad (4.29)$$

4. Refractive index and extinction coefficient

When the temperature is lower than T_m, the target is in the solid phase. For the incident laser with a wavelength of $\lambda_l = 808$ nm, the target has a refractive index n_R of 2.685 and an extinction coefficient X_R of 8.45; hence, the liquid/solid reflectivity R is 0.87, and the liquid/solid absorption coefficient α_R is 1.33×10^6 cm^{-1}. For the incident laser with $\lambda_l = 1064$ nm, $n_R = 1.24$ and $X_R = 10.42$; hence $R = 0.956$ and $\alpha_R = 1.23 \times 10^6$ cm^{-1}.

When the temperature satisfies $T_m \leq T \leq 0.8\ T_{cr}$, the refractive index and extinction coefficient of the liquid metal layer are

$$\begin{cases} n_R = \sqrt{0.5\left(A_R + \sqrt{A_R^2 + B_R^2}\right)} \\ \chi_R = \sqrt{0.5\left(-A_R + \sqrt{A_R^2 + B_R^2}\right)} \end{cases} \tag{4.30}$$

where A_R and B_R are

$$\begin{cases} A_R = 1 - c^2\mu_0\gamma\sigma/\left(\gamma^2 + \omega_l^2\right) \\ B_R = (1 - A_R)\gamma/\omega_l \end{cases} \tag{4.31}$$

When the temperature is greater than $0.8T_{cr}$, the refractive index and extinction coefficient of the D-T layer are

$$\begin{cases} n_d = \sqrt{0.5\left(A_d + \sqrt{A_d^2 + B_d^2}\right)} \\ \chi_d = \sqrt{0.5\left(-A_d + \sqrt{A_d^2 + B_d^2}\right)} \end{cases} \tag{4.32}$$

where A_d and B_d are

$$\begin{cases} A_d = 1 - \sigma\gamma/[\varepsilon_0(\gamma^2 + \omega_l^2)] \\ B_d = (1 - A_d)\gamma/\omega_l \end{cases} \tag{4.33}$$

where c, μ_0, and ε_0 are the speed of light, vacuum magnetic permeability, and vacuum permittivity, respectively, and the laser frequency is $\omega_l = 2\pi c/\lambda_l$. The collision frequency of the target for the infrared laser absorption process is determined by the Drude model as $\gamma(T) = n_e e^2/[m_e\delta(T)]$, in which the electron number density $n_e = 3\rho(T)/m_a$.

5. Reflectivity and absorption coefficient

The reflectivity of the surface of the target is determined by the complex refractive index $n_c = n_R + i\chi$ as

$$R = \frac{(n_R - 1)^2 + \chi^2}{(n_R + 1)^2 + \chi^2} \tag{4.34}$$

where the real part n_R of the complex refractive index is generally referred to as the refractive index and the imaginary part X is the extinction coefficient.

As shown in Fig. 4.2a, when there is no dielectric transition, the reflectivity on the ablated surface of the target is

$$R(T_s) = \frac{(n_R - 1)^2 + \chi_R^2}{(n_R + 1)^2 + \chi_R^2} \tag{4.35}$$

As shown in Fig. 4.2b, the reflectivity on the outer surface of the D-T layer, i.e., the ablated surface of the target, is

$$R_s(T_s) = \frac{(n_d - 1)^2 + \chi_d^2}{(n_d + 1)^2 + \chi_d^2} \tag{4.36}$$

The reflectivity at the interface between the D-T layer and the liquid layer is

$$R_d(0.8T_{cr}) = \frac{(n_d - n_R)^2 + \chi_R^2}{(n_d + n_R)^2 + \chi_R^2} \tag{4.37}$$

For lasers with wavelengths of 1064 and 808 nm, the reflectivity values at the interface between the D–T layer and the liquid layer are 76.3 and 76.7%, respectively.

The absorption coefficient of the target is expressed as

$$\alpha_R = \frac{4\pi \chi_R}{\lambda_l} \tag{4.38}$$

Based on Eqs. (4.24) to (4.38), the density, specific heat capacity, thermal conductivity, and absorption coefficient of the target for lasers with wavelengths of 808 and 1064 nm, respectively, are obtained, as shown in Figs. 4.3 and 4.4, respectively.

4.1.3 Laser Parameters and Heat Source

1. Laser parameters

It is assumed that the laser intensity follows a Gaussian distribution, that is, $I_0(r, t) = I_p(t)\exp(-r^2/r)$, where $I_p(t)$ is the laser intensity at the center of the spot. In addition, the dimensionless laser power is defined as $h(t) = P_0(t)/P_{peak}$, where $P_0(t)$ is the instantaneous laser power and P_{peak} is the peak power. Obviously, the instantaneous laser power is

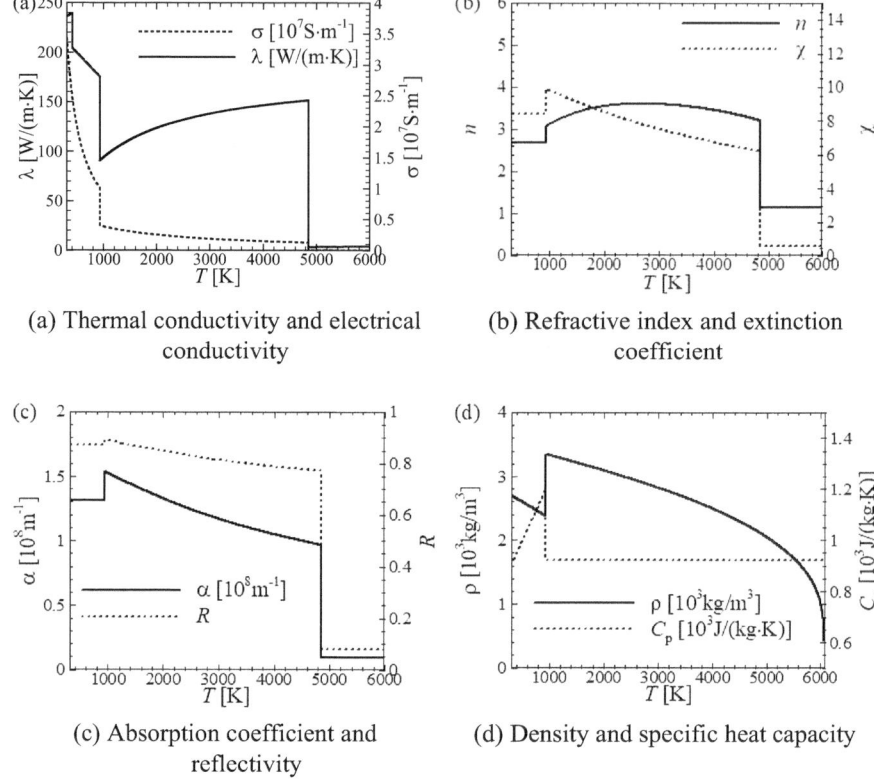

(a) Thermal conductivity and electrical conductivity

(b) Refractive index and extinction coefficient

(c) Absorption coefficient and reflectivity

(d) Density and specific heat capacity

Fig. 4.3 Physical properties of Al (wavelength = 808 nm)

$$P_0(t) = \int_0^{r_p} I_0(r, t) 2\pi r dr = \pi \left(1 - e^{-1}\right) r_p^2 I_p(t) \approx 2 r_p^2 I_p(t) \tag{4.39}$$

Therefore,

$$I_0(r, t) = I_{\text{peak}} h(t) \exp\left(-\frac{r^2}{r_p^2}\right) \tag{4.40}$$

The laser intensity at the center of the spot is

$$I_p(t) = I_{\text{peak}} h(t) \tag{4.41}$$

where the peak laser intensity is

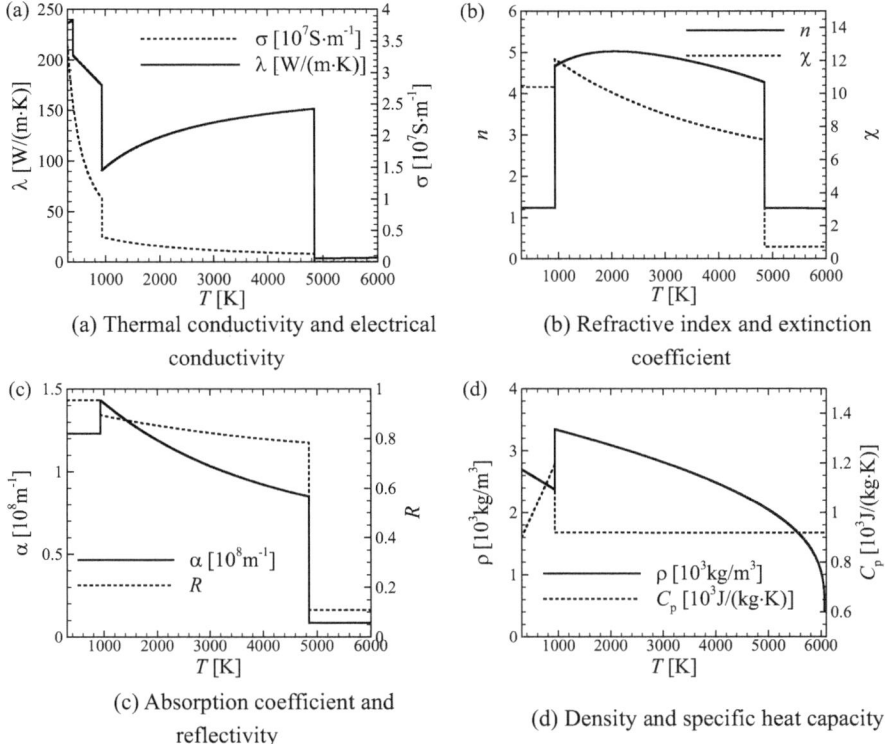

(a) Thermal conductivity and electrical conductivity

(b) Refractive index and extinction coefficient

(c) Absorption coefficient and reflectivity

(d) Density and specific heat capacity

Fig. 4.4 Physical properties of Al (wavelength = 1064 nm)

$$I_{\text{peak}} \approx 0.5 \frac{P_{\text{peak}}}{r_p^2} \tag{4.42}$$

The laser energy can be obtained by integrating the laser power over time, that is,

$$E_p = \int_0^{\tau_p} P_0(t)\mathrm{d}t = P_{\text{peak}} \int_0^{\tau_p} h(t)\mathrm{d}t \tag{4.43}$$

Therefore, the peak laser power is

$$P_{\text{peak}} = \frac{E_p}{\int_0^{\tau_p} h(t)\mathrm{d}t} \tag{4.44}$$

Obviously, when the laser power does not change with time during the pulse time, that is, when the function $h(t)$ can be represented as a unit step function, the peak laser

Fig. 4.5 Dimensionless power of a certain model of a Nd:YAG laser

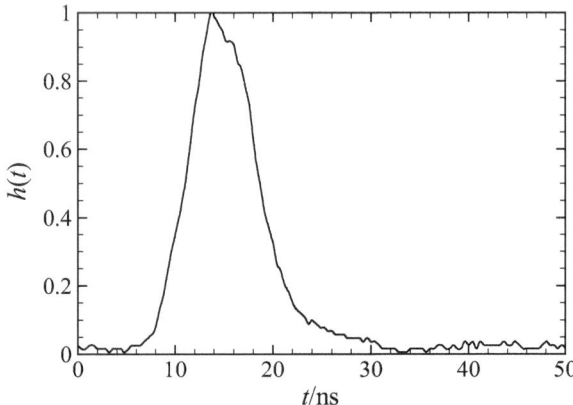

intensity is $I_{peak} = \frac{E_p}{2\tau_p r_p^2}$. In general, however, the laser power constantly changes during the pulse time.

A Nd:YAG laser (model SpitLight 600, InnoLas, Germany) is chosen here as an example, its laser parameters are as follows: wavelength λ_l, 1064 nm; beam diameter, 6 mm; pulse width τ_p, ~ 8 ns; repetition rate, 0–20 Hz; and single-pulse energy E_p, 600 mJ. To more accurately calculate the laser ablation, the time-varying characteristics of the laser power output of this model were measured experimentally, as shown in Fig. 4.5. Its dimensionless power can be fitted as

$$
h(t) = \begin{cases} 0.168t & 0 \le t \le \tau_{peak} \\ 2.991 - 0.796t + 0.108t^2 - 5.127 \times 10^{-3}t^3 & \tau_{peak} < t \le \tau_1 \\ 9.656 - 1.844t + 0.123t^2 - 2.831 \times 10^{-3}t^3 & \tau_1 < t \le \tau_p \\ 0.0 & t > \tau_p \end{cases} \quad (4.45)
$$

where the time t is expressed in units of ns, and the time parameters are $\tau_{peak} = 5.95$ ns, $\tau_1 = 9.8$ ns, and $\tau p = 16.3$ ns.

Based on the fitting Eq. (4.45), a P_{peak} of 7.2×10^7 W can be calculated. Furthermore, using Eq. (4.42) and assuming a focal spot radius r_p of 1.5 mm, we obtain $I_{peak} = 1.6$ GW/cm^2. In addition, the fluence is $F = E_p/(\pi r) = 8.5$ J/cm^2.

2. Plasma absorption and shielding of laser

As shown in Fig. 4.1, after the plasma is generated near the ablated surface, it moves in the direction opposite to the incident laser, that is, toward the negative direction of the x-axis. The length of the plasma expansion calculation region is set to δ_p. Due to plasma absorption, the laser intensity is attenuated according to Beer's law. At position x ($-\delta_p \le x \le s(t)$) in the plasma expansion region, the instantaneous laser intensity is given by

$$I(x, r, t) = I(-\delta_p, r, t) \exp\left(-\int_{-\delta_p}^{x} \beta \mathrm{d}x\right) \tag{4.46}$$

where $I(-\delta_p, r, t)$ can be considered as the initial laser intensity before the shielding and absorption of plasma, that is, $I_0(r, t) = I(-\delta_p, r, t)$, which can be determined by Eq. (4.40). After partial absorption by the plasma, the intensity of the laser reaching the ablated surface of the target is $I_s(r, t) = I(s(t), r, t)$.

In the plasma absorption region, the local internal heat source generated by pulsed laser deposition can be expressed as

$$S_{\text{laser}}(x, r, t) = \beta I(x, r, t) \tag{4.47}$$

There are many mechanisms of laser absorption in plasma, which can be categorized into classical collisional absorption and anomalous absorption. Classical collisional absorption refers to the absorption caused by the interaction between electrons and ions or neutral particles and is also known as the inverse bremsstrahlung absorption process. Anomalous absorption is the partial conversion of laser energy into plasma wave energy through various non-collision mechanisms and then the conversion of wave energy into the irregular motion energy of plasma through Landau damping, wave breaking, or other dissipation mechanisms. Anomalous absorption includes resonance absorption, various nonlinear parametric stable absorption, and anomalous collisional absorption. When a laser beam irradiates an ionized gas, because the photon energy is much lower than the atomic ionization energy, photon ionization can generally be ignored.

In this book, the first, second, and third ionizations of Al are considered. Therefore, the Al plasma contains five compositions: Al, Al^+, Al^{2+}, Al^{3+}, and e^-. Assuming that only the inverse bremsstrahlung absorption mechanism is considered for the absorption of laser in the plasma, the absorption coefficient can be expressed as

$$\beta = \beta_{\text{e-Al}}^{\text{IB}} + \beta_{\text{e-I}}^{\text{IB}} \tag{4.48}$$

$\beta_{\text{e-Al}}^{\text{IB}}$ is the electron-neutral-particle inverse bremsstrahlung absorption coefficient and is given by

$$\beta_{\text{e-Al}}^{\text{IB}} = \left[1 - \exp\left(-\frac{h\nu_l}{k_B T}\right)\right] n_e n_{\text{Al}} Q_{\text{e-Al}} \tag{4.49}$$

$\beta_{\text{e-I}}^{\text{IB}}$ is the electron–ion inverse bremsstrahlung absorption coefficient and is given by

$$\beta_{\text{e-I}}^{\text{IB}} = \left(1 - e^{-\frac{h\nu_l}{k_B T_e}}\right) \frac{4e^6 \lambda_l^3}{3hc^4 m_e} \sqrt{\frac{2\pi}{3m_e k_B T_e}} n_e \left(n_{\text{Al}^+} + 4n_{\text{Al}^{2+}} + 9n_{\text{Al}^{3+}}\right) \tag{4.50}$$

where the absorption coefficients are calculated using parameters in Gaussian units, $h = 6.6262 \times 10^{-27}$ erg·s, $e = 4.8032 \times 10^{-10}$ statcoulomb, $m_e = 9.1094 \times 10^{-28}$ g, $c = 2.9979 \times 10^{10}$ cm/s, $\lambda_l = 1064 \times 10^{-7}$ cm, $k_B = 1.3807 \times 10^{-16}$ erg/K. T is the translational temperature, T_e is the electron temperature, n_{Al+}, $n_{Al}{}^{2+}$, and n_e are the particle number densities (cm^{-3}) of ions and electrons, and $Q_{e\text{-}Al} \approx 10^{-36}$ cm^5 characterizes the average electron-neutral-particle collision cross section.

Evidently, the plasma absorption coefficient β and the plasma absorption region scale δ_p change with the motion of the plasma. Therefore, to calculate the laser absorption and shielding by the plasma, it is necessary to couple the plasma expansion process in the calculation.

It is assumed that the Al plasma is electrically neutral and contains five compositions (electrons, ions, and atoms): Al, Al$^+$, Al^{2+}, Al^{3+}, and e$^-$. To simplify the solution of the plasma flow field, it is assumed that the temperature T in the plasma absorption region is approximately equal to T_e. In addition, it is assumed that the various compositions of the plasma are in chemical equilibrium; thus, the mass fraction c_s of each composition can be obtained from the temperature and the total density or pressure. The Al plasma flow field satisfies the fluid dynamics governing equations as follows:

$$\frac{\partial}{\partial t}\begin{bmatrix} \rho \\ \rho u \\ E \end{bmatrix} + \frac{\partial}{\partial x}\begin{bmatrix} \rho u \\ \rho u^2 + p \\ (E+p)u \end{bmatrix} = \begin{bmatrix} 0 \\ 0 \\ S_{\text{laser}} \end{bmatrix} \tag{4.51}$$

The total energy E per unit volume of Al plasma is composed of internal energy and kinetic energy, which is given by

$$E = \frac{3}{2}\rho\overline{R}T + \sum_{s \neq e} \frac{\rho_s R_s g_1^{(s)} \Theta_{el,1}^{(s)} \exp\left(-\Theta_{el,1}^{(s)}/T_e\right)}{\sum\limits_{i=0}^{j^s} g_i^{(s)} \exp\left(-\Theta_{el,i}^{(s)}/T_e\right)} + \frac{1}{2}\rho u^2 \tag{4.52}$$

where $\overline{R} = R_0/\overline{M}$, $\overline{M} = \left[\sum C_s/M_s\right]^{-1}$, $R_0 = 8.3145$ J/(mol K), and $\Theta_{el,i}^{(s)}$ is the characteristic temperature of electron excitation corresponding to the i-th electronic energy level of composition s, that is, $\Theta_{el,i}^{(s)} = \varepsilon_{el,i}^{(s)}/k_B$, and $g_i^{(s)}$ is the degeneracy of the i-th electronic energy level of composition s.

The equation of state is

$$p = \sum_{s=1}^{4} \rho_s R_s T + \rho_e R_e T_e \tag{4.53}$$

3. Laser energy deposition in the propellant

For a 1D problem, the laser intensity on the ablated surface of the target is represented by the value at the spot center, $I_s(0, t)$, as shown in Fig. 4.2. When there is no dielectric transition, the internal heat source corresponding to the laser energy absorbed by the propellant can be expressed as

$$S(x, t) = (1 - R_s)I_s(0, t)\alpha \exp\left(-\int_{s(t)}^{x} \alpha dx\right) \tag{4.54}$$

where R_s is the reflectivity on the ablated surface and α is the laser absorption coefficient of the propellant.

As shown in Fig. 4.2b, after the dielectric transition occurs, there exists a D–T layer outside the liquid metal layer. The laser is reflected at the interface between the D–T layer and the liquid metal layer. The internal heat source of the laser in the dielectric transition region ($s(t) \leq x \leq d$) can still be expressed by Eq. (4.54).

The internal heat source of the laser in the solid/liquid region ($d(t) \leq x \leq \delta$) is

$$S(x, t) = (1 - R_d)I_d\alpha \exp\left(-\int_{d(t)}^{x} \alpha dx\right) \tag{4.55}$$

where $d(t)$ is the position of the interface between the D–T layer and the liquid metal layer, and R_d and I_d are the reflectivity and laser intensity at the interface between the dielectric transition layer and the liquid metal layer, respectively. I_d is calculated as follows:

$$I_d = (1 - R_s)I_s(0, t)\exp\left(-\int_{s(t)}^{d(t)} \alpha dx\right) \tag{4.56}$$

4.2 Numerical Calculation Methods

4.2.1 Solution of the Temperature Field of the Target

In the process of solving the temperature of the non-Fourier differential equation (Eq. 4.5) established by the enthalpy method, it is not necessary to distinguish the different ablation stages of the target or to divide the computational domain into multiple regions. Therefore, the calculations are simpler and easier to perform.

The initial conditions of the partial differential equation (Eq. 4.5) are

$$\begin{cases} T(x, t)|_{t=0} = T_0 \\ \frac{\partial T}{\partial t}|_{t=0} = 0 \end{cases} \tag{4.57}$$

At the back face of the solid phase ($x = \delta$, $i = ni$, or $\xi = 1$), the boundary conditions are

$$-\lambda \frac{\partial T}{\partial x}\bigg|_{x=\delta} = 0; \quad \frac{\lambda}{s(t)} \frac{\partial T}{\partial \xi}\bigg|_{\xi=1} = 0 \qquad (4.58)$$

At the gas–liquid interface boundary ($x = s(t)$, $i = 1$, or $\xi = 0$), according to the law of conservation of energy, we have

$$\frac{1}{2} \frac{\partial H}{\partial t} \delta x_{1/2} = \frac{1}{2} S \delta x_{1/2} - q|_{1/2} - (L_m + L_v)\rho v_s \qquad (4.59)$$

From the heat conduction differential equation, we can obtain a system of discrete equations in the following form:

$$\mathbf{AH} + \mathbf{BT} = \mathbf{RHS} \qquad (4.60)$$

where the column vectors H and T represent the enthalpy and temperature of each discrete point, respectively.

By solving the above system of equations, the temperature value of the next time layer can be calculated from the current temperature. The steps are as follows: (1) Let the temperature T^n of the nth time layer be the initial value, and assume the estimated (guessed) temperature of the $(n + 1)$th time layer is $T_{\text{guess}} = T^n$; (2) Substitute the guessed temperature T_{guess} into the linear system of equations (Eq. 4.60), and solve for the estimated enthalpy H_{guess} at the $(n + 1)$th time layer using the TDMA; (3) Obtain a new guessed temperature T_{guess} through iteration from using the enthalpy H_{guess}; and (4) Repeat steps (2) to (3) until the relative error between the guessed temperature values from two consecutive iterations is less than the specified error; then, the temperature at time $n + 1$ is $T^{n+1} = T_{\text{guess}}$.

4.2.2 Solving the Plasma Plume Field

The velocity, temperature, and particle number density at the ablation boundary are determined by the ablated mass flux. Linear interpolation is performed at the downstream exit boundary. In the calculation, the plasma region has a length of 200 μm and is meshed using 1000 cells. The target region has a length of 5 μm and is meshed using 400 meshes. In the process of solving the system of governing equations, the AUSM$^+$-up scheme is used for difference discretization of the inviscid flux, and the fourth-order Runge–Kutta method is used for calculation in the temporal direction. In theory, the solving process can be applied to achieve second-order or higher accuracy in both the spatial and temporal directions.

After obtaining the flux Q^{n+1} in the next time layer, various physical quantities can be obtained. The procedure is as follows: (1) Calculate the parameters ρ, u, and

E_t from the flux Q^{n+1} to obtain the internal energy e_t per unit mass; (2) Assume that $c_s^*(s) = \{1, 0, 0, 0, 0\}$ to find $\rho_s^* = c_s^*(s)\rho$. The particle number density of each composition is $n_s^* = \rho_s^* M_s / N_A \lim_{x \to \infty}$. Therefore, the total particle number density is $n_T^* = \sum n_s^*$; (3) Note that the internal energy per unit mass e_t is obtained in step (1) and that e_t is a function of the partial density and temperature of each composition, i.e., $e_t = e_t(\rho_1, \rho_2, \rho_1, \rho_2, \rho_2, T)$. The fractional density ρ_s^* of each composition obtained in step (2) can be substituted into the expression for e_t to obtain the temperature T^*; (4) Based on the total particle number density n_T^* and temperature T^* and using with the method for solving chemical equilibrium compositions, a new set of particle number density values for different compositions can be obtained n_s^{**}; thus, the partial density $\rho_s^{**} = n_s^{**} M_s / N_A$ and mass fraction $c_s^{**}(s) = \rho_s^{**} / \rho$ of each composition are obtained; and (5) Repeat steps (2) to (4) until all physical quantities converge. Finally, each composition has a partial density of $\rho_s = \rho_s^{**}$, a particle number density of $n_s = n_s^{**}$, a temperature of $T = T^*$, and a pressure of $p = n_T^* k_B T^*$.

4.3 Numerical Simulation Results and Analysis

4.3.1 Influence of the Laser Fluence and Plasma Shielding Effect

Figure 4.6 shows the ablation depth of the target under different laser fluences and a laser wavelength λ of 1064 nm. The experimental results in the figure are from the literature [5, 6]. A comparison of the experimental and numerical calculation results in Fig. 4.6 shows that the two are in good agreement. There is a fluence threshold in the laser ablation of Al: the laser ablation depth and rate increase suddenly and substantially at a laser fluence of approximately 5.2 J/cm^2. After considering the shielding and absorption of the laser by the plasma, the laser energy reaching the target is relatively low, resulting in a decrease in the corresponding ablation depth of the target. A higher laser fluence results in higher density and temperature of the plasma generated by laser ablation and hence stronger absorption of the laser by the plasma. Therefore, the shielding and absorption of the plasma are more pronounced, as shown in Fig. 4.6. In addition, considering the absorption by the plasma, the fluence threshold for laser ablation still remains at approximately 5.2 J/cm^2. This is because when the plasma is first generated during laser ablation, the absorption of the laser by the plasma is weak, hence the impact of plasma absorption on the laser fluence threshold is small.

Figure 4.7 shows the variation in the temperature of the ablated surface of the target with time under different laser fluences. When the laser fluence is less than the laser ablation fluence threshold, i.e., $F < 5.2$ J/cm^2, the temperature of the ablated surface of the target increases from the initial temperature to the solid–liquid transition temperature T_m. Then, after the temperature is maintained at T_m for a period

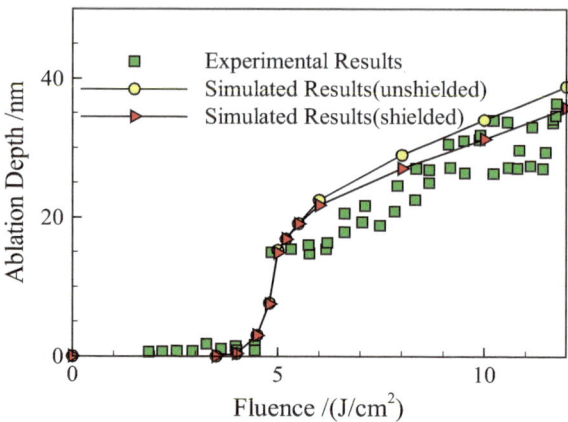

Fig. 4.6 Ablation depth under different laser fluences (wavelength = 1064 nm)

of time, it continues to increase and subsequently decreases as the laser intensity decreases. During the laser ablation process, the peak temperature of the ablated surface of the target is less than $0.8T_{cr}$ and its occurrence is delayed relative to the occurrence of the peak laser intensity. As shown in Fig. 4.7, a higher laser fluence leads to a faster increase in the temperature of the ablated surface of the target and a higher peak temperature. When the laser fluence is greater than the laser ablation fluence threshold, the temperature of the ablated surface of the target rises faster and remains unchanged for a period of time when it reaches approximately $0.8T_{cr}$. After considering the plasma shielding effect, the numerical simulation results for the laser ablation of Al are slightly different, as shown in Fig. 4.7b. Figure 4.7a and b both provide the intensity of the laser reaching the ablated surface of the target. As shown in Fig. 4.7b, when the plasma shielding effect is considered, the intensity of the laser reaching the ablated surface of the target decreases in the later stage of ablation.

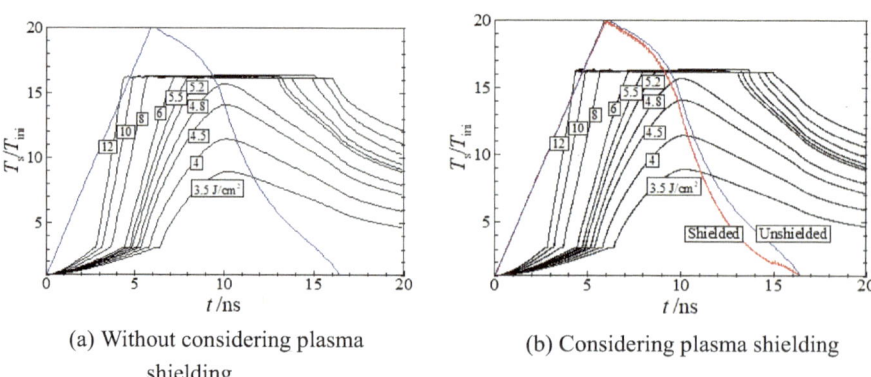

(a) Without considering plasma shielding

(b) Considering plasma shielding

Fig. 4.7 Variation characteristics of the ablated surface temperature of the target with time under different laser fluences

As shown in Fig. 4.8, during the laser ablation process, the D–T layer of the Al target dynamically changes over time. As shown in the figure, a higher laser fluence leads to a larger thickness d_{D-T} of the D–T layer, and when the laser fluence F is in the range of 8–12 J/cm^2, the maximum thickness of the D–T layer is approximately between 60 and 450 nm.

Figures 4.9 and 4.10 show the ablated mass flux and recession velocity of the ablated surface under different laser fluences. A higher laser fluence leads to a faster increase in the ablated mass flux and in the recession velocity of the ablated surface and higher peaks. The ablated mass flux can reach more than 6000 kg/(m^2 s), and the recession velocity of the ablated surface can reach more than 3 m/s.

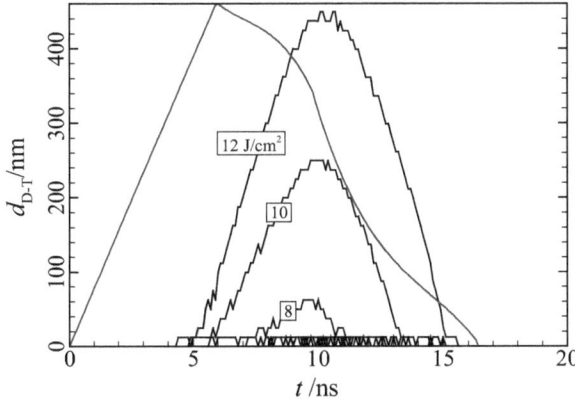

Fig. 4.8 D–T layer thickness under different laser fluences (wavelength = 1064 nm)

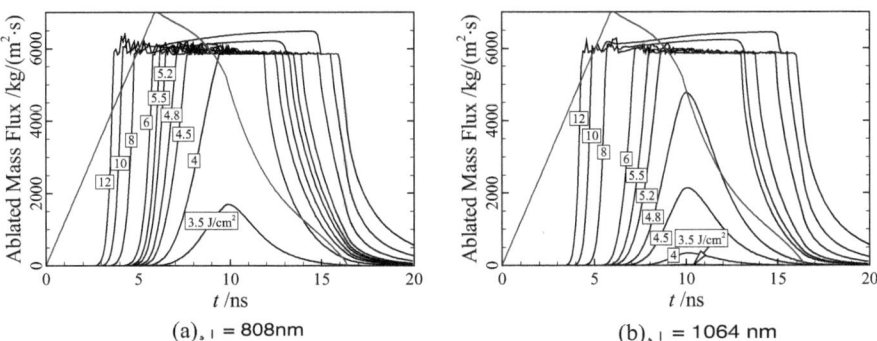

Fig. 4.9 Ablated mass fluxes under different laser fluences

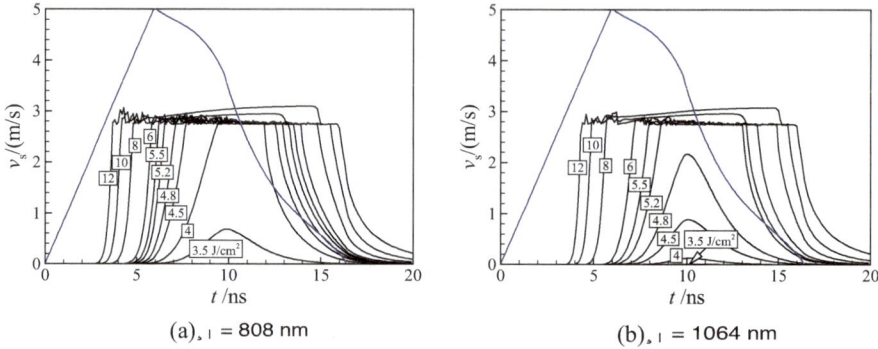

Fig. 4.10 Recession velocities of the ablated surface under different laser fluences

4.3.2 Influence of the Laser Wavelength

It has been reported in the literature [7] that the laser wavelength may have a great impact on the laser ablation process of Al. Therefore, the ablation characteristics of Al under lasers with wavelengths of 808 and 1064 nm, respectively, are comparatively studied. The physical parameters of Al corresponding to lasers with wavelengths of 808 and 1064 nm are shown in Figs. 4.3 and 4.4 of Sect. 4.2.2, respectively. Figure 4.11 shows the ablation depths of ablated Al under different laser fluences. The experimental results for the ablation depth under the laser with a wavelength of 1064 nm are also shown in Fig. 4.11. The ablation depths of Al ablated by lasers with wavelengths of 808 and 1064 nm are of the same order, but the ablation depth of Al was greater under the laser with a shorter wavelength.

In addition, the laser wavelength also has a great impact on the laser fluence threshold of Al. A shorter laser wavelength corresponds to a lower laser fluence

Fig. 4.11 Ablation depths under different laser fluences

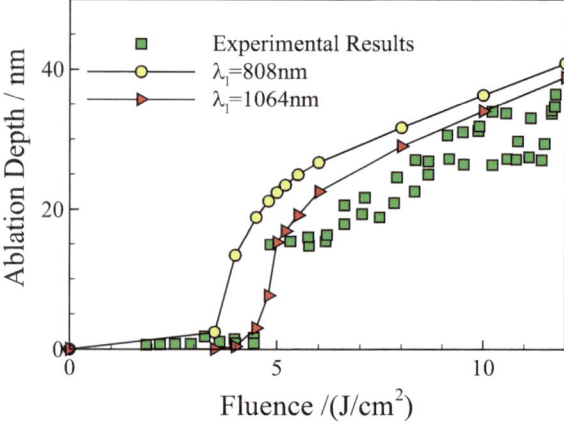

threshold of Al. The calculation results in Fig. 4.11 show that the fluence threshold F_{th} of Al ablation for a laser with a wavelength of 808 nm is approximately 3.5–4.5 J/cm^2.

As shown in Figs. 4.9 and 4.10, under the same laser fluence, a shorter laser wavelength corresponds to a faster increase in the ablated mass flux and the recession velocity of the ablated surface. Under a high fluence, the ablated mass flux and the recession velocity of the ablated surface quickly reach their peak values and remain unchanged for a period of time. Therefore, the laser wavelength has little impact on the peak values of ablated mass flux and ablated surface recession velocity, but a shorter wavelength leads to a longer duration of the peak value.

4.3.3 Influence of the Background Gas Pressure

In numerical calculations, the background gas pressure may affect the calculation results for laser ablation of propellants. As shown in Fig. 4.12, the ablation depths corresponding to different laser fluences are calculated for background gas pressures of 5.0×10^{-3} and 4.14×10^{4} Pa, respectively. From this figure, it is found that a lower background gas pressure leads to a larger ablation depth, and the background gas pressure has little impact on the laser fluence threshold. Figures 4.13, 4.14, and 4.15 present the ablated surface temperature, ablated mass flux, and ablated surface recession velocity under different background gas pressures, respectively. Clearly, the influence of the background gas pressure is not significant.

Figures 4.16 and 4.17 show the variations in the temperature and velocity distributions with time in the 1D flow process of Al plasma under different background gas pressures. At 20 ns, under background pressures of 5.0×10^{-3} and 4.14×10^{4} Pa, the Al plasma reaches temperatures of 25,000 and 7000 K, respectively, and velocities of 7000 and 3000 m/s, respectively. Under a lower background pressure, the Al plasma has a higher peak temperature and faster velocity expansion.

Fig. 4.12 Influence of the background gas pressure on the ablation depth

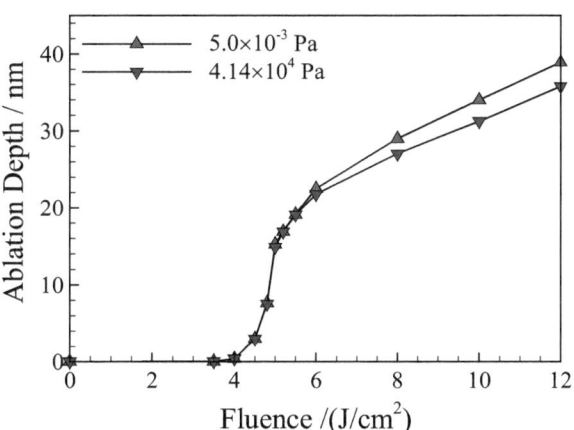

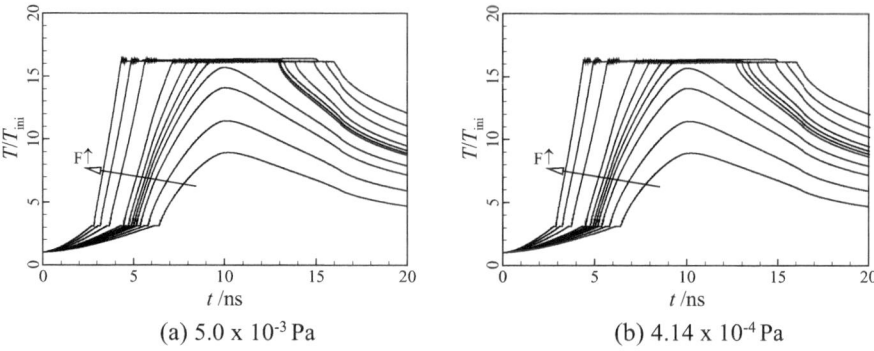

Fig. 4.13 Influence of the background gas pressure on the temperature of the ablated surface (wavelength = 1064 nm)

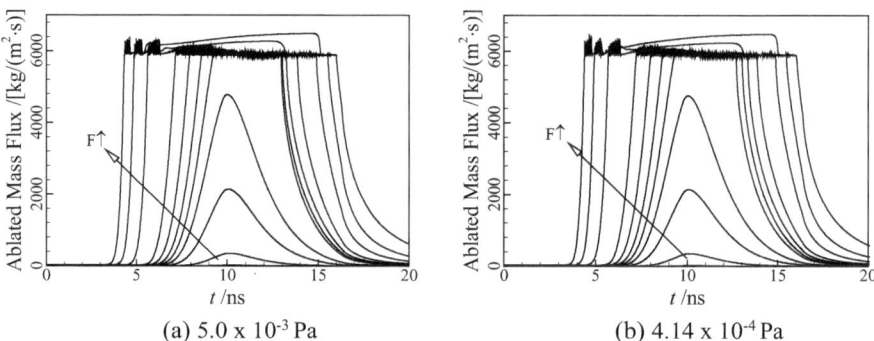

Fig. 4.14 Influence of the background gas pressure on the ablated mass flux (wavelength = 1064 nm)

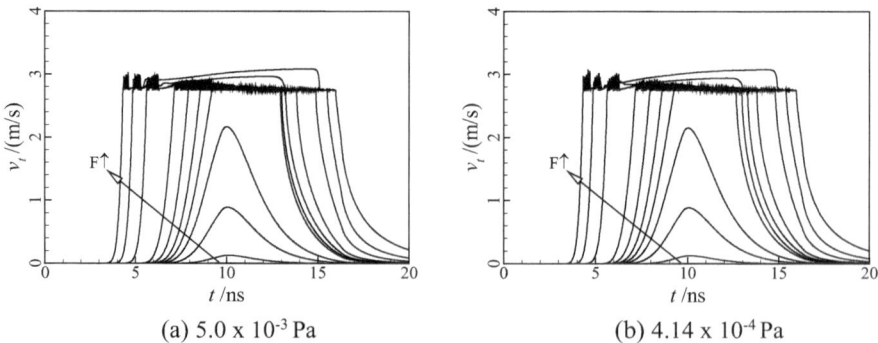

Fig. 4.15 Influence of the background gas pressure on the recession velocity of the ablated surface (wavelength = 1064 nm)

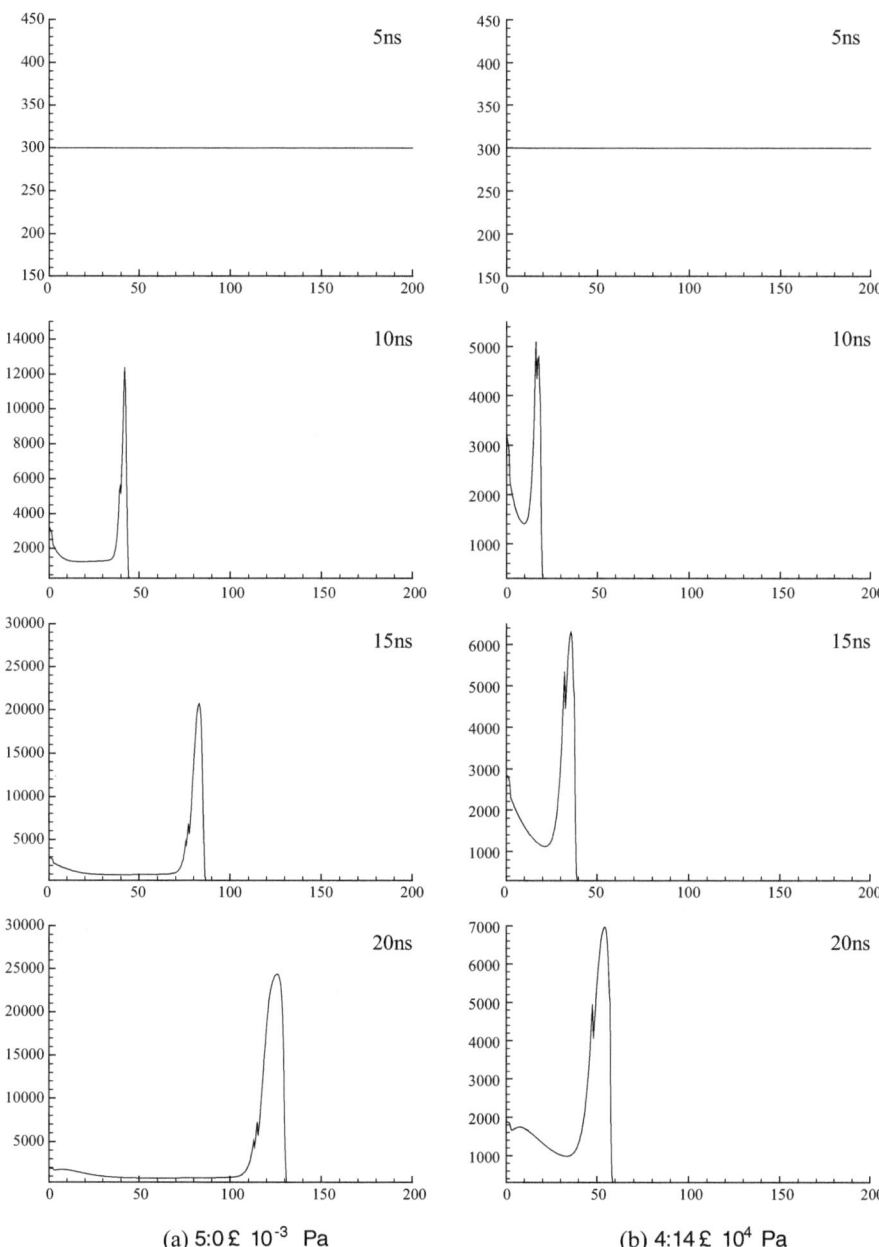

(a) 5:0 £ 10^{-3} Pa   (b) 4:14 £ 10^4 Pa

Fig. 4.16 Influence of the background gas pressure on the plasma temperature distribution (wavelength = 1064 nm)

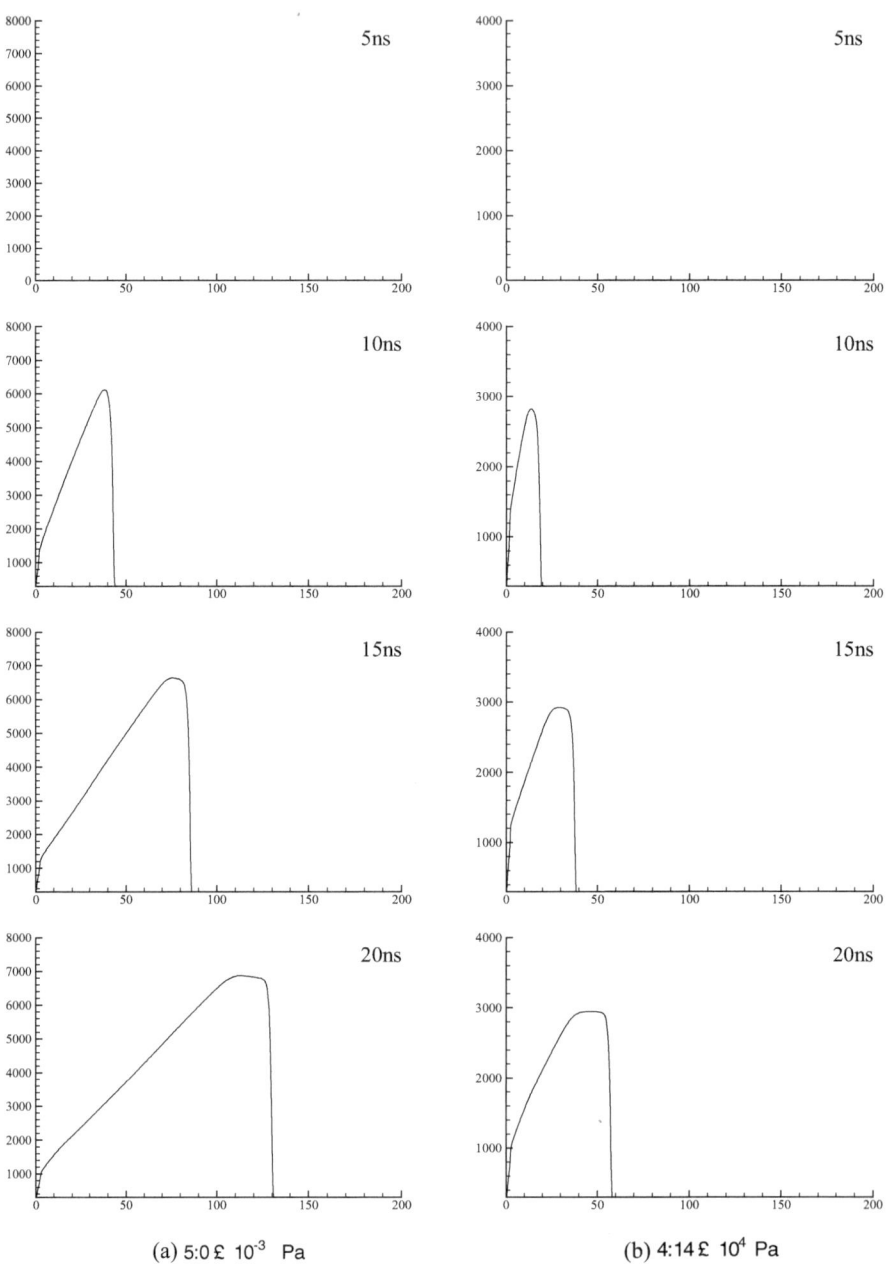

Fig. 4.17 Influence of the background gas pressure on the plasma velocity distribution (wavelength = 1064 nm)

References

1. Ziemer JK, Choueiri EY. Is the gas-fed PPT an electromagnetic accelerator an investigation using measured performance. In: AIAA 99-2289; 1999.
2. Ziemer JK, Cubbin EA, Choueiri EY. Performance characterization of a high efficiency gas-fed pulsed plasma thruster. In: AIAA-97-2925; 1997.
3. Ziemer JK. Performance scaling of gas-fed pulsed plasma thrusters. Princeton University; 2001.
4. Ziemer JK, Petr RA. Performance of gas fed pulsed plasma thrusters using water vapour propellant. In: AIAA 2002-4273; 2002.
5. Porneala C, Willis DA. Time-resolved dynamics of nanosecond laser induced phase explosion. J Phys D: Appl Phys. 2009;42(155503):1–7.
6. Porneala C, Willis DA. Observation of nanosecond laser-induced phase explosion in aluminium. Appl Phys Lett. 2006;89(211121):1–3.
7. Lu Q, Mao SS, Mao X, Russo RE. Theory analysis of wavelength dependence of laser-induced phase explosion of silicon. J Appl Phys. 2008;104(083301):1–7.
8. Gragossian A, Tavassoli SH, Shokri B. Laser ablation of aluminium from normal evaporation to phase explosion. J Appl Phys. 2009;105(103304):1–7.
9. Zhuang H-Z, Zou X-W, Jin Z-Z, Tian D-C. Metal-nonmetal transition of fluid CS along the liquid-vapour coexistence curve. Phys B. 1998;253:68–72.
10. Morel V, Bultel A, Cheron BG. The critical temperature of aluminium. Int J Thermophys. 2009;30:1853–63.
11. Brandt R, Neuer G. Electrical resistivity and thermal conductivity of pure aluminium and aluminium alloys up to and above the melting temperature1. Int J Thermophys. 2007;28(5):1429–45.

Part II
Discharge

Chapter 5
Numerical Simulation of the PPT Discharge Process Based on Electromechanical Models

Zero-dimensional (0D) and one-dimensional (1D) models have low computational complexity and short computation time. In addition, these models can predict the macroscopic performance parameters of a pulsed plasma thruster (PPT), such as the specific impulse, impulse bit, and thrust efficiency. However, these models depend highly on experimental data and empirical parameters. Therefore, it is essential to establish a numerical model that can quickly and accurately predict the performance parameters of a PPT without relying on experimental results. In an electromechanical model, a PPT is equivalent to an electromechanical device in which dynamic and circuit elements interact. It is believed that all the ablated mass of a propellant is concentrated in a very thin current sheet and accelerated and ejected in the form of a slug under the Lorentz force, thereby generating thrust. This type of model can objectively reflect the multiphysical field coupling discharge characteristics of the PPT operation process.

In this chapter, an electromechanical model of the PPT discharge process is established based on a fixed mass current sheet. On this basis, the assumption that the ablated mass of the propellant is constant is abandoned, and a new electromechanical model based on the mass accumulation on the current sheet is established to calculate the ablated mass of the solid propellant during the PPT operation process. This model enables the prediction of the performance parameters of the PPT and provides information on the ablation characteristics of the propellant during the PPT operation process. During the calculation process, there is no need to provide the ablated mass of the pulsed propellant, and the advantages of the original model in terms of its low computational complexity and short computation time are maintained. Finally, an electromechanical model of the PPT considering an additional magnetic field is established. The electromechanical models established in this chapter for the PPT discharge process under different operating conditions can lay a foundation for the study of thruster discharge mechanisms and provide accurate and reliable numerical simulation analysis tools for the optimization design and performance evaluation of PPTs.

© The Author(s) 2025
J. Wu et al., *Numerical Simulation of Pulsed Plasma Thruster*,
https://doi.org/10.1007/978-981-97-7958-1_5

5.1 Electromechanical Model Based on Invariable Mass Current Sheet

An electromechanical model considers a PPT system as a simplified electrome-chanical device and a circuit as a discrete, movable inductance, resistance, and capacitance (LRC) circuit. Typically, Kirchhoff's voltage law is used to describe the dynamic characteristics of the circuit, where the inductance and current are func-tions of time, and the resistance is considered constant during the plasma acceler-ation process. A dynamic system is idealized as a current sheet with an invariable mass. The current sheet is accelerated by the Lorentz force and ejected from the thruster. This dynamic process is described comprehensively using Newton's second law of motion. Figure 5.1 shows a schematic of an electromechanical system of a parallel-plate PPT.

Figure 5.2 shows the circuit model of the parallel-plate electrode PPT. In this figure, L_c, L_{pe}, and L_e are the inductances of the capacitor, parallel-plate electrode, and wire and capacitor lead, respectively, and R_c, R_e, R_{pe}, and R_p are the resis-tances of the capacitor, wire and capacitor lead, parallel-plate electrode, and plasma, respectively. Referring to Fig. 5.2, according to Kirchhoff's law and Faraday's law of electromagnetic induction, we have

$$V_e(t) = \mathrm{IR}_T(t) + \frac{\mathrm{d}}{\mathrm{d}t}[\lambda_{\mathrm{PPT}}(t)] \tag{5.1}$$

where $R_T(t) = R_c + R_e + R_{pe} + R_p(t)$ and $\lambda_{\mathrm{PPT}}(t)$ is the magnetic flux passing through the entire circuit. The total magnetic flux is composed of the magnetic flux $\lambda_c(t)$ generated by the capacitor inductance, the magnetic flux $\lambda_e(t)$ generated by the wire and capacitor lead inductance, and the magnetic flux $\lambda_{pe}(t)$ passing through the parallel-plate electrode channel.

$$\lambda_{\mathrm{PPT}}(t) = \lambda_c(t) + \lambda_6(t) + \lambda_{p6}(t) \tag{5.2}$$

Fig. 5.1 Schematic diagram of the electromechanical system of a parallel-plate electrode PPT

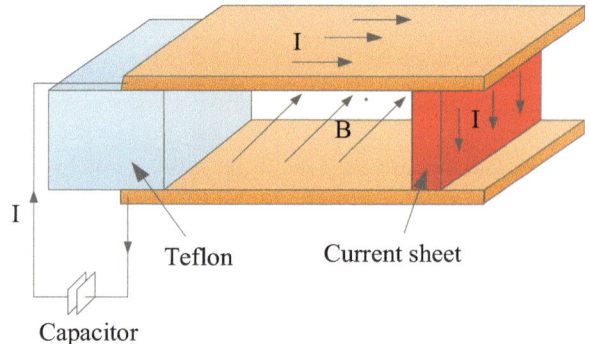

Fig. 5.2 A parallel-plate
electrode PPT circuit model

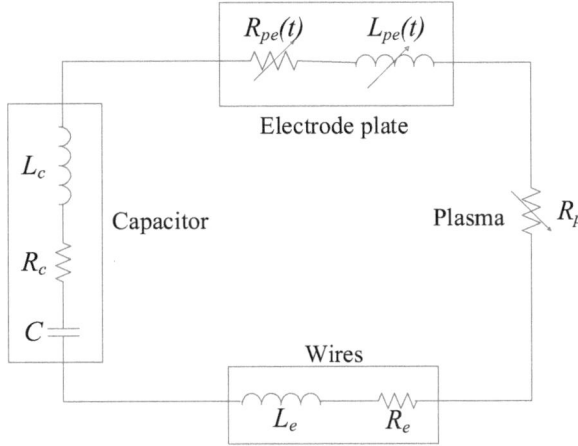

Considering the self-inductance of the capacitor, wire, and capacitor lead, the above equation is rewritten as

$$\lambda_{\mathrm{PPT}}(t) = L_c I(t) + L_e I(t) + \iint\limits_{\text{electrodes}} B_{\mathrm{ind}}(x, y)\mathrm{d}a \tag{5.3}$$

where L_c, L_e, B_{ind}, and A are the self-inductance of the capacitor, the self-inductance of the wire and the capacitor lead, the strength of the self-induced magnetic field across the current sheet, and the area vector of the plasma sheet, respectively.

5.1.1 Strength of the Self-Induced Magnetic Field

A parallel-plate electrode can be approximated as a single-turn solenoid composed of thin sheets with a quasi-infinite width ($w \gg h$) and perfect conduction (i.e., with an infinite conductivity σ), as shown in Fig. 5.3. Additionally, it is assumed that each thin sheet has a uniform current per unit width, K.

Fig. 5.3 Perfectly
conducting single-turn
solenoid

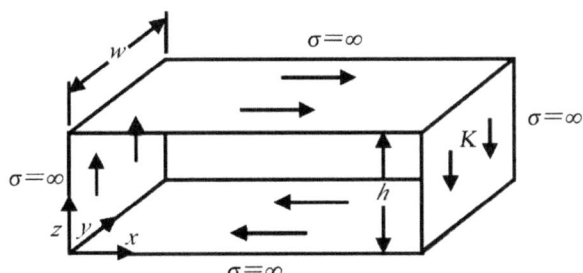

Fig. 5.4 Current sheet

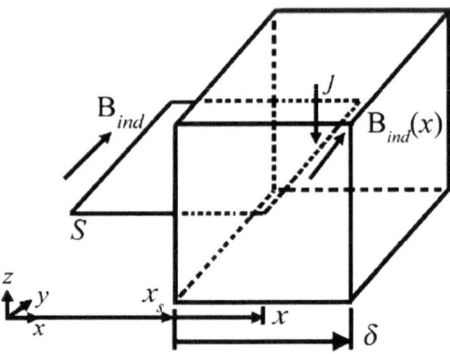

Based on Ampere's continuity condition and the boundary condition of a perfect conductor, the magnetic field behind the current sheet is obtained as

$$B_{ind} = \mu_0 K \hat{y} = \mu_0 \frac{I}{w} \hat{y} \tag{5.4}$$

By applying Ampere's law to the surface S of the current sheet and using Eq. (5.4) as the boundary condition (Fig. 5.4), the magnetic field penetrating the current sheet is obtained as

$$B_{ind}(x, t) = \mu_0 \frac{I(t)}{w} \left[1 - \frac{x - x_s(t)}{\delta} \right] \hat{y} \tag{5.5}$$

The magnetic field in front of the current sheet is assumed to be equal to zero. As a result, the complete expression for the self-induced magnetic field intensity is

$$B_{ind}(x, t) = \begin{cases} \mu_0 \frac{I(t)}{w} \hat{y}, & 0 < x < x_s(t) \\ \mu_0 \frac{I(t)}{w} \left[1 - \frac{x - x_s(t)}{\delta} \right] \hat{y}, & x_s(t) < x < x_x(t) + \delta \\ 0, & x > x_x(t) + \delta \end{cases} \tag{5.6}$$

5.1.2 Inductor Model

Substituting the self-induced magnetic field intensity (Eq. (5.6)) of the parallel-plate electrode into Eq. (5.3) gives

$$\lambda_{PPT}(t) = L_c I(t) + L_e I(t)$$

$$+ \int_0^{x_s(t)} \int_0^h \mu_0 \frac{I(t)}{w} dy dx + \int_{x_s(t)}^{x_s(t)+\delta} \int_0^h \mu_0 \frac{I(t)}{w} \left[1 - \frac{x - x_s(t)}{\delta} \right] dy dx \quad (5.7)$$

Integrating the last two terms on the right-hand side gives

$$\lambda_{PPT}(t) = L_c I(t) + L_e I(t) + \left[\mu_0 \frac{h}{w} x_s(t) + \mu_0 \frac{\delta}{2} \frac{h}{w} \right] I(t) \quad (5.8)$$

The term in the brackets of Eq. (5.8) is the self-inductance of the parallel-plate electrode

$$L_{pe}(x_s(t)) = \frac{\lambda_{pe}(x_s(t))}{I(t)} = \mu_0 \frac{h}{w} x_s(t) + \mu_0 \frac{\delta}{2} \frac{h}{w} \quad (5.9)$$

Assuming that the thin sheet has an infinitesimally small thickness, i.e., $\delta = 0$, Eq. (5.9) reduces to

$$L_{pe}(x_s(t)) = \mu_0 \frac{h}{w} x_s(t) \quad (5.10)$$

5.1.3 Dynamic Model

The motion of the current sheet follows Newton's second law

$$\frac{d}{dt}[m(t)\dot{x}(t)] = \sum F(t) \quad (5.11)$$

where $m(t)$ is the mass of the current sheet and $\sum F(t)$ is the sum of the forces acting on the current sheet.

Assuming that the force acting on the current sheet is the Lorentz force, we have

$$F_L(t) = \iiint_{\substack{current \\ sheet}} j \times B dV$$

$$= \iiint_{\substack{current \\ sheet}} \mu_0 \frac{[I(t)]^2}{\delta w^2} \left[1 - \frac{x - x_s(t)}{\delta} \right] dx dy dz = \frac{1}{2} \mu_0 \frac{h}{w} [I(t)]^2 \quad (5.12)$$

where j is the current density passing through the current sheet.

$$j = -\frac{I}{w\delta} \tag{5.13}$$

Substituting Eq. (5.12) into Eq. (5.11) yields the dynamic equation of the electromechanical model

$$\frac{d}{dt}[m(t)\dot{x}_s(t)] = \frac{1}{2}\mu_0\frac{h}{w}[I(t)]^2\hat{x} \tag{5.14}$$

Assuming that all the propellant gas is concentrated on the surface of the propellant at time $t = 0$ and that there is no mass accumulation during the process of the current sheet accelerating downstream of the discharge channel, we have $m(t) = m_0$. Therefore,

$$m_0\ddot{x}_s(t) = \frac{1}{2}\mu_0\frac{h}{w}[I(t)]^2 \tag{5.15}$$

5.1.4 Plasma Resistance Model

Assuming that the plasma undergoes first-order ionization and is fully ionized,

$$R_p = \frac{h}{\sigma_p w\delta} \tag{5.16}$$

where σ_p is the plasma conductivity given by the Spitzer–Harm conductivity model.

$$\sigma_p = 1.53 \times 10^{-2}\frac{T_e^{\frac{3}{2}}}{\ln\Lambda} \tag{5.17}$$

where Λ is the ratio of the Debye length to the collision parameter

$$\Lambda = \frac{\lambda_D}{b_0} = 1.24 \times 10^7\left(\frac{T_e^3}{n_e}\right)^{\frac{1}{2}} \tag{5.18}$$

where T_e is the electron temperature and n_e is the electron number density.

The thickness of the current sheet is approximately equal to the magnetic field diffusion depth

$$\delta = \sqrt{\frac{\tau}{\sigma_p\mu_0}} \tag{5.19}$$

where τ is the characteristic pulse time and μ_0 is the vacuum magnetic permeability, which has a value of $4\pi \times 10^{-7}$.

From Eqs. (5.16), (5.17), (5.18), and (5.19), the plasma resistance is obtained as

$$R_p = 8.08 \frac{h}{T_e^{\frac{3}{4}} w} \sqrt{\frac{\mu_0 \ln\left[1.24 \times 10^7 \left(\frac{T_e^3}{n_e}\right)^{\frac{1}{2}}\right]}{\tau}} \tag{5.20}$$

5.1.5 Circuit Model

During the discharge process of the capacitor, its voltage can be written as

$$V_0 - \frac{1}{C} \int_0^t I(t)dt = I(t)\left(R_c + R_e + R_{pe} + R_p\right)$$

$$+ \left[L_c + L_e + \mu_0 \frac{h}{w} x_s(t) + \mu_0 \frac{\delta}{2} \frac{h}{w}\right] \dot{I}(t) + \mu_0 \frac{h}{w} \dot{x}_s(t) I(t) \tag{5.21}$$

In summary, a coupled nonlinear second-order integral–differential system of equations for the electromechanical model can be obtained as follows:

$$
\begin{cases}
V_0 - \frac{1}{C} \int_0^t I(t)dt = I(t)\left(R_c + R_e + R_{pe} + R_p\right) \\
\qquad + \left[L_c + L_e + \mu_0 \frac{h}{w} x_s(t) + \mu_0 \frac{\delta}{2} \frac{h}{w}\right] \dot{I}(t) + \mu_0 \frac{h}{w} \dot{x}_s(t) I(t) \\
m_0 \ddot{x}_s(t) = \frac{1}{2} \mu_0 \frac{h}{w} [I(t)]^2 \\
R_p = 8.08 \frac{h}{T_e^{\frac{3}{4}} w} \sqrt{\dfrac{\mu_0 \ln\left[1.24 \times 10^7 \left(\frac{T_e^3}{n_e}\right)^{\frac{1}{2}}\right]}{\tau}}
\end{cases} \tag{5.22}
$$

By solving the above system of equations, relevant parameters such as the voltage across the capacitor terminals, circuit current, and current sheet displacement and velocity of the PPT during the discharge process can be obtained. Other PPT performance parameters can be calculated from these parameters.

5.2 Electromechanical Model Based on Variable Mass Current Sheet

5.2.1 Model of Time-Varying Ablated Mass

Traditional electromechanical models assume that the ablated mass of the propellant is fully generated at the beginning of the calculation and remains constant during the plasma acceleration process. As a result, the ablated mass is constant throughout the entire PPT operation process and is given by the experimental measurements. In the actual operation of a PPT, the propellant is ablated gradually, and the ablated mass of the propellant increases gradually over time. Therefore, the assumption that the ablated mass of the propellant is constant in the electromechanical model differs significantly from the actual operation of the PPT. In this book, we abandon this assumption and consider the actual operation of the thruster, accounting for the fact that the ablated mass of the propellant gradually accumulates during the discharge process. The ablated mass is calculated by Eq. (2.15) in Sect. 2.1 of Chap. 2.

If the ablated mass of the propellant changes with time, the equation of motion for the current sheet should be rewritten as

$$\frac{\mathrm{d}}{\mathrm{d}t}[m(t)\dot{x}(t)] = F(t) \tag{5.23}$$

That is

$$m(t)\ddot{x}(t) + \dot{m}(t)\dot{x}(t) = \frac{1}{2}L'_{\mathrm{pe}}[I(t)]^2 \tag{5.24}$$

Then, the electromechanical model changes accordingly to

$$\begin{cases} V_0 - \frac{1}{C}\int_0^t I(t)\mathrm{d}\tau = I(t)\left(R_C + R_e + R_p\right) + \frac{\mathrm{d}}{\mathrm{d}t}\left[\left(L_C + L_e + L_{\mathrm{pe}}\right)I(t)\right] \\ m(t)\ddot{x}(t) + \dot{m}(t)\dot{x}(t) = \frac{1}{2}L'_{\mathrm{pe}}[I(t)]^2 \\ R_p = 8.08\frac{h}{T_e^{\frac{3}{4}}w}\sqrt{\dfrac{\mu_0\ln\left[1.24\times10^7\left(\frac{T_e^3}{n_e}\right)^{\frac{1}{2}}\right]}{\tau}} \end{cases} \tag{5.25}$$

where $m(t)$ is the ablated mass of the propellant in each time step and is calculated using Eq. (2.15) above. Then, $m(t) = m(t_0) + m(t)$ is the cumulative ablated mass of the propellant at time t, in which $m(t_0)$ is the cumulative ablated mass of the propellant in the previous time step relative to time t.

5.2.2 Model Validation

The Lincoln Experimental Satellite (LES-6) PPT is a mature parallel-plate PPT with years of flight experience. A large number of theoretical and experimental studies have been carried out on the LES-6 PPT, resulting in a large amount of experimental data and some research outcomes. In this book, the LES-6 PPT is used to verify the reliability of the improved electromechanical model. The relevant electrical parameters and structural parameters of the LES-6 PPT are shown in Table 5.1, and the propellant parameters are presented in Table 5.2. The experimental measurements and simulation results of the discharge voltage and current are shown in Figs. 5.5 and 5.6, respectively. The experimental and simulated performance parameters of the thruster are listed in Table 5.3.

Table 5.1 LES-6 PPT parameters

Initial voltage (V)	1360	Electrode width (mm)	10
Capacitance capacity (μF)	2	Electrode length (mm)	6
Capacitance resistance (mΩ)	30	Characteristic pulse time (μs)	0.4
Initial inductance (nH)	34	Plasma temperature (eV)	1.5
Electrode spacing (mm)	30	Electron density (m^{-3})	10^{21}

Table 5.2 PTFE parameters used in the numerical simulations

Parameter	Value	Unit
Solid thermal conductivity k_s	$\left(5.023 + 6.11 \times 10^{-2}T\right) \times 10^{-2}$	W/m/K
Molten thermal conductivity k_m	$\left(87.53 - 0.14T + 5.82 \times 10^{-5}T^2\right) \times 10^{-2}$	W/m/K
Solid density ρ_s	$\left(2.119 + 7.92 \times 10^{-4}T - 2.105 \times 10^{-6}T^2\right) \times 10^3$	kg/m^3
Molten density ρ_m	$\left(2.07 - 7 \times 10^{-4}T\right) \times 10^3$	kg/m^3
Reference density ρ_i	1933	kg/m^3
Solid specific heat C_s	$514.9 + 1.563T$	J/kg/K
Molten specific heat C_m	$904.2 + 0.653T$	J/kg/K
Surface absorption coefficient ε	0.92	
Specific depolymerization energy E_p	$1.774 \times 10^6 - 279.2T$	J/kg
Activation energy E_A	3.473	MJ/kg
Depolymerization frequency factor A_p	3.1×10^{19}	s^{-1}
Depolymerization activation temperature B_p	41,769	K

Fig. 5.5 Experimental results of discharge voltage and current of the LES-6 PPT

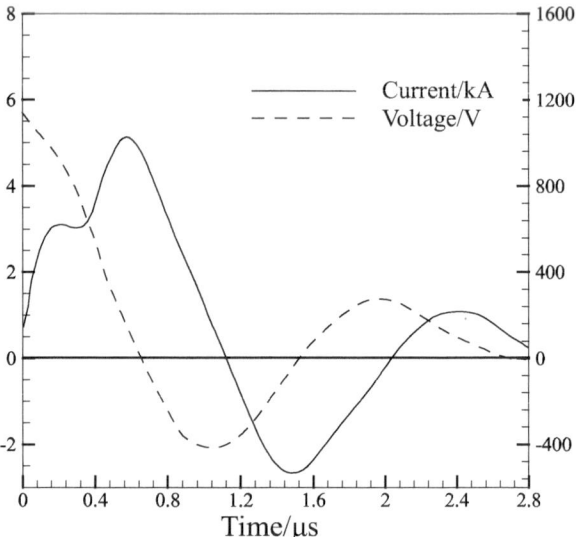

Fig. 5.6 Simulation results of discharge voltage and current LES-6 PPT

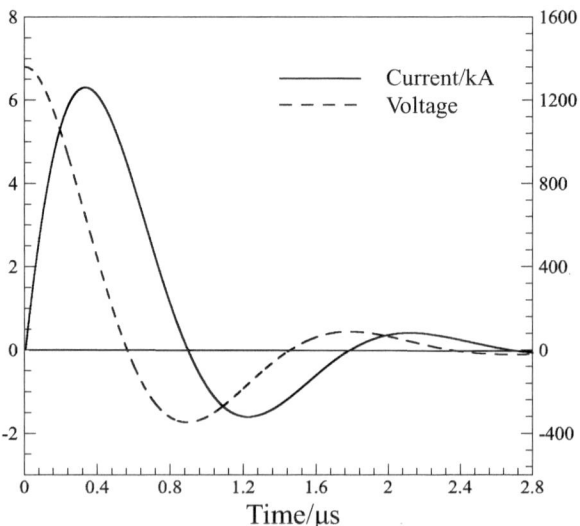

Table 5.3 Experimental and simulation results for the LES-6 PPT performance parameters

	Experiment	Simulation
Exit velocity (m/s)	3000	2925
Specific impulse (s)	300	298
Impulse bit (μN s)	31.2	33.6
Ablated mass (μg)	10	13.5

Figures 5.5 and 5.6 show that the simulation results are in good agreement with the experimental results in terms of the discharge voltage and current. The plasma resistance is regarded as a constant in the new electromechanical model, while the plasma resistance of the thruster varies with time and the circuit parameters during the actual operation of the thruster. Therefore, the calculated waveforms of the discharge voltage and discharge current are slightly different from the experimental measurements. However, these differences are within an acceptable margin of error. Table 5.3 presents a comparison between the experimental and simulation results for the performance parameters of the LES-6 PPT, including the plasma exit velocity, specific impulse, impulse bit, and single-pulse ablated mass of the propellant. It is observed from this table that the performance parameters of the LES-6 PPT during the operation of the thruster obtained from numerical simulations using the improved electromechanical model agree well with the experimental measurements, thereby validating the reliability of the model.

Figure 5.7 shows the simulation results in terms of the energy distribution of the LES-6 PPT. It is observed that 96.67% of the discharge energy is converted to ohmic heat, and only about 3.3% of the energy is eventually converted to the kinetic energy of the plasma. Vondra et al. [1] experimentally studied the discharge energy distribution of the LES-6 PPT, and their results indicated that only approximately 3% of the energy was ultimately converted to the kinetic energy of the plasma, which is consistent with the simulation results using the improved electromechanical model in our study, further validating the reliability of the model.

Figure 5.8 shows the variations in the position and velocity of the current sheet in the discharge channel over time. After the discharge starts, the plasma micelles are gradually accelerated by the Lorentz force. As the discharge current intensity gradually increases, the acceleration of the current sheet also gradually increases. During the period when the discharge current and the discharge voltage are in the

Fig. 5.7 Simulation results of the energy distribution of the LES-6 PPT

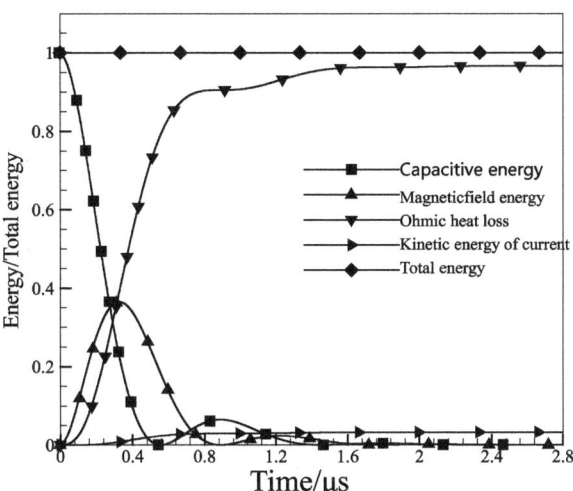

Fig. 5.8 Variations in the position and velocity of the current sheet in the discharge channel over time

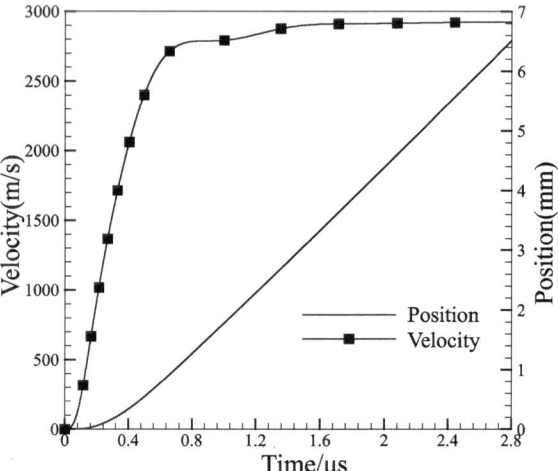

opposite direction, the current sheet continues to accelerate. However, its acceleration gradually decreases until the discharge voltage and the discharge current are in the same direction. Then, the acceleration of the plasma gradually increases again. As the discharge energy gradually dissipates, the acceleration gradually decreases, and the velocity of the current sheet stops increasing before reaching the exit of the discharge channel. The Lorentz force primarily accelerates the current sheet in a short period immediately after the discharge starts. In the second half of the discharge process, the conversion efficiency of discharge energy into kinetic energy is very low.

The improved electromechanical model not only simulates the operation process of the PPT and calculates its discharge waveform and macroscopic performance parameters but also reflects the temperature distribution of Teflon and the specific ablation process of the propellant during the operation of the thruster. Figure 5.9 shows the variations in the heat flux and temperature on a Teflon surface over time. It is observed that the heat flux is concentrated in a short period of time after the start of discharge, accounting for the vast majority of the total discharge energy, and fluctuates over time. Given the heat flux on the Teflon surface, the temperature of the ablated surface rises rapidly to over 1300 K after the start of discharge and fluctuates slightly as the heat flux changes. Figure 5.10 shows the temperature distribution of Teflon at different positions relative to the ablated surface at different time points. After the discharge begins, the temperature of the propellant on the ablated Teflon surface and in the surrounding area rises sharply, rapidly exceeding the phase transition temperature of Teflon. As the heat is transferred toward the lower-temperature region, the temperature gradually decreases in the vicinity of the ablated surface and gradually increases in the interior of the Teflon. The heat flux reaches its maximum at 0.3 μs, and when the amorphous region of the Teflon has a depth of approximately 0.25 μm, the ablation process of the propellant is mainly concentrated in the vicinity of the ablated surface within a short period of time after the start of discharge.

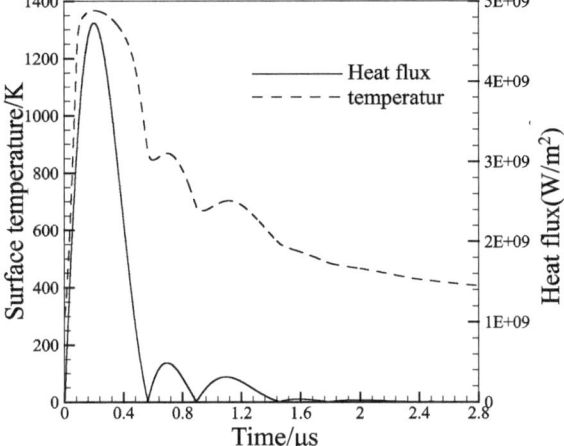

Fig. 5.9 Variations in the heat flux and temperature on a Teflon surface over time

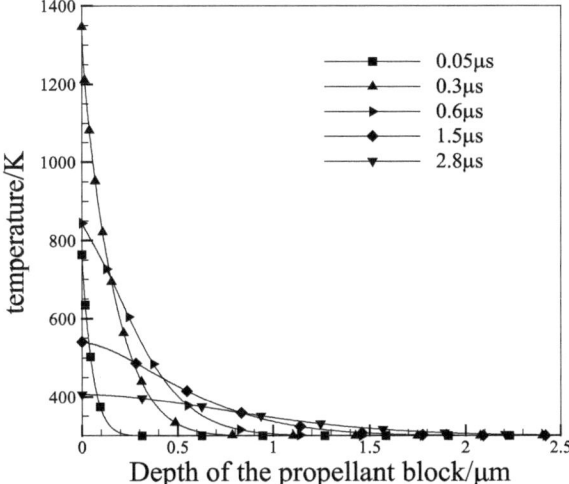

Fig. 5.10 Temperature distribution of Teflon at different time points

Figure 5.11 shows the curves of the single-pulse ablated mass and the ablated mass flux of the propellant as a function of time, reflecting the specific changes in the ablated mass of the propellant during the discharge process. The propellant ablation process mainly occurs within the first 0.6 μs after discharge starts, and the total ablated mass of the propellant during the entire discharge process is 13.5 μg. Figure 5.12 presents the variation in the length of the Teflon propellant over time. As the Teflon propellant is ablated and consumed, the length of the propellant block decreases continuously. During a single-pulse operation of the thruster, the propellant block has a total ablation length of 0.02 μm.

Fig. 5.11 Variations in the mass flux and ablated mass over time

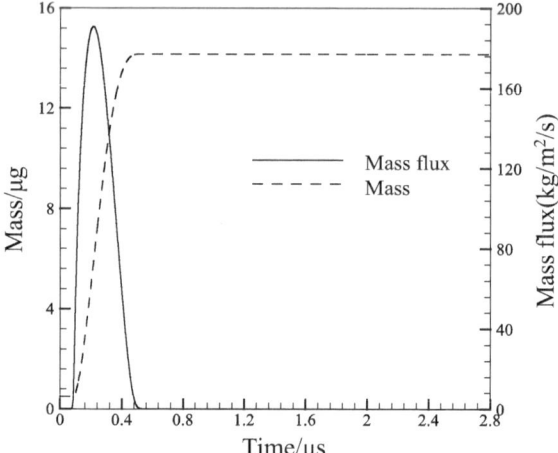

Fig. 5.12 Variation in the propellant length over time

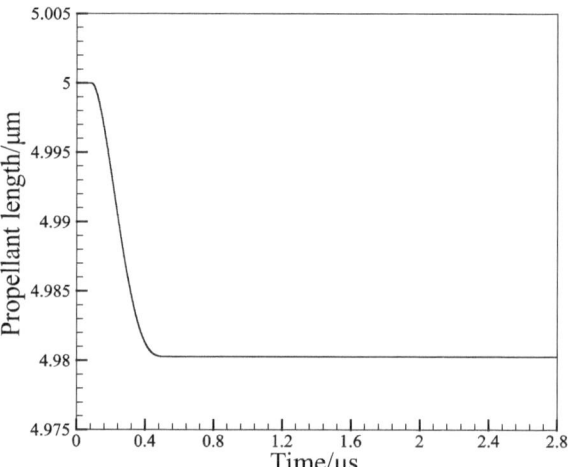

5.3   Improved Electromechanical Model Considering an Additional Magnetic Field

5.3.1   Additional Magnetic Field Model

Figure 5.13 shows a schematic diagram of a parallel-plate PPT with an applied magnetic field. In the figure, B_{induce} represents the self-induced magnetic field, B_{applied} represents the applied magnetic field, l is the plate length, h is the spacing between the plates, and w is the plate width. In the electromechanical model of the PPT, the

Fig. 5.13 Schematic diagram of a PPT with an applied magnetic field

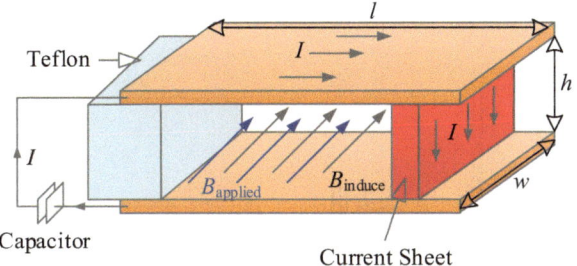

Fig. 5.14 Equivalent circuit diagram of a PPT with an applied magnetic field

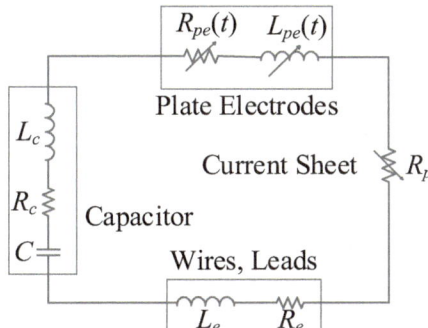

discharge acceleration process of the thruster is simplified as a resistor-inductor-capacitor (RLC) circuit, and the equivalent circuit is shown in Fig. 5.14. Discharge circuit is mainly composed of the capacitor and its resistance R_c and inductance L_c, the resistance R_e and inductance L_e of the wires and leads, the resistance $R_{pe}(t)$ and inductance $L_{pe}(t)$ of the plates, and the resistance $R_p(t)$ of the current sheet.

The plasma plume generated by the laser PPT (LPPT) during the laser ablation stage passes through the circular hole in the middle of the ceramic separator plate and then enters the discharge acceleration channel between the anode and cathode plates. At this moment, the plume has a roughly cylindrical shape, with a size that does not exceed the diameter of the circular hole in the middle of the ceramic separator plate. However, the plasma plume is distributed throughout the entire discharge channel during discharge. At the same time, the plasma plume entering the discharge acceleration channel has a certain initial velocity. Therefore, the plasma plume of the LPPT is assumed to be a thin current sheet with an initial velocity, and the height and width of the current sheet are equal to the plate spacing and plate width of the thruster, respectively. During the discharge acceleration process, the current sheet is accelerated and ejected from the thruster by the Lorentz force and aerodynamic force. A schematic diagram of the LPPT with an applied magnetic field is shown in Fig. 5.15. The equivalent circuit diagram of the LPPT is similar to that of the PPT, as shown in Fig. 5.14.

In the equivalent circuit, R_c, R_e, L_c, and L_e are determined by the design parameters of the thruster and are constant, while $R_p, R_{pe}(t)$, and $L_{pe}(t)$ change with different

Fig. 5.15 Schematic diagram of an LPPT with an applied magnetic field

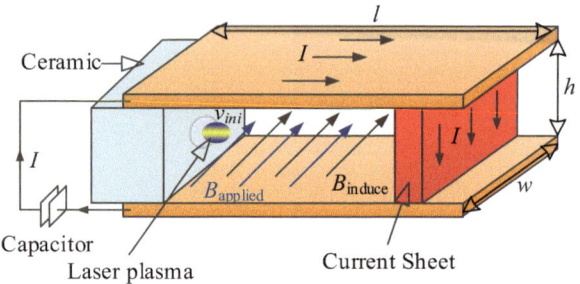

operating states. R_p is calculated by Eq. (5.26), that is,

$$R_p = 8.08 \frac{h}{T_e^{\frac{3}{4}} w} \sqrt{\frac{\mu_0 \ln\left[1.24 \times 10^7 \left(\frac{T_e^3}{n_e}\right)^{\frac{1}{2}}\right]}{\tau}} \tag{5.26}$$

where T_e is the electron temperature, μ_0 is the vacuum magnetic permeability, n_e is the electron number density, and τ is the characteristic pulse time.

From Faraday's law of electromagnetic induction, the equivalent circuit equation can be written as

$$V_c(t) = I(t)R_{\text{total}}(t) + \frac{d\phi_{\text{total}}(t)}{dt} \tag{5.27}$$

where $V_c(t)$ denotes the voltage across the capacitor, $I(t)$ represents the circuit current, $R_{\text{total}}(t) = R_c + R_e + R_{pe}(t) + R_p$ is the total circuit resistance, $\phi_{\text{total}}(t) = \phi_c(t) + \phi_e(t) + \phi_{pe}(t)$ represents the total magnetic induction flux of the circuit, in which $\phi_c(t)$ is the magnetic induction flux caused by the capacitor inductance, $\phi_e(t)$ is the magnetic induction flux caused by the inductance of the wires and leads, and $\phi_{pe}(t)$ is the magnetic induction flux passing through the parallel-plate channel.

The sum of the magnetic induction flux caused by the inductance of the capacitor and the inductance of the wires and the leads is

$$\phi_c(t) + \phi_e(t) = L_c I(t) + L_e I(t) \tag{5.28}$$

The magnetic induction flux passing through the parallel-plate channel consists of the magnetic induction flux of the self-induced magnetic field and the applied magnetic field, that is,

$$\phi_{pe}(t) = \iint\limits_{\text{electrodes}} B_{\text{induce}}(t, x, y)dA + \iint\limits_{\text{electrodes}} B_{\text{applied}}(t, x, y)dA \tag{5.29}$$

where $\boldsymbol{B}_{\text{induce}}(t, x, y)$ is the self-induced magnetic field intensity, $\boldsymbol{B}_{\text{applied}}(t, x, y)$ is the applied magnetic field intensity, and A denotes the area vector of the current sheet.

Assuming that the parallel-plate electrodes are thin planes with a quasi-infinite width ($w \gg h$) and that the current density is uniform, according to Ampere circuital theorem, we have

$$
\iint\limits_{\text{electrodes}} \boldsymbol{B}_{\text{induce}}(t, x, y)\mathrm{d}A = \int_0^{x(t)} \int_0^h \mu_0 \frac{I(t)}{w} \mathrm{d}y\mathrm{d}x
$$

$$
+ \int_{x(t)}^{x(t)+\delta} \int_0^h \mu_0 \frac{I(t)}{w} \left[\frac{\delta + x(t) - x}{\delta} \right] \mathrm{d}y\mathrm{d}x
$$

$$
= \left[\mu_0 \frac{h}{w} x(t) + \mu_0 \frac{\delta}{2} \frac{h}{w} \right] I(t) \tag{5.30}
$$

where $x(t)$ is the distance between the current sheet and the ablated surface of the propellant and δ is the thickness of the current sheet.

Assuming that the applied magnetic field is uniformly distributed in space and time, we have

$$
\iint\limits_{\text{electrodes}} \boldsymbol{B}_{\text{applied}}(t, x, y)\mathrm{d}A = \int_0^{x(t)+\delta} \int_0^h \boldsymbol{B}_{\text{applied}}(t, x, y)\mathrm{d}y\mathrm{d}x
$$

$$
= h B_{\text{applied}}[x(t) + \delta] \tag{5.31}
$$

Therefore, from Eqs. (5.28), (5.30), and (5.31), we have

$$
\phi_{\text{total}}(t) = L_c I(t) + L_e I(t) + \mu_0 \frac{h}{w} x(t) I(t)
$$

$$
+ \mu_0 \frac{\delta}{2} \frac{h}{w} I(t) + h B_{\text{applied}}[x(t) + \delta] \tag{5.32}
$$

Considering the small thickness of the current sheet, we set $\delta = 0$ and then simplify the above equation to

$$
\phi_{\text{total}}(t) = \left[L_c + L_e + \mu_0 \frac{h}{w} x(t) \right] I(t) + h B_{\text{applied}} x(t) \tag{5.33}
$$

Therefore, we have

$$V_c(t) = V_0 - \frac{1}{C}\int_0^t I(\tau)d\tau = I(t)R_{\text{total}}(t) + I(t)\mu_0\frac{h}{w}\dot{x}(t)$$

$$+ \dot{I}(t)\left[L_c + L_e + \mu_0\frac{h}{w}x(t)\right] + hB_{\text{applied}}\dot{x}(t) \tag{5.34}$$

where V_0 is the initial voltage of the capacitor, C is the capacitance of the capacitor, and $\dot{x}(t)$ is the velocity of the current sheet.

According to Newton's second law, we have

$$\frac{d}{dt}[m(t)\dot{x}(t)] = F(t) = F_v + F_s + F_{\text{initial}} \tag{5.35}$$

where $m(t)$ is the mass of the current sheet and $F(t)$ represents the resultant force on the current sheet, including the volume force F_v, surface force F_s, and F_{initial}, which is the equivalent force of the initial momentum.

The current sheet can be considered a quasi-charge-neutral and inviscid fluid. Therefore, only the Lorentz force is considered the volume force, including the Lorentz force F_{induce} generated by the self-induced magnetic field and the Lorentz force F_{applied} generated by the applied magnetic field; only the aerodynamic force F_{gas} is considered the surface force. Therefore, Eq. (5.35) can be rewritten as

$$\frac{d}{dt}[m(t)\dot{x}(t)] = F_{\text{induce}} + F_{\text{applied}} + F_{\text{gas}} + F_{\text{initial}} \tag{5.36}$$

The Lorentz force generated by the self-induced magnetic field is

$$F_{\text{induce}} = \iiint_{\substack{\text{current} \\ \text{sheet}}} J(t)B_{\text{induce}}(t, x, y)dv$$

$$= \int_0^h \int_0^w \int_{x(t)}^{x(t)+\delta} \frac{I(t)}{w\delta} \cdot \frac{\mu_0 I(t)}{w}\left[\frac{\delta + x(t) - x}{\delta}\right]dxdydz$$

$$= \frac{1}{2}\mu_0\frac{h}{w}[I(t)]^2 \tag{5.37}$$

where $J(t)$ is the current density across the current sheet.

Since the applied magnetic field is uniformly distributed in space and time, the Lorentz force generated by the applied magnetic field is as follows:

$$F_{\text{applied}} = \int_0^h \int_0^w \int_{x(t)}^{x(t)+\delta} \frac{I(t)}{w\delta}B_{\text{applied}}dxdydz = hB_{\text{applied}}I(t) \tag{5.38}$$

The aerodynamic force on the current sheet can be expressed as

$$F_{\text{gas}} = hwn_e kT_e \tag{5.39}$$

where k is the Boltzmann constant.

According to the momentum theorem, the force equivalent to the initial momentum of the current sheet can be expressed as

$$F_{\text{initial}} = \frac{\text{d}}{\text{d}t}[m(t)v_{\text{ini}}(t)] \tag{5.40}$$

where $v_{\text{ini}}(t)$ is the velocity of the current sheet as it enters the discharge channel.

Therefore, the governing equation of motion can be written as

$$\frac{\text{d}}{\text{d}t}[m(t)\dot{x}(t)] = \frac{1}{2}\mu_0 \frac{h}{w}[I(t)]^2 + hB_{\text{applied}}I(t)$$
$$+ hwn_e kT_e + \frac{\text{d}}{\text{d}t}[m(t)v_{\text{ini}}(t)] \tag{5.41}$$

The propellant mass of the LPPT is supplied by short-pulse laser ablation, and almost no additional mass is generated during the discharge acceleration stage. Therefore, this model assumes that the current sheet mass enters the discharge channel in its entirety immediately at the beginning of discharge and remains unchanged during the whole discharge process. This mass is set to m_0, i.e., the ablated mass of the single laser pulse, in the model. Moreover, it is assumed that all the propellants entering the discharge channel have the same initial velocity, which is set to v_{ini} in the model. Therefore, Eq. (5.41) can be simplified as

$$m_0\ddot{x}(t) = \frac{1}{2}\mu_0 \frac{h}{w}[I(t)]^2 + hB_{\text{applied}}I(t) + hwn_e kT_e \tag{5.42}$$

Combining Eqs. (5.26), (5.34), and (5.42) yields the electromechanical model of the LPPT with an applied magnetic field, namely

$$\begin{cases} V_0 - \frac{1}{C}\int_0^t I(t)\text{d}\tau = \dot{I}(t)\left[L_c + L_e + \mu_0\frac{h}{w}x(t)\right] \\ \quad + I(t)R_{\text{total}}(t) + I(t)\mu_0\frac{h}{w}\dot{x}(t) + hB_{\text{applied}}\dot{x}(t) \\ m_0\ddot{x}(t) = \frac{1}{2}\mu_0\frac{h}{w}[I(t)]^2 + hwn_e kT_e + hB_{\text{applied}}I(t) \\ R_p = 8.08\frac{h}{T_e^{\frac{3}{4}}w}\sqrt{\dfrac{\mu_0\ln\left[1.24\times10^7\left(\frac{T_e^3}{n_e}\right)^{\frac{1}{2}}\right]}{\tau}} \end{cases} \tag{5.43}$$

where the initial conditions are $x(0) = 0$, $\dot{x}(0) = v_{\text{ini}}$, and $I(0) = 0$.

By solving the above system of equations, the discharge voltage, discharge current, and current sheet motion parameters during the LPPT operation process can be calculated, and then these parameters can be used to calculate the relevant performance parameters of the LPPT.

By setting the time when the current sheet is ejected from the plate to be t^*, that is, $x(t^*) = l$, the velocity v_{out} when the current sheet is ejected from the plate is

$$v_{out} = \dot{x}(t^*) \tag{5.44}$$

The specific impulse I_{sp} can be expressed as

$$I_{sp} = \frac{\dot{x}(t^*)}{g} \tag{5.45}$$

The impulse bit generated by the self-induced magnetic field (hereinafter referred to as the impulse bit of the self-induced magnetic field) is

$$I_{bit\text{-}induce} = \int_0^{t^*} F_{induce} d\tau = \frac{\mu_0 h}{2w} \int_0^{t^*} [I(\tau)]^2 d\tau \tag{5.46}$$

The impulse bit generated by the applied magnetic field (hereinafter referred to as the impulse bit of the applied magnetic field) is

$$I_{bit\text{-}applied} = \int_0^{t^*} F_{applied} d\tau = h B_{applied} \int_0^{t^*} I(\tau) d\tau \tag{5.47}$$

The impulse bit generated by the aerodynamic force (hereinafter referred to as the aerodynamic impulse bit) is

$$I_{bit\text{-}gas} = \int_0^{t^*} F_{gas} d\tau = h w n_e k T_e t^* \tag{5.48}$$

The impulse bit generated by the initial velocity of the current sheet (hereinafter referred to as the initial velocity impulse bit) is

$$I_{bit\text{-}initial} = m_0 v_{ini} \tag{5.49}$$

The impulse bit I_{bit} can be expressed as

$$I_{bit} = m_0 v_{ini} + \frac{\mu_0 h}{2w} \int_0^{t^*} [I(\tau)]^2 d\tau + hwn_e kT_e t^* + hB_{applied} \int_0^{t^*} I(\tau) d\tau \qquad (5.50)$$

The thrust efficiency η_{th} can be expressed as

$$\eta_{th} = \frac{I_{bit}^2}{2m_0(E_C + E_{laser})} = \frac{I_{bit}^2}{m_0(CV_0^2 + 2E_{laser})} \qquad (5.51)$$

where E_c is the discharge energy and E_{laser} is the single-pulse laser energy.

With Eqs. (5.43)–(5.51), the discharge characteristics and propulsion performance of the LPPT can be obtained. Moreover, by setting the initial velocity of the propellant and laser energy to zero, the electromechanical model can be used for the simulation of parallel-plate PPTs with an applied magnetic field.

5.3.2 Model Validation

1. Model validation when the applied magnetic field is zero

The LES-6 PPT and LES-8/9 PPT are two well-developed types of parallel-plate PPTs with a background in space flight applications. To date, a large number of experimental and theoretical studies have been carried out on these two types of PPTs. Furthermore, the discharge energies of these two types of PPTs differ greatly, so the applicability of the present model can be better demonstrated by using their experimental data for model validation. In this chapter, these two types of PPTs are chosen as the research objects to validate the created model when the applied magnetic field is zero. The parameters selected for the simulation process are shown in Table 5.4. The experimental and simulation results are compared in Table 5.5. As shown in the table, the simulation results and experimental results are in high agreement, indicating that the model can effectively perform in situations when the applied magnetic field is zero.

2. Model validation when the applied magnetic field is nonzero

In this chapter, the model with an applied magnetic field is validated using the relevant parameters of the TMU PPT, as shown in Table 5.6. The magnetic field settings in the simulation are consistent with those in the experiment, except that an external magnetic field is applied to the front 17.5 mm of the thruster plate. A comparison of the simulation results and experimental results is shown in Fig. 5.16. It is observed that although the simulation results are generally consistent with the experimental results, there are some differences. For example, there is a significant difference between the simulation and experimental results for the performance parameters when the applied magnetic field is 0.3 T. This difference may be caused by the improper setting of the plasma resistance; the results would match better if the plasma resistance is reduced

Table 5.4 Relevant parameters of the LES-6 PPT and LES-8/9 PPT

Thruster	LES-6 PPT	LES-8/9 PPT
Initial voltage (V)	1360	1538
Capacitance (μF)	2	17
Capacitance resistance (mΩ)	30	30
Initial inductance (nH)	34	35
Plate spacing (mm)	30	25.4
Plate width (mm)	10	25.4
Plate length (mm)	6	25.4
Characteristic pulse time (μs)	0.4	1.0
Plasma temperature (eV)	1.5	5.0
Electron number density (m^{-3})	1e21	1e21
Single-pulse mass (μg)	10	28.5

Table 5.5 Comparison of the experimental and simulation results

Thruster	LES-6 PPT		LES-8/9 PPT	
	Experiment	Simulation	Experiment	Simulation
Impulse bit (μN s)	32	32.6	300	298.8

in this case. The simulation and experimental results are in overall good agreement despite some differences between them, suggesting that the model can effectively predict the PPT performance when the applied magnetic field is nonzero. Moreover, since the discharge energy of the TMU PPT is 125.0 J, the good applicability of the model in this chapter is further verified.

Table 5.6 Relevant parameters of the TMU PPT

Initial voltage (V)	2500	Capacitance resistance (mΩ)	30		
Capacitance (μF)	40	Initial Inductance (nH)	120		
Plate width (mm)	15	Strength of the applied magnetic field (T)	0	0.15	0.30
Plate length (mm)	60	Single-pulse mass (μg)	320	120	60
Plate spacing (mm)	50	Plasma resistance (mΩ)	18	8	4

Fig. 5.16 Discharge current curves under different capacitances

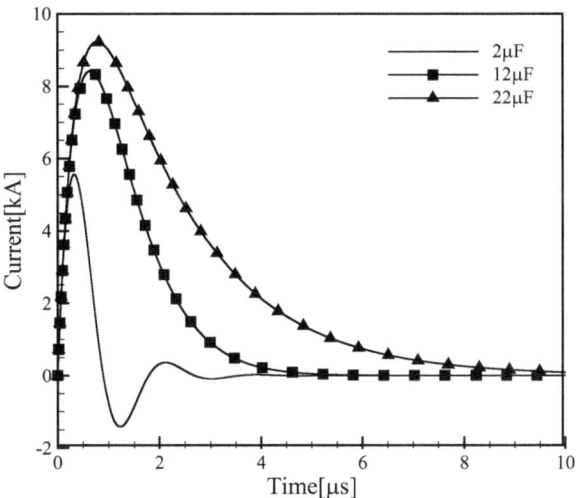

5.4 Numerical Simulation Results and Analysis

5.4.1 Influence of the Electrical Parameters on the PPT Performance

The electrical parameters of the PPT mainly include the circuit capacitance (primarily the capacitance of the capacitor), initial discharge voltage at both ends of the capacitor, circuit resistance, and inductance. The capacitance and initial discharge voltage of the capacitor determine the discharge energy of a single-pulse operation of the PPT. Changing the capacitance and initial discharge voltage of the capacitor provides two main ways to control the operating energy of the thruster and has a great impact on the performance of the thruster. The electrical parameters of the circuit affect the specific discharge process and energy conversion process of the thruster, considerably affecting the efficiency of the thruster. The numerical simulation of the operation process of the thruster is performed by changing one electrical parameter at a time while keeping the other parameters constant to obtain the variation patterns of the discharge waveform, performance parameters, and propellant ablation characteristics of the thruster with different electrical parameters.

1. Analysis of the influence of the capacitor capacitance on the PPT performance

When the other operating parameters are constant, increasing the capacitance of the capacitor means increasing the discharge energy per operation. Figures 5.16, 5.17, 5.18, 5.19, and 5.20 show the simulation results of the PPT obtained by changing the capacitance of the capacitor, while keeping the other parameters constant.

Fig. 5.17 Discharge voltage
curves under different
capacitances

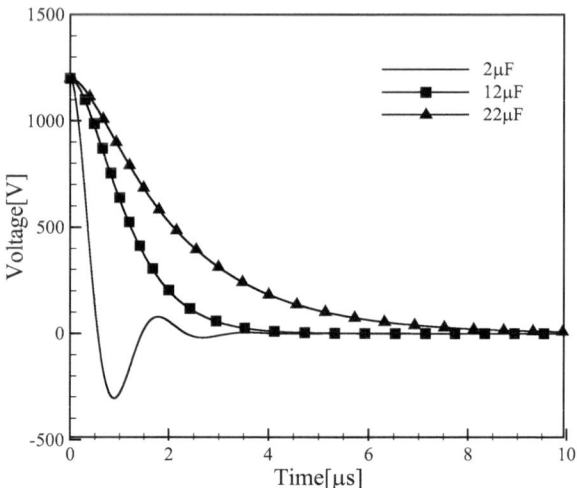

Fig. 5.18 Ablated surface
temperature of the propellant
under different capacitances

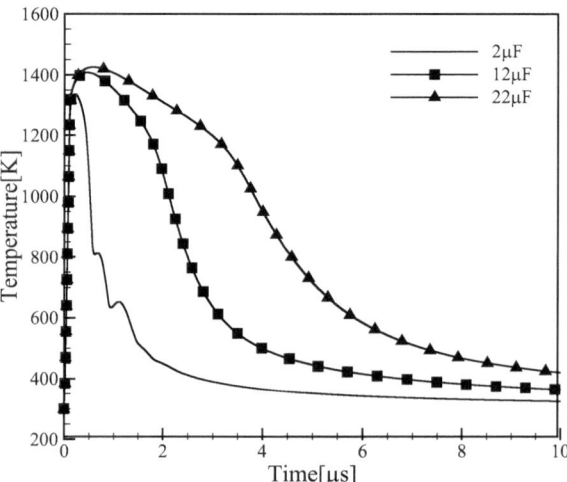

Figures 5.16 and 5.17 present the discharge current curve and discharge voltage
curve under different capacitances, respectively. It is observed that when the capacitance of the capacitor increases, the discharge cycle of the PPT increases, the peak
discharge current increases, the reverse current decreases, and the rate of current
change decreases, effectively reducing the impact of the current on the capacitor
and improving its service life. When the capacitance increases to a certain value,
the reverse current disappears. As the capacitance increases, the rate of increase in
the thruster specific impulse gradually decreases. Figure 5.18 shows the variation
curve of the ablated surface temperature of the propellant over time under different

Fig. 5.19 Variation in the ablated mass of the propellant with capacitance

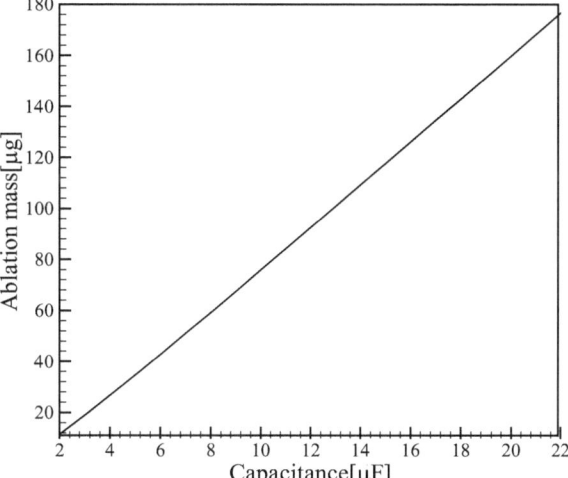

Fig. 5.20 Variations in the specific impulse and impulse bit of the PPT with capacitance

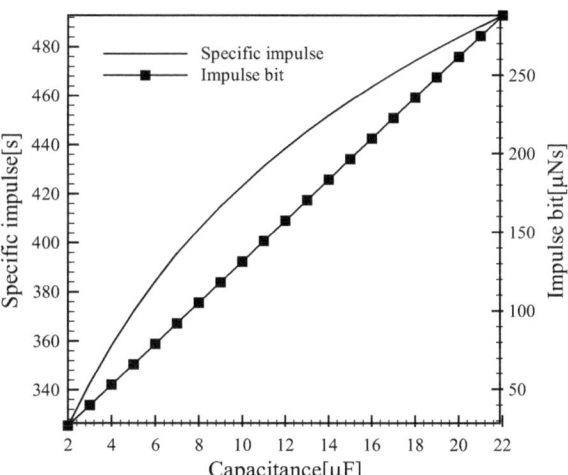

capacitances. It is observed that increasing the capacitance can reduce the temperature fluctuation of the ablated surface of the propellant, increase the peak ablated surface temperature of the propellant, and greatly prolong the duration during which the propellant surface temperature is higher than the melting temperature. This is mainly because the increase in the capacitance reduces the discharge waveform fluctuation and increases the discharge energy. An increase in the peak ablated surface temperature of the propellant and an increase in the duration of the temperature above the melting temperature increase the ablated mass of the propellant, as shown in Fig. 5.19. Figure 5.20 shows the variation curves of PPT's specific impulse and impulse bit with an increase in capacitance. As observed in this figure, both the specific impulse and impulse bit of the thruster increase with increasing capacitance. In summary, when the other operating parameters are constant, increasing the capacitance can effectively weaken the oscillation characteristics of the circuit and improve the overall performance of the thruster. However, increasing the capacitance will inevitably increase the volume and mass of the capacitor, thereby increasing the overall volume and mass of the thruster system. Therefore, in the actual PPT design process, the capacitance should be reasonably selected by comprehensively considering the requirements of the flight mission on the performance, volume, and mass of the thruster.

2. Analysis of the influence of the initial discharge voltage on the PPT performance

When the other operating parameters are constant, increasing the initial discharge voltage means increasing the discharge energy per operation. Figures 5.21, 5.22, 5.23, 5.24, and 5.25 show the simulation results of the PPT by changing the initial discharge voltage, while keeping the other parameters unchanged.

Figures 5.21 and 5.22 present the discharge current curves and discharge voltage curves under different initial discharge voltages, respectively. It is observed in these figures that as the initial discharge voltage increases, the peak discharge current increases significantly while the discharge cycle remains unchanged. Increasing the initial discharge voltage also intensifies the oscillation of the discharge waveform, which increases the impact on the capacitor. Therefore, capacitor failure is more likely, which is unfavorable for the service life of the capacitor. Figure 5.23 presents the variations in the ablated surface temperature of the propellant over time under different initial discharge voltages. As shown in this figure, increasing the initial discharge voltage increases the ablated surface temperature of the propellant and prolongs the duration during which the surface temperature of the propellant exceeds the melting temperature. As the initial discharge voltage increases, the increment of

Fig. 5.21 Discharge current curves under different initial voltages

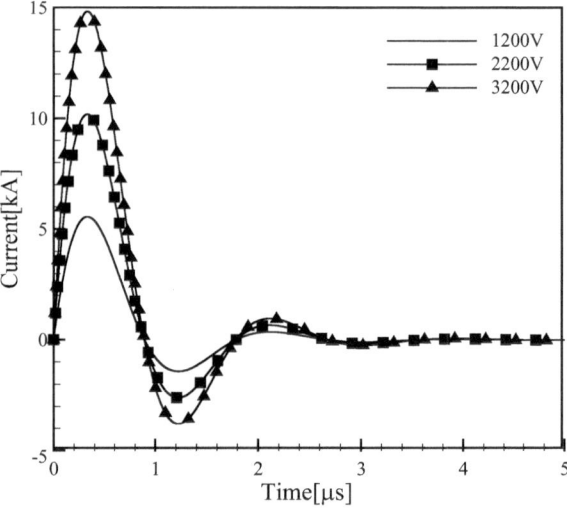

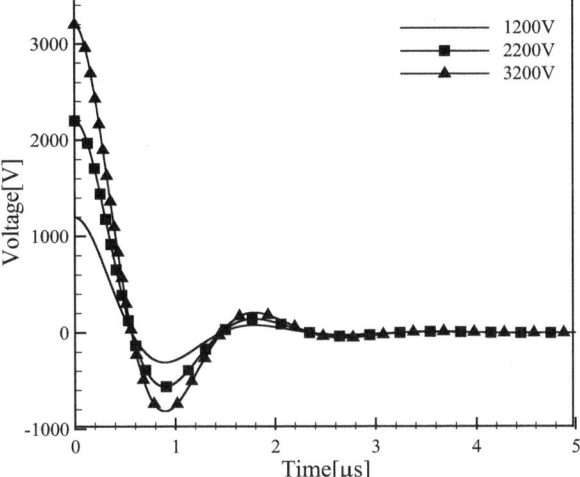

Fig. 5.22 Discharge voltage curves under different initial voltages

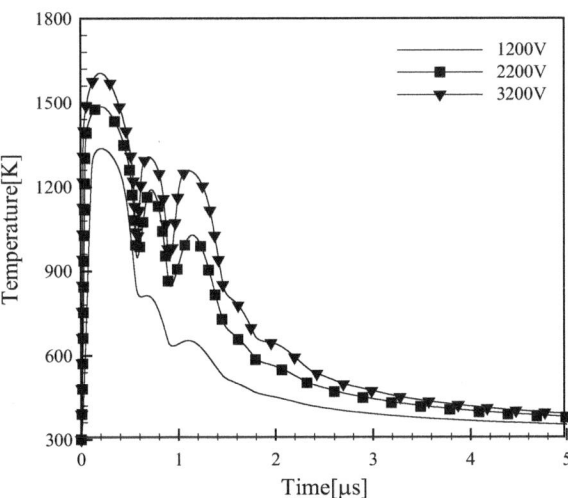

Fig. 5.23 Ablated surface temperature of the propellant under different initial voltages

the ablated surface temperature of the propellant gradually decreases. The increase in the ablated surface temperature of the propellant and the prolongation of the duration of temperature exceeding the melting temperature result in an increase in the single-pulse ablated mass of the propellant, as shown in Fig. 5.24. Since the ablation rate of the solid propellant is very sensitive to the propellant temperature, the ablation rate of the propellant can be effectively increased by increasing the propellant temperature. Therefore, the increment of the single-pulse ablated mass of the propellant increases with increasing initial discharge voltage. Figure 5.25 shows the variations in the specific impulse and impulse bit of the PPT with the initial discharge voltage. As the initial discharge voltage increases, the specific impulse of the thruster gradually

Fig. 5.24 Variation in the ablated mass of the propellant with the initial voltage

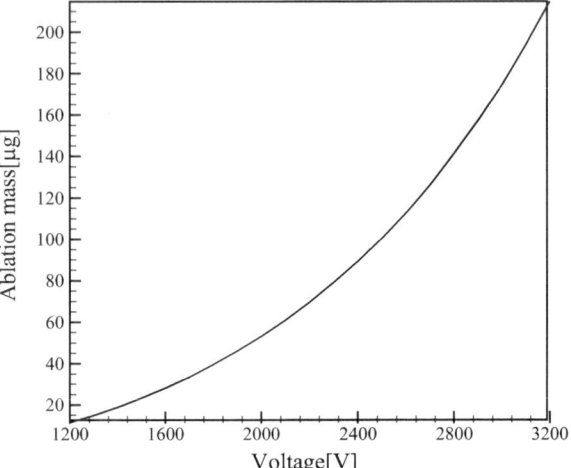

Fig. 5.25 Variations in the specific impulse and impulse bit of the PPT with initial voltage

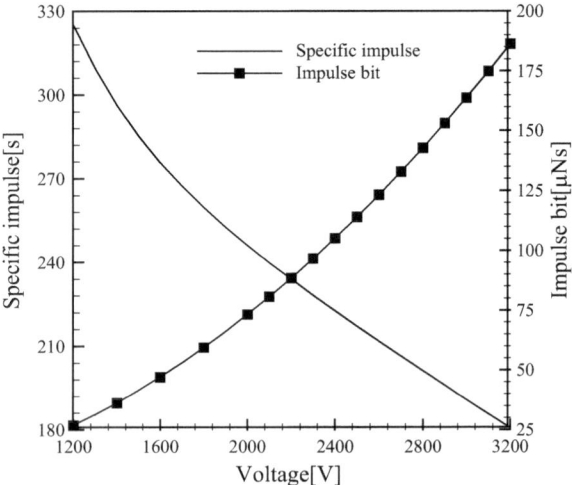

decreases, while the impulse bit of the thruster gradually increases. This is mainly caused by the excessive initial discharge voltage resulting in an overly high ablated mass of the propellant. Increasing the initial discharge voltage increases the ablated mass and impulse bit of the thruster but reduces the specific impulse of the thruster. In addition, increasing the initial discharge voltage increases the peak discharge current and current oscillation, thus reducing the service life of the capacitor.

3. Analysis of the influence of different capacitances and initial voltages on the PPT performance under the same discharge energy

Keeping the single-pulse discharge energy at 15 J, the operation process of the PPT is simulated by changing different combinations of the capacitance and initial discharge voltage of the capacitor to study the influence of different energy release modes under the same discharge energy on the overall performance of the thruster. The calculation results are shown in Figs. 5.26, 5.27, 5.28, 5.29, and 5.30.

According to $E = 1/2\, C_0 V$, to keep the discharge energy of the thruster constant, the initial discharge voltage must be reduced while simultaneously increasing the capacitance of the capacitor. As shown in Figs. 5.26 and 5.27, increasing the capacitance increases the discharge cycle of the thruster, reduces the peak discharge current, and weakens the oscillation of the discharge waveform. The service life of the capacitor is a key factor affecting that of the thruster. The weakening of the discharge

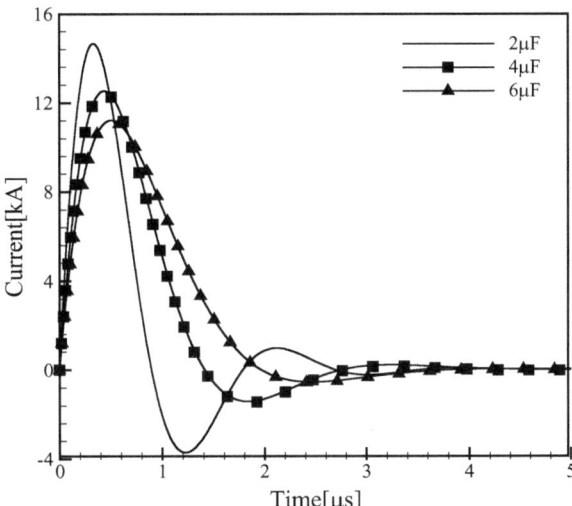

Fig. 5.26 Discharge current waveforms under the same discharge energy and different capacitances, and initial voltages

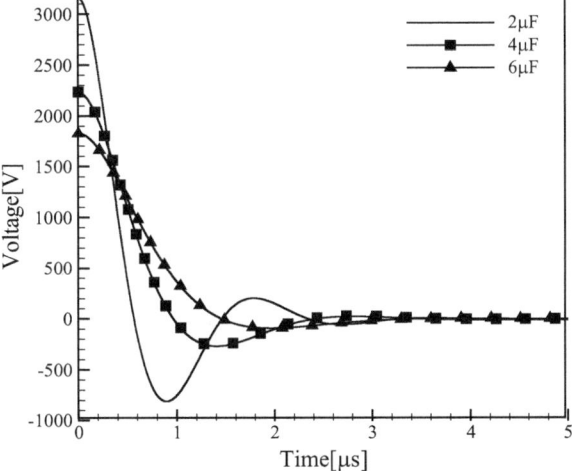

Fig. 5.27 Discharge voltage waveforms under the same discharge energy and different capacitances and initial voltages

Fig. 5.28 Ablated surface temperature of the propellant under the same discharge energy and different capacitances and initial voltages

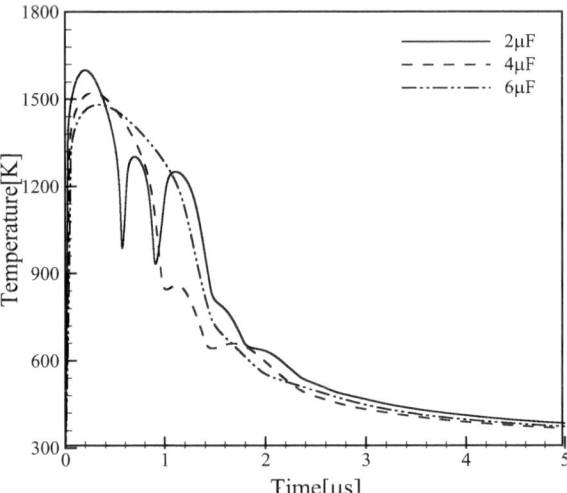

Fig. 5.29 Ablated mass of the propellant under the same discharge energy and different capacitances and initial voltages

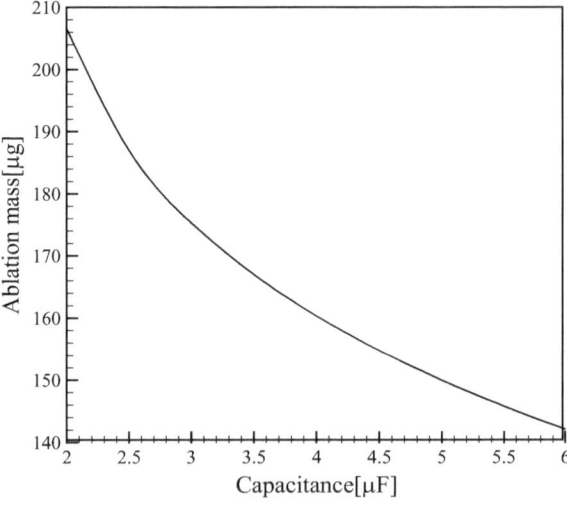

waveform oscillation is helpful for extending the service life of the capacitor and thus that of the thruster. As shown in Fig. 5.28, while the discharge energy of the thruster is constant, as the capacitance of the capacitor increases, the peak ablated surface temperature of the propellant decreases, and the rate of change of the ablated surface temperature of the propellant stabilizes, thereby causing a gradual reduction in the single-pulse ablated mass of the propellant. The eclipse quality gradually decreases. Figure 5.30 shows the variations in the specific impulse and the impulse bit of the thruster with the capacitance of the capacitor when the discharge energy of the thruster is constant. It is observed that the specific impulse of the PPT increases as the

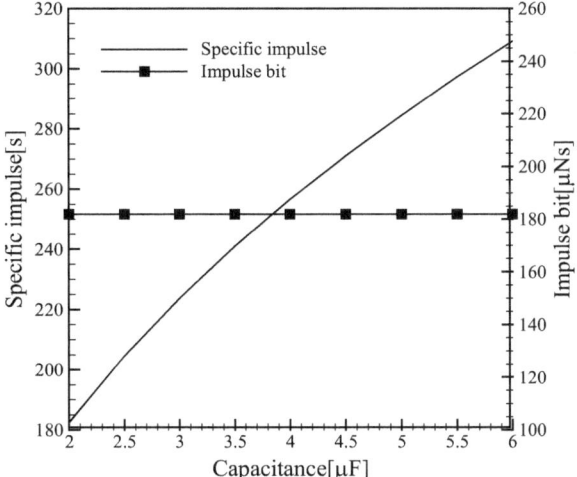

Fig. 5.30 Specific impulse and impulse bit under the same discharge energy and different capacitances and initial voltages

capacitor capacitance increases. Additionally, since the single-pulse ablated mass of the propellant decreases as the capacitance increases, the impulse bit of the thruster does not change with the capacitance under the same discharge energy. Therefore, under the same thruster discharge energy, the overall performance of the thruster can be improved by selecting a larger capacitance. However, increasing the capacitance will increase the mass and volume of the capacitor, accounting for a major part of the overall mass and volume of the thruster. Therefore, an increase in the mass and volume of the capacitor will significantly increase the overall mass and volume of the thruster. Accordingly, in the actual PPT design process, after determining the discharge energy level of the thruster, it is necessary to reasonably select the capacitor capacitance and initial discharge voltage based on the performance and space requirements of the capacitor. Under the same thruster discharge energy, choosing a large capacitor capacity and a small initial discharge voltage within a reasonable range can effectively improve the overall performance of the thruster and prolong its service life.

4. Analysis of the influence of the circuit resistance on the PPT performance

The circuit resistance of the PPT is an important factor that causes discharge energy loss and affects the thruster performance. As shown in Figs. 5.31 and 5.32, increasing the circuit resistance reduces the peak discharge current while increasing the damping of the oscillating circuit, thereby weakening the oscillation of the discharge waveform.

As shown in Fig. 5.33, increasing the circuit resistance has little impact on the ablated surface temperature of the propellant; the peak value and duration of the ablated surface temperature of the propellant above the melting temperature only slightly decrease with increasing circuit resistance. As shown in Fig. 5.34, the single-pulse ablated mass of the propellant decreases with increasing circuit resistance. The main reason is that the increase in circuit resistance increases the ohmic heat

Fig. 5.31 Discharge current curves under different circuit resistances

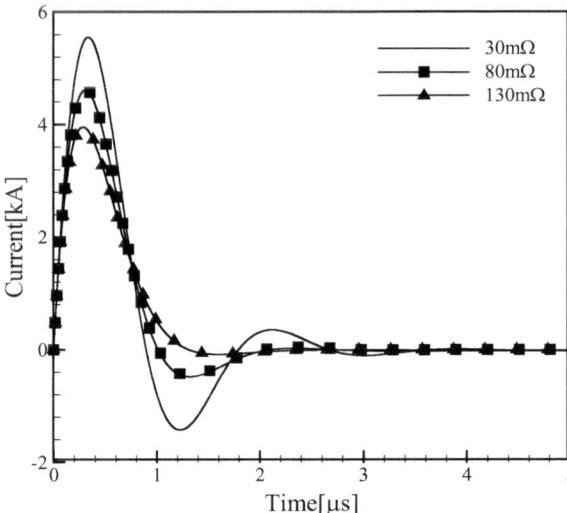

Fig. 5.32 Discharge voltage curves under different circuit resistances

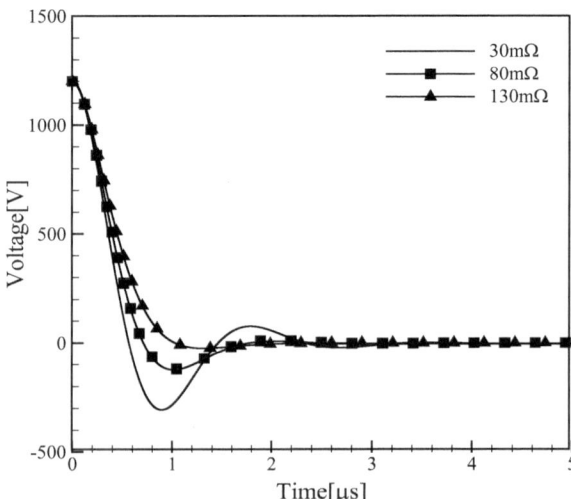

loss from the circuit, reducing the energy available for ablating the propellant and accelerating the plasma. Figure 5.35 shows the variations in the specific impulse and impulse bit of the thruster with the circuit resistance. It is observed that both the specific impulse and impulse bit of the thruster decrease with increasing circuit resistance. The circuit resistance increases the system energy loss, severely affecting the overall performance of the thruster. Therefore, in the actual design process of a PPT, the circuit resistance should be minimized to achieve a high overall thruster performance.

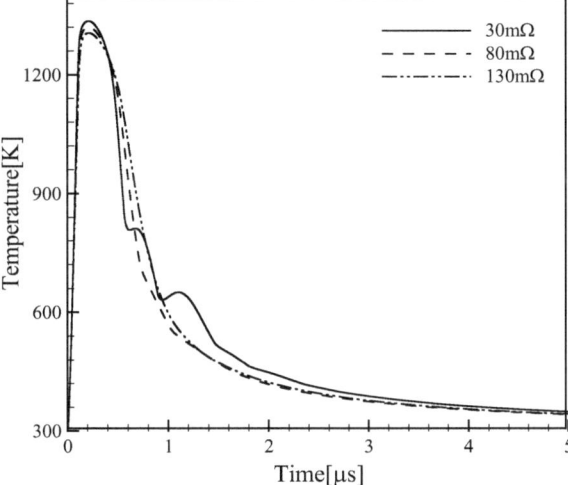

Fig. 5.33 Ablated surface temperature of the propellant under different circuit resistances

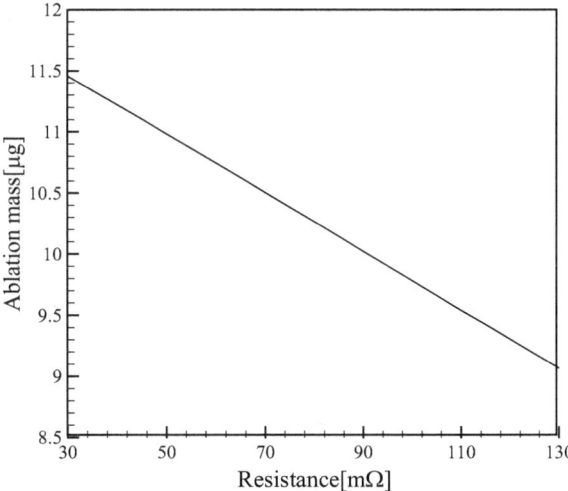

Fig. 5.34 Ablated mass of the propellant under different circuit resistances

5. Analysis of the influence of the plasma temperature on the PPT performance

The plasma temperature has a great impact on the plasma resistance, and it also has an important influence on the propellant ablation and plasma acceleration processes. Figures 5.36 and 5.37 show that as the plasma temperature increases, the peak discharge current increases and the plasma resistance decreases, thus the oscillation of the discharge waveform increases.

Figure 5.38 shows the variations in the ablated surface temperature of the propellant under different plasma temperatures. The peak ablated surface temperature of the propellant does not change with increasing plasma temperature, but the duration

Fig. 5.35 Specific impulse and impulse bit under different circuit resistances

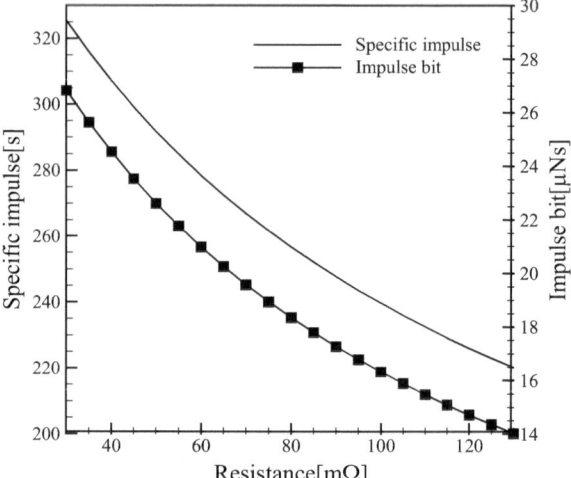

Fig. 5.36 Discharge current curves under different plasma temperatures

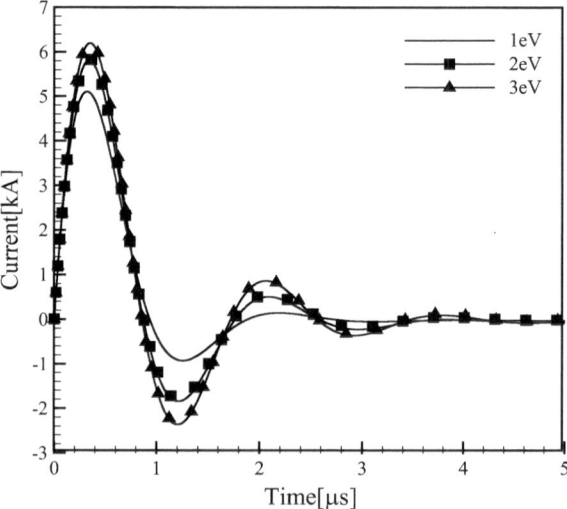

of the ablated surface temperature of the propellant being higher than the melting temperature is prolonged. As a result, the single-pulse ablated mass of the propellant increases slightly, as shown in Fig. 5.39. It is observed in Fig. 5.40 that increasing the plasma temperature increases the specific impulse and impulse bit of the thruster, effectively improving the overall performance of the PPT. In the actual operation process of a PPT, increasing the plasma temperature increases the degree of ionization of the ablated propellant, which can effectively improve the utilization rate of the propellant.

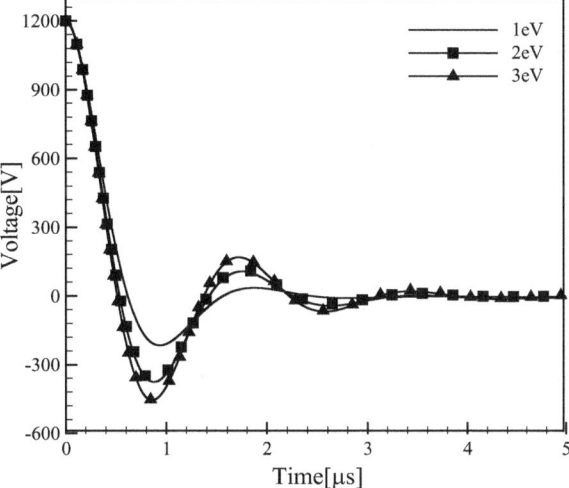

Fig. 5.37 Discharge voltage curves under different plasma temperatures

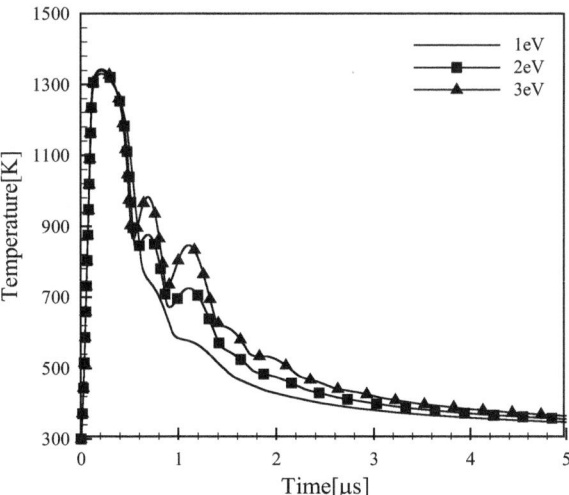

Fig. 5.38 Ablated surface temperature of the propellant under different plasma temperatures

5.4.2 Influence of the Structural Parameters on the PPT Performance

1. Analysis of the influence of the plate spacing on the PPT performance

Figures 5.41, 5.42, 5.43, 5.44, and 5.45 show the simulation results of the thruster operation process obtained by increasing the plate spacing while keeping other operating parameters constant. The plate spacing is one of the main structural parameters of the PPT. Different plate spacings directly affect the plasma resistance of the thruster, the distribution of arcs on the ablated surface of the propellant, and the

Fig. 5.39 Ablated mass of the propellant under different plasma temperatures

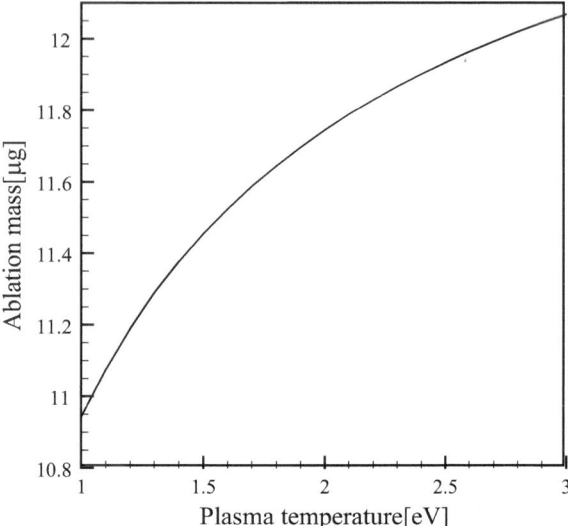

Fig. 5.40 Specific impulse and impulse bit under different plasma temperatures

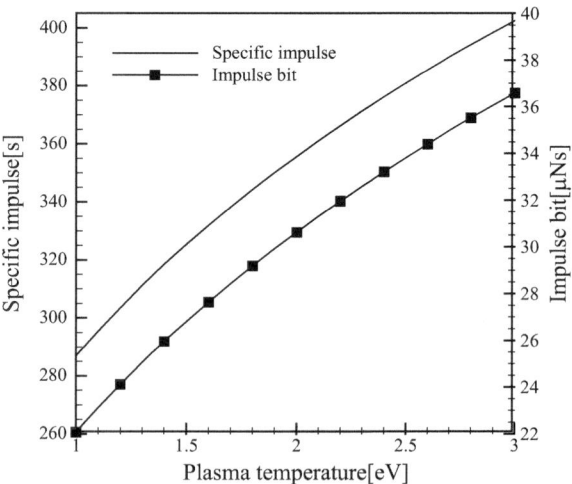

ablation process of the propellant, which have a great impact on the overall performance and ignition reliability of the thruster. When other operating parameters are held constant, increasing the plate spacing increases the plasma resistance, and the area of the ablated surface of the propellant also increases with increasing plate spacing. As a result, the fluence per unit area on the ablated surface of the propellant decreases. As shown in Figs. 5.41 and 5.42, as the plate spacing increases, the peak discharge current gradually decreases, and the fluctuations in the discharge current and discharge voltage gradually decrease. This is mainly because the thruster circuit resistance increases with increasing plate spacing. As shown in Fig. 5.43, as the

plate spacing increases, the peak ablated surface temperature of the propellant gradually decreases, and the duration of the ablated surface temperature exceeding the
melting temperature gradually decreases. Therefore, the single-pulse ablated mass
of the propellant also significantly decreases with increasing plate spacing, as shown
in Fig. 5.44.

Fig. 5.41 Discharge current
under different plate spacings

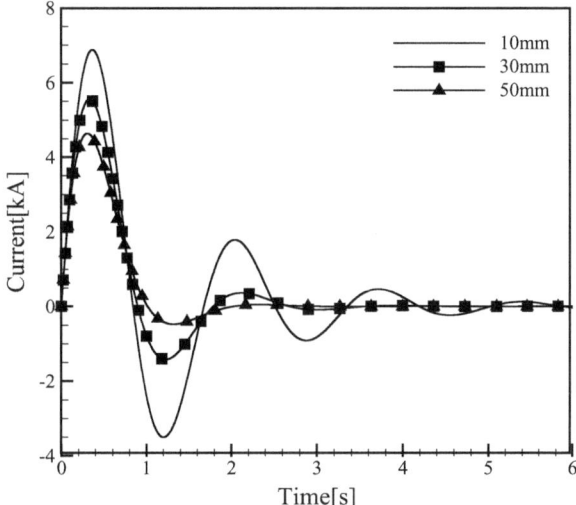

Fig. 5.42 Discharge voltage
under different plate spacings

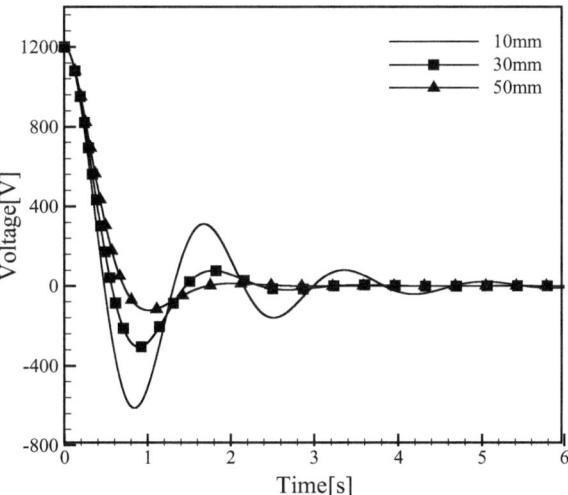

Fig. 5.43 Ablated surface
temperature under different
plate spacings

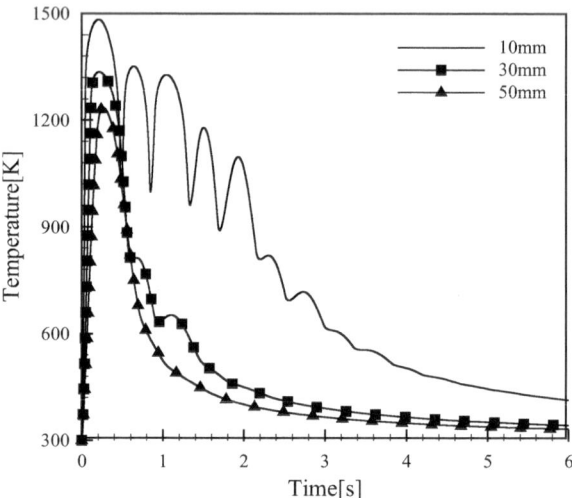

Fig. 5.44 Ablated mass
under different plate spacings

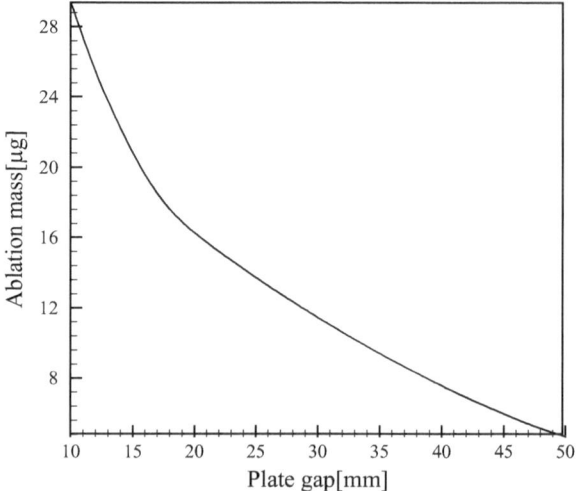

Figure 5.45 shows the variations in the specific impulse and impulse bit of the
thruster with the plate spacing. It is observed that the specific impulse and impulse
bit of the thruster both increase with increasing plate spacing. This is mainly because
increasing the plate spacing effectively reduces the single-pulse ablated mass of the
thruster propellant; under the same discharge energy, a higher thruster performance
can be obtained with a lower ablated mass of the propellant. As the plate spacing
gradually increases, the specific impulse increment per unit plate spacing gradually
increases, while the impulse bit increment per unit plate spacing gradually decreases.
Therefore, increasing the plate spacing when the plate spacing is large can effectively

Fig. 5.45 Specific impulse and impulse bit under different plate spacings

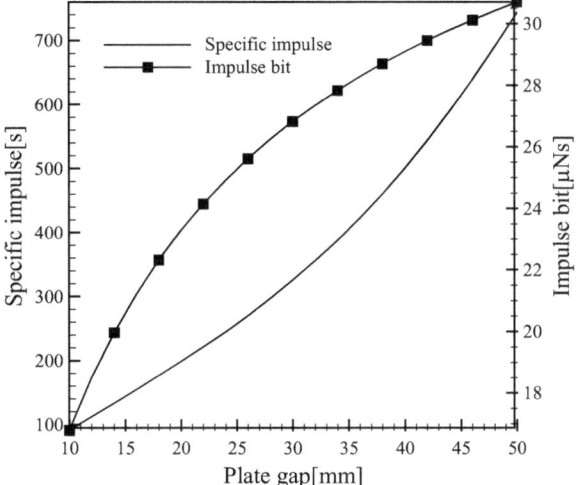

improve the specific impulse of the thruster, and increasing the plate spacing when the plate spacing is small can effectively improve the impulse bit of the thruster.

2. Analysis of the influence of the plate width on the PPT performance

With the other operating parameters held constant, increasing the plate width reduces the plasma resistance and thus lowers the resistance of the whole circuit. Increasing the plate width increases the area of the ablated surface of the propellant, which decreases the fluence at the ablated surface. As shown in Figs. 5.46 and 5.47, as the plate width increases, the peak discharge current gradually decreases, the circuit impedance gradually increases, and the oscillation of the discharge curve weakens. It is observed in Fig. 5.48 that as the plate width increases, the peak ablated surface temperature of the propellant gradually decreases, and so does the duration of this temperature exceeding the melting temperature. Therefore, the single-pulse ablated mass of the propellant decreases with increasing plate width, as shown in Fig. 5.49. Figure 5.50 presents the variations in the specific impulse and impulse bit of the thruster with plate width. It is observed that as the plate width increases, the specific impulse of the thruster increases while the impulse bit of the thruster decreases. This is mainly attributed to a decrease in the single-pulse ablated mass of the propellant due to a decrease in the fluence per unit of the ablated surface.

3. Analysis of the influence of different aspect ratios on the PPT performance under the same exposure area

First, the aspect ratio is defined as the ratio of the plate spacing to the plate width. The spacing and width of the plate determine the area of the ablated surface of the propellant. Under the same area of the ablated surface of the propellant, different aspect ratios have significantly different impacts on the operation process and overall performance of the thruster. Figures 5.51 and 5.52 show that as the aspect ratio of

Fig. 5.46 Discharge current curves under different plate widths

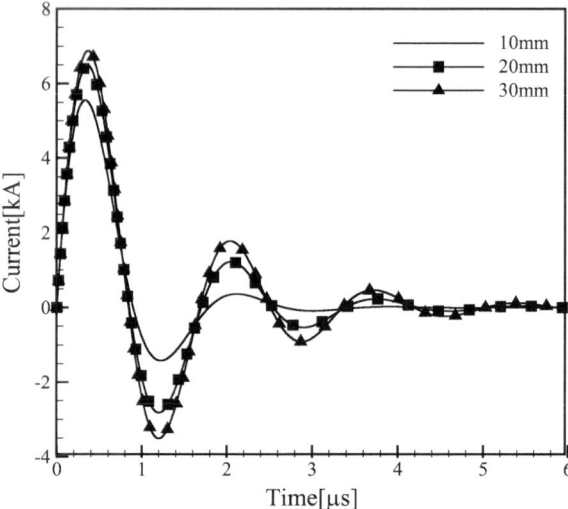

Fig. 5.47 Discharge voltage curves under different plate widths

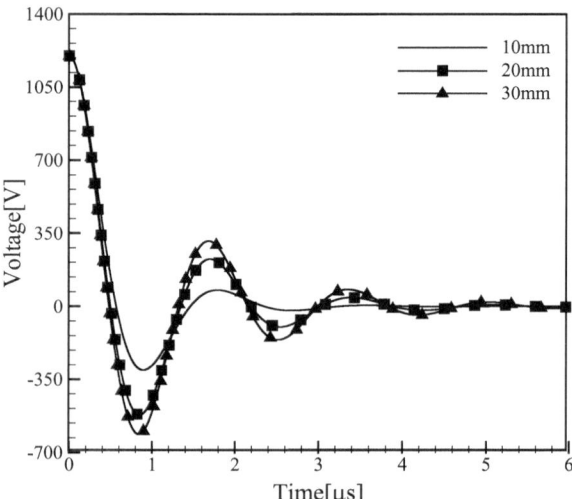

the plate increases, the peak discharge current gradually decreases, the circuit resistance gradually increases, and the degree of oscillation of the discharge waveform weakens. As shown in Fig. 5.53, as the aspect ratio of the plate increases, the peak ablated surface temperature of the propellant remains almost unchanged, but the duration of the ablated surface temperature of the propellant being higher than the melting temperature decreases. Figure 5.54 presents the curve of the variations in the single-pulse ablated mass of the propellant with the aspect ratio of the plate. It is observed that as the aspect ratio of the plate increases, the single-pulse ablated mass of the propellant gradually decreases. Figure 5.55 shows the variations in the specific

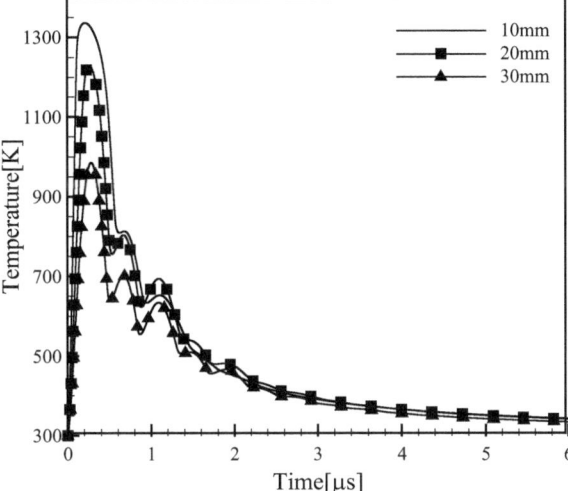

Fig. 5.48 Ablated surface temperature of the propellant under different plate widths

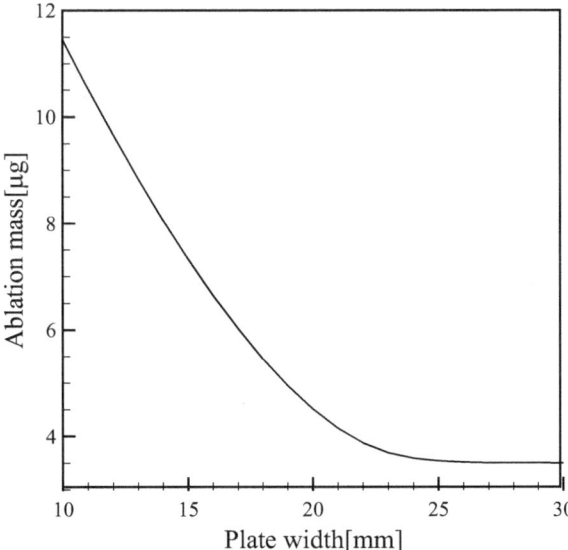

Fig. 5.49 Ablated mass of the propellant under different plate widths

impulse and impulse bit of the thruster with the aspect ratio of the plate. As shown in the figure, the specific impulse and impulse bit of the thruster both increase as the aspect ratio of the plate increases, suggesting that increasing the aspect ratio of the plate can improve the overall performance of the thruster.

Fig. 5.50 Specific impulse
and impulse bit under
different plate widths

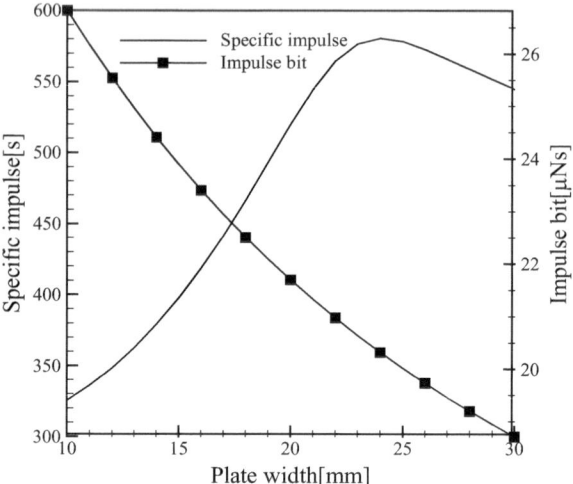

Fig. 5.51 Discharge current
of the propellant with the
same exposure area and
different plate aspect ratios

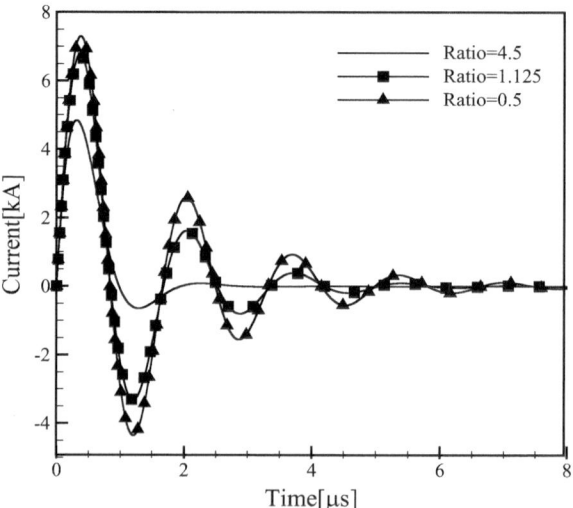

5.4.3 Influence of the Applied Magnetic Field on the PPT Performance

1. Influence of the applied magnetic field intensity on the LPPT performance

In this section, a constant accelerating magnetic field is applied along the entire plate length of the thruster. Figure 5.56 shows the variations in the impulse bit, impulse bit growth rate, and impulse bit relative growth rate with the strength of the applied magnetic field. It is observed that as the strength of the applied magnetic

Fig. 5.52 Discharge voltage of the propellant with the same exposure area and different plate aspect ratios

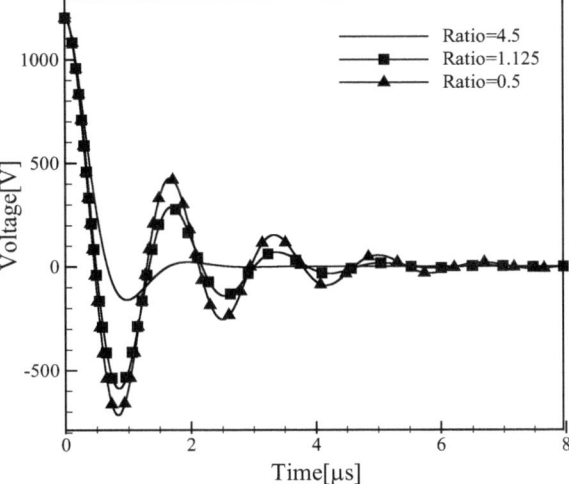

Fig. 5.53 Ablated surface temperature of the propellant with the same exposure area and different plate aspect ratios

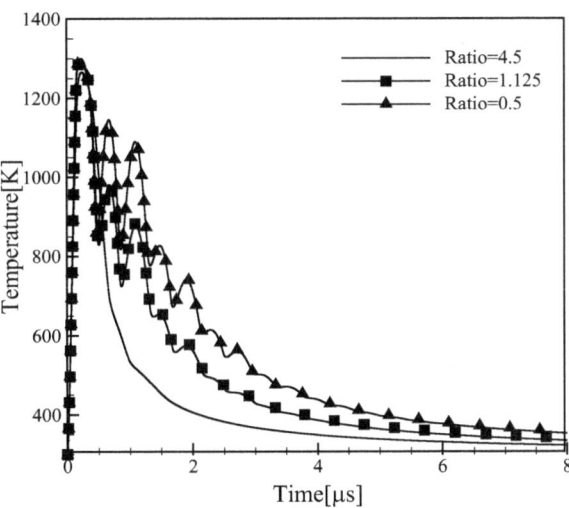

field increases, both the impulse bit and the relative growth rate of the impulse bit first increase and then decrease, reaching a maximum at approximately 1.0 T, while the growth rate of the impulse bits gradually decreases. When the strength of the applied magnetic field is changed, the change in the impulse bit is dominated by the impulse bit of the self-induced magnetic field and is also affected by the impulse bit of the applied magnetic field. Therefore, it is necessary to focus on the analysis of the influence of the strength of the applied magnetic field on the impulse bits of the self-induced magnetic field and the applied magnetic field influence. As shown in Fig. 5.57, as the strength of the applied magnetic field increases, the impulse bit

Fig. 5.54 Ablation mass of the propellant with the same exposure area and different plate aspect ratios

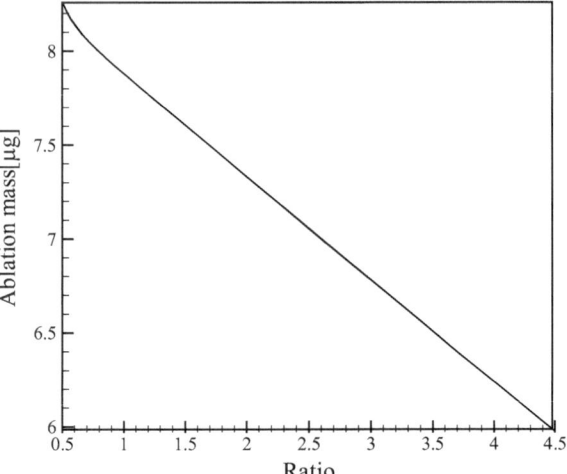

Fig. 5.55 Specific impulse and impulse bit of the propellant with the same exposure area and different plate aspect ratios

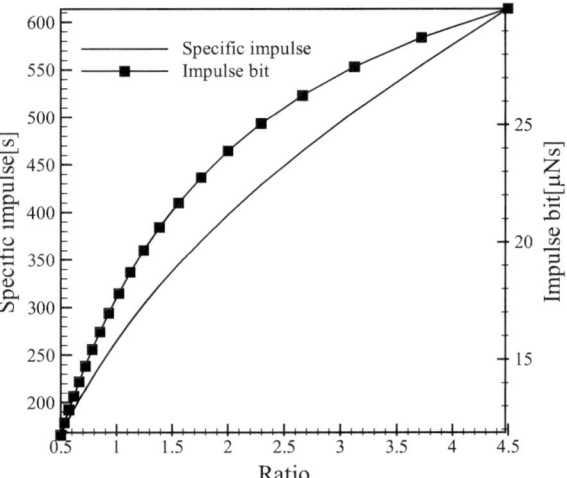

of the self-induced magnetic field gradually decreases, while the impulse bit of the applied magnetic field first increases and then decreases.

When the applied magnetic field is weak (0.0–1.0 T), the discharge current and the square waveform of the discharge current under different strengths of the applied magnetic field are shown in Fig. 5.58. It is observed that as the strength of the applied magnetic field increases, the peak discharge current and the area of the impulse bit of the self-induced magnetic field decrease; therefore, the impulse bit of the self-induced magnetic field decreases. According to Eq. (5.47), the impulse bit generated by the applied magnetic field, on the one hand, is proportional to the strength of the applied magnetic field and, on the other hand, is proportional to the area enclosed

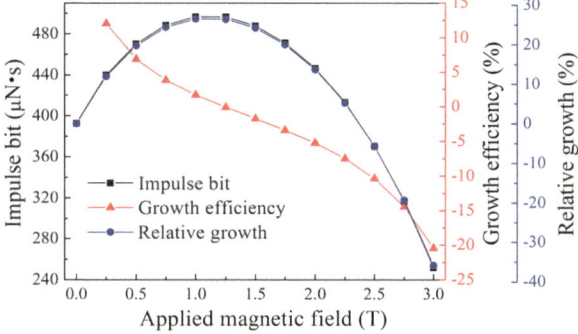

Fig. 5.56 Variations in the impulse bit, growth rate of the impulse bit, and relative growth rate of the impulse bit with the strength of the applied magnetic field

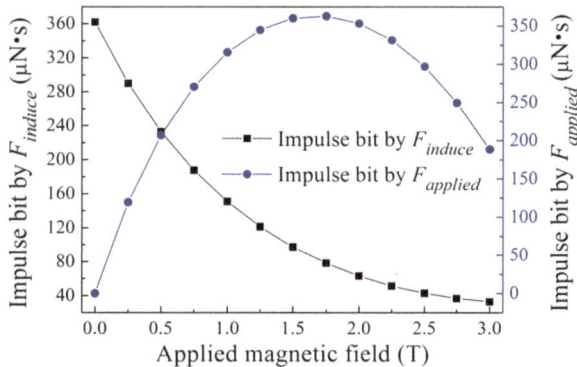

Fig. 5.57 Variations in the impulse bit of the self-induced magnetic field and the impulse bit of the applied magnetic field with the strength of the applied magnetic field

by the discharge current curve and the straight line with an ordinate of 0 (hereinafter referred to as the area of the impulse bit of the applied magnetic field). When the area is greater than zero, a positive impulse bit is generated; when the area is less than zero, a negative impulse bit is generated. As shown in Fig. 5.58, at this time, the area of the impulse bit of the applied magnetic field is always greater than zero, thus generating a positive impulse bit. The area of the impulse bit of the applied magnetic field decreases as the strength of the applied magnetic field increases, thus reducing the positive impulse bit to some extent. However, increasing the strength of the applied magnetic field multiple times leads to multiple increases in the positive impulse bit generated by the applied magnetic field. At this time, increasing the positive impulse bit by increasing the strength of the applied magnetic field plays a dominant role; therefore, an increase in the applied magnetic field increases the positive impulse bit generated by it. Although the increase in the impulse bit generated by the applied magnetic field continues until the applied magnetic field reaches approximately 1.75 T, after the applied magnetic field exceeds 1.0 T, the increase in the impulse bit of the applied magnetic field is less than the decrease in the impulse bit of the self-induced magnetic field caused by the increase in the strength of the applied magnetic field. Therefore, after the strength of the applied magnetic field is

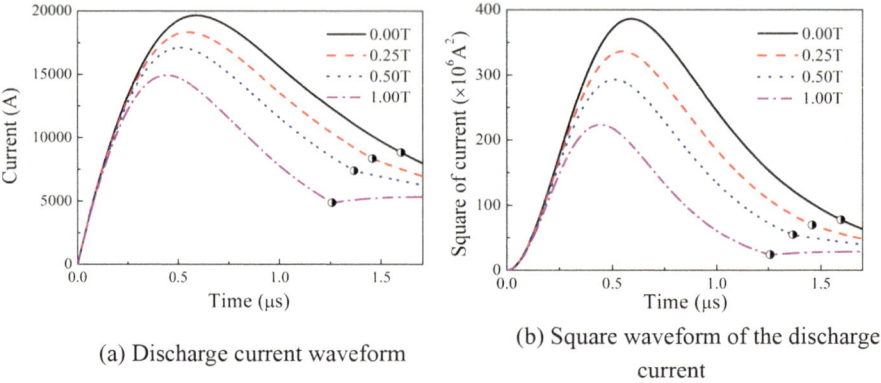

(a) Discharge current waveform

(b) Square waveform of the discharge
current

Fig. 5.58 Discharge current and its square waveform under different applied magnetic field
intensities (0.0–1.0 T)

greater than 1.0 T, the impulse bit decreases as the strength of the applied magnetic
field increases.

When the applied magnetic field is strong (1.5–3.0 T), the discharge current and
the square waveform of the discharge current under different strengths of the applied
magnetic field are presented in Fig. 5.59. Figure 5.59a shows that as the strength of
the applied magnetic field increases, the forward peak discharge current decreases,
the reverse current gradually appears before the current sheet ejects from the plate,
and the reverse peak discharge current gradually increases. As the strength of the
applied magnetic field increases, the increase in the reverse peak results in an increase
in the negative impulse bit generated by the applied magnetic field. As shown in
Fig. 5.59b, the area of the impulse bit of the self-induced magnetic field decreases as
the strength of the applied magnetic field increases. When the negative effect of the
applied magnetic field is greater than the positive effect, the impulse bit decreases
as the strength of the applied magnetic field increases.

When the strength of the applied magnetic field increases from 0.0 to 3.0 T,
the relative growth rate of the impulse bit ranges from −35.8 to 26.5%, a range of
62.3%, indicating that the strength of the applied magnetic field has a great impact
on the performance of the thruster. Therefore, the use of an applied magnetic field
of a certain strength is conducive to improving the thruster performance. However,
the strength of the applied magnetic field should not be too large; otherwise, it will
adversely affect the thruster performance.

2. Influence of the position and length of the applied magnetic field on the LPPT
   performance

To study the influence of the position of the applied magnetic field on the LPPT
performance, the length of the applied magnetic field of the thruster is kept constant,
and only the position of the left boundary of the applied magnetic field is changed.
The left boundary starts from the leftmost end of the plate and is increased by 0.1 mm
each time until the right boundary of the applied magnetic field reaches the rightmost

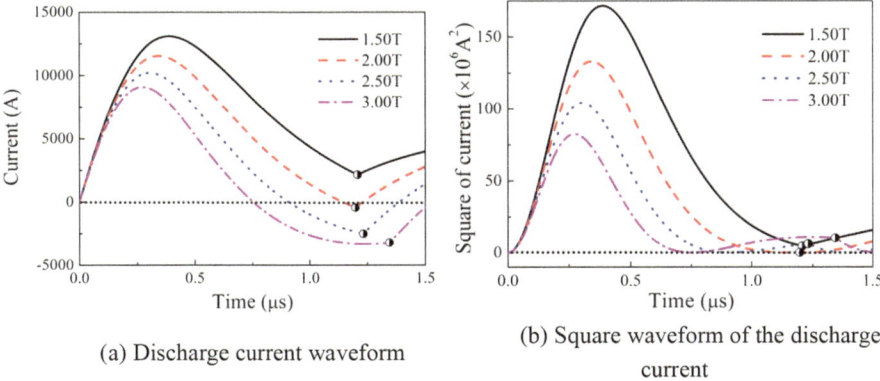

(a) Discharge current waveform

(b) Square waveform of the discharge current

Fig. 5.59 Discharge current and its square waveform under different applied magnetic field intensities (1.5–3.0 T)

end of the plate. Figure 5.60 shows the trends of the variations in the impulse bit with the position of the left boundary of the applied magnetic field when the length of the applied magnetic field is 1 and 5 mm, respectively (5 calculation points between the two points are omitted for clarity). It is found in Fig. 5.60 that as the position of the left boundary of the applied magnetic field gradually moves to the right, the impulse bit first increases and then decreases, reaching a maximum at a certain position between the plates. Therefore, it is most effective to apply an external magnetic field at some position between the plates. This is different from the simulation results for the LES-6 PPT and LES-8/9 PPT, where the external magnetic field is most effective when applied at the leftmost end of the plate.

Since the optimal position for applying an external magnetic field in the LPPT is somewhere between the plates, to study the influence of the length of the applied magnetic field on the thruster performance, the same approach as that used to study

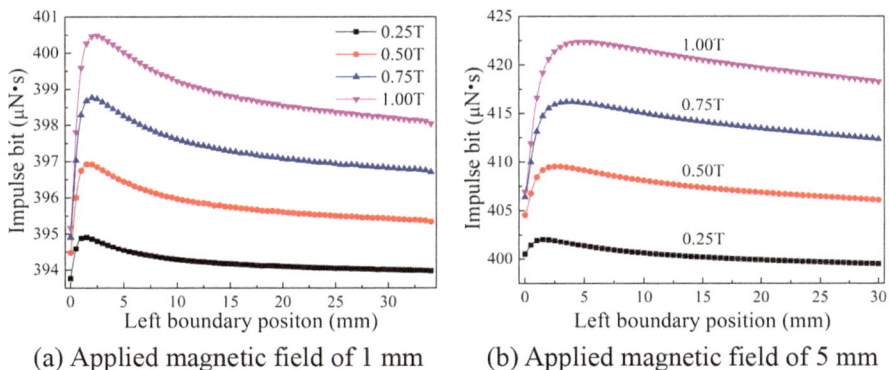

(a) Applied magnetic field of 1 mm

(b) Applied magnetic field of 5 mm

Fig. 5.60 Variations in the impulse bit with the position of the left boundary of the applied magnetic field

the influence of the position of the applied magnetic field is adopted; that is, under any length of the applied magnetic field, the left boundary is moved starting from the leftmost end of the plate and increased by 0.1 mm each time until the right boundary of the applied magnetic field reaches the rightmost end of the plate. Finally, the maximum impulse bit for each length of the applied magnetic field is extracted to obtain the variation in the maximum impulse bit with the length of the applied magnetic field, as shown in Fig. 5.61a. It is observed that as the length of the applied magnetic field increases, the maximum impulse bit first increases and then decreases slightly, suggesting that a larger length of the applied magnetic field does not necessary result in better performance. The variation in the position of the left boundary of the maximum impulse bit with the length of the applied magnetic field is shown in Fig. 5.61b. It is observed from the figure that despite slight fluctuations, the position of the left boundary generally shows a trend of increasing first and then decreasing as the length of the applied magnetic field increases.

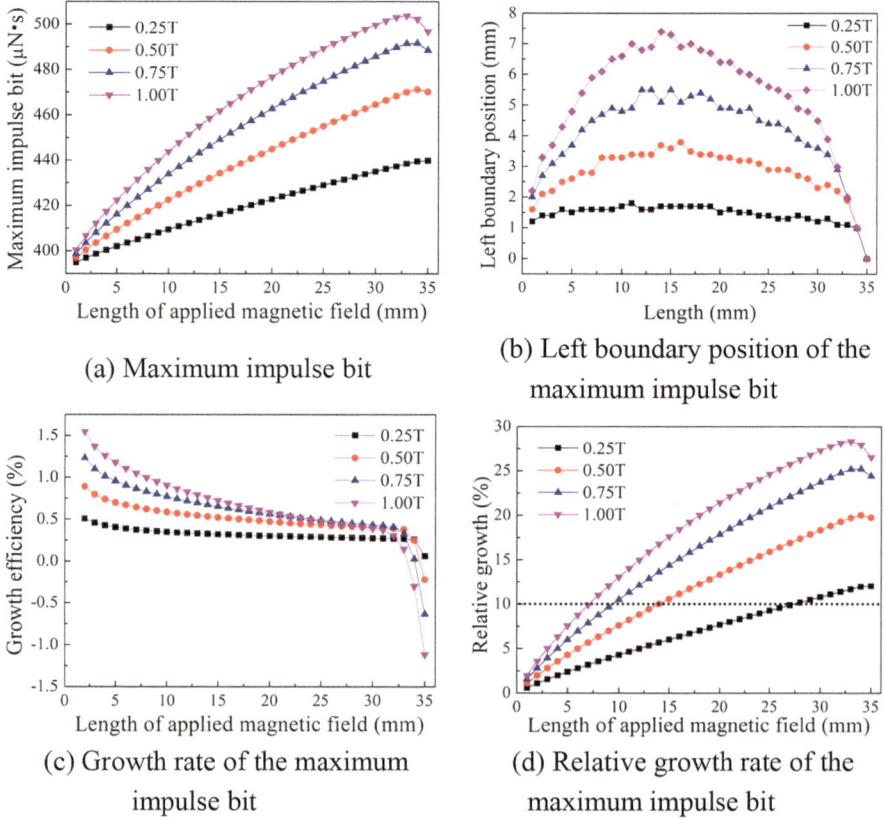

(a) Maximum impulse bit

(b) Left boundary position of the maximum impulse bit

(c) Growth rate of the maximum impulse bit

(d) Relative growth rate of the maximum impulse bit

Fig. 5.61 Influence of the length of the applied magnetic field on the thruster performance

A Fig. 5.61c shows the variation trend of the growth rate of the maximum impulse bit with the length of the applied magnetic field. It is observed that as the length of the applied magnetic field increases, the growth rate of the maximum impulse bit keeps decreasing, and that the growth rate of the maximum impulse bit is negative when the length is large. The variation in the relative growth rate of the maximum impulse bit with the length of the applied magnetic field is shown in Fig. 5.61d. It is observed that as the length of the applied magnetic field increases, the relative growth rate of the maximum impulse bit first increases and then decreases slightly. Moreover, considering that the system mass of the thruster increases when the length of the applied magnetic field increases, it is necessary to consider the optimal length of the applied magnetic field to achieve a balance between the system mass and the performance of the thruster. Here, the length of the applied magnetic field corresponding to a relative growth rate of 10% of the maximum impulse bit is considered as the optimal length of the applied magnetic field. The dotted line in Fig. 5.61d represents the dividing line corresponding to a relative growth rate of 10% of the impulse bit. As shown in Fig. 5.61d, when the applied magnetic field is 0.25, 0.50, 0.75, and 1.00 T, the optimal length of the applied magnetic field is approximately 28, 14, 10, and 8 mm, respectively, indicating that an increase in the strength of the applied magnetic field shortens the optimal length of the applied magnetic field.

Since an applied magnetic field inevitably increases the size and weight of the thruster system, the size, position, and length of the applied magnetic field should be optimally designed according to the needs of the specific thruster to reduce the added weight of the system while improving the thruster performance.

Reference

1. Vondra RJ, Thomassen K, Solbes A. Analysis of solid Teflon pulsed plasma thruster. J Spacecraft. 1970;7(12):1402–6.

Chapter 6
Numerical Simulation of the PPT Discharge Process Based on Magnetohydrodynamic Models

The pulsed plasma thruster (PPT) discharge process involves complex physical phenomena such as electromagnetism, flow, heat transfer, ionization, and recombination, and there is strong nonlinear coupling between the dynamic behaviors at different spatiotemporal scales. Although electromechanical models can clearly describe the change characteristics of the discharge parameters and the propulsion performance characteristics of the thruster, they fail to reflect the spatiotemporal characteristics of the plasma and its interaction with the electromagnetic field. To gain a deeper understanding of the physical process of PPT discharge plasma acceleration and fully elucidate the internal operating mechanism and energy conversion pattern of the thruster, it is necessary to carry out a detailed and comprehensive study on the PPT discharge process by means of microscopic dynamic theoretical analysis and numerical simulation.

The PPT discharge process involves the interaction between the electromagnetic field and plasma, and is usually investigated through theoretical study and numerical simulation using magnetohydrodynamic (MHD) methods. MHD simulation can provide detailed information on propellant heating and plasma flow. Therefore, considering complex and key physical processes such as propellant ablation and plasma flow, this chapter focuses on the development of a polytetrafluoroethylene (PTFE) two-phase ablation model and an MHD model in the form of generalized Lagrange multiplier (GLM) in combination with a circuit model, a thermochemical model and a transport model to establish a physical model of the PPT discharge process. Considering a three-dimensional (3D) MHD numerical simulation method, a computer calculation program is developed and validated using multiple cases, providing an important tool for PPT discharge process simulation and analysis.

© The Author(s) 2025   141
J. Wu et al., *Numerical Simulation of Pulsed Plasma Thruster*,
https://doi.org/10.1007/978-981-97-7958-1_6

6.1 Physical Model

6.1.1 Discharge Circuit Model

PPTs use the electrical energy stored in capacitors to ablate propellant through pulse discharge, causing it to decompose and ionize to generate plasma, which is accelerated to obtain pulsed thrust. The discharge process during the PPT operation can be considered a resistor-inductor-capacitor (RLC) circuit discharge process. According to Kirchhoff's law [1],

$$L_0 \frac{d^2 q_c}{dt^2} + R_0 \frac{dq_c}{dt} + \frac{q_c}{C} = V_{sh} + V_{pl} \tag{6.1}$$

where q_c is the charge stored in the capacitor, L_0 is the circuit inductance, R_0 is the circuit resistance, C is the capacitor capacitance, V_{sh} is the potential drop across the electrode sheath, and V_{pl} is the potential drop across the plasma. The sheath is a non-electrically neutral region formed by contact between the plasma and solid wall, with a thickness of several Debye lengths. The time required to establish the sheath is very short. Assuming that the sheath is in a quasi-steady state, the potential drop of the sheath can be calculated as [2]

$$V_{sh} = -\frac{\kappa T_e}{e} \ln\left(\frac{J_{th}}{J_y}\right) \tag{6.2}$$

where J_y is the current density perpendicular to the electrode surface and J_{th} is the random electron current density.

$$J_{th} = \frac{1}{4} e n_e \left(\frac{8\kappa T_e}{\pi m_e}\right)^{1/2} \tag{6.3}$$

In magnetohydrodynamics, the relationship between the current density J and the electric field strength E is described by the generalized Ohm's law, which is expressed as follows [3]:

$$J = \sigma_e \left(E + V \times B - \frac{1}{e n_e} J \times B\right) \tag{6.4}$$

where σ_e is the electrical conductivity, V is the plasma velocity, B is the magnetic induction intensity, and the term $J \times B$ represents the Hall effect electromotive force, which can be ignored when the electron cyclotron frequency is much lower than the plasma collision frequency. Using the generalized Ohm's law, the electric field strength E is integrated along path l from the anode to the cathode surface passing through the high-density plasmoid to calculate the plasma potential drop.

$$V_{pl} = \int \left(\frac{J}{\sigma_e} - V \times B + \frac{1}{en_e} J \times B \right) \cdot dl \tag{6.5}$$

After calculating the change in the charge of the capacitor according to Eq. (6.1), the discharge current of the PPT can be determined as

$$I(t) = -\frac{dq_c}{dt} \tag{6.6}$$

6.1.2 Two-Phase Ablation Model

In the MHD numerical study of the PPT discharge process, the specific physicochemical changes in the PTFE ablation process are rarely considered. A simplified one-dimensional (1D) ablation model of PTFE was established in Chap. 2. To accurately understand the influence of PTFE ablation characteristics on discharge, a two-phase ablation model of PTFE in 3D space is established in this chapter.

1. Ablation Heat Transfer Equation

As discussed in Sect. 2.1, when the temperature of PTFE reaches 600 K, a phase transition occurs, and PTFE becomes an amorphous body with very high viscosity. At the same time, the polymer molecules begin to depolymerize and decompose, yielding products consisting of small-molecule monomers [4]. Figure 6.1 shows a schematic diagram of the ablation process of PTFE in 3D Cartesian coordinates. Heat transfer processes in the crystalline and amorphous regions are considered separately, and the corresponding heat transfer differential equations are given as follows:

$$\rho_c c_c \frac{\partial T_c}{\partial t} = \frac{\partial}{\partial x}\left(k_c \frac{\partial T_c}{\partial x}\right) + \frac{\partial}{\partial y}\left(k_c \frac{\partial T_c}{\partial y}\right) + \frac{\partial}{\partial z}\left(k_c \frac{\partial T_c}{\partial z}\right) \tag{6.7}$$

$$\rho_a c_a \frac{\partial T_a}{\partial t} = \frac{\partial}{\partial x}\left(k_a \frac{\partial T_a}{\partial x}\right) + \frac{\partial}{\partial y}\left(k_a \frac{\partial T_a}{\partial y}\right) + \frac{\partial}{\partial z}\left(k_a \frac{\partial T_a}{\partial z}\right) + Q_p \tag{6.8}$$

where P, C, T, and K represent the density, specific heat capacity, temperature and thermal conductivity of PTFE, respectively. The subscripts c and a denote the crystal and amorphous bodies, respectively. Q_P represents the heat generated by the decomposition of the unit volume of PTFE and can be expressed using the Arrhenius reaction law as follows:

$$Q_p = -A_p \rho_a H_p \exp\left(-\frac{B_p}{T_a}\right) \tag{6.9}$$

where A_P is the frequency factor, H_P is the pyrolysis heat, and B_P is the activation temperature.

2. Ablated Surface and Phase Interface Treatments

On an ablated surface, the net incoming heat flux is determined by the external incoming heat flux q, surface radiation loss q_{rad}, and the energy transferred by the ablated mass, that is,

$$\left(-k_a \frac{\partial T_a}{\partial x} \cos\varphi - k_a \frac{\partial T_a}{\partial y} \cos\alpha - k_a \frac{\partial T_a}{\partial z} \cos\beta\right)\bigg|_{x=s} = q - q_{rad} - \dot{m} h_s \quad (6.10)$$

where s represents the recession distance of the ablated surface, h_s is the specific enthalpy of the ablation product, and ψ, α, and β are the angles between the normal vector of the surface and the x, y, and z coordinate axes, respectively, which satisfy

$$\cos\varphi = \frac{1}{\sqrt{1 + s_y'^2 + s_z'^2}}, \cos\alpha = -\frac{s_y'}{\sqrt{1 + s_y'^2 + s_z'^2}}, \cos\beta = -\frac{s_z'}{\sqrt{1 + s_y'^2 + s_z'^2}}$$

$$(6.11)$$

Propellant ablation occurs in a very small region of micrometer-scale thickness near the ablated surface, and the recession velocity and ablated mass flux on the ablated surface can be calculated by the following equation based on the PTFE decomposition rate:

$$\dot{m}(y, z, t) = \rho_0 \frac{\partial s}{\partial t} \cos\varphi = A_p \int_s^{s+\ell} \rho_a \exp\left(-\frac{B_p}{T_a}\right) dx \quad (6.12)$$

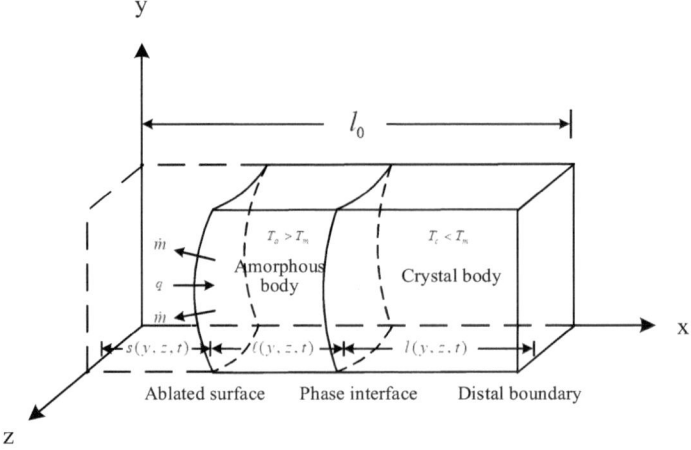

Fig. 6.1 Schematic diagram of the ablation process

where P_0 is the reference density of PTFE and l denotes the length of the amorphous region.

At the phase interface, the temperature of the propellant is equal to the phase transition temperature T_m, and the velocity of the phase interface can be derived from the heat flow equilibrium.

$$
\left(-k_a \frac{\partial T_a}{\partial x} \cos \varphi - k_a \frac{\partial T_a}{\partial y} \cos \alpha - k_a \frac{\partial T_a}{\partial z} \cos \beta \right) \Bigg|_{x=s+\ell}
$$
$$
= \left(-k_c \frac{\partial T_c}{\partial x} \cos \varphi - k_c \frac{\partial T_c}{\partial y} \cos \alpha - k_c \frac{\partial T_c}{\partial z} \cos \beta \right) \Bigg|_{x=s+\ell} + \rho_m h_m \left(\dot{s} + \dot{\ell} \right) \cos \varphi
$$

(6.13)

where s is replaced by $s + l$, P_m is the average density at the phase interface, and h_m is the latent heat of phase transition. The position of the phase interface can be obtained by integrating the velocity of the phase interface over time.

6.1.3 Governing System of MHD Equations

Plasma is a conducting fluid, and the study of the motion of conducting fluids in an electromagnetic field is called magnetohydrodynamics. When a conducting fluid moves in an electromagnetic field, a current is generated inside the fluid. This current interacts with the magnetic field to generate the Lorentz force, which alters the motion of the fluid and leads to changes in the electromagnetic field. To study this complex problem, the electromagnetic phenomenon and mechanical phenomenon within the fluid must be investigated simultaneously, and the electromagnetic and MHD equations, i.e., the governing system of MHD equations, [5] should be solved in a coupled manner.

1. System of MHD Equations

Changes in the electromagnetic field follow Maxwell's electromagnetic field theory, including Faraday's law of electromagnetic induction, Ampere's law, Gauss's theorem and the magnetic flux continuity law, which are, respectively, expressed as follows [6]:

$$
\nabla \times \boldsymbol{E} = -\frac{\partial \boldsymbol{B}}{\partial t}
$$

(6.14)

$$
\nabla \times \boldsymbol{H} = \boldsymbol{J} + \frac{\partial \boldsymbol{D}}{\partial t}
$$

(6.15)

$$
\nabla \cdot \boldsymbol{D} = \rho_e
$$

(6.16)

$$\nabla \cdot \boldsymbol{B} = 0 \qquad (6.17)$$

where H is the magnetic field intensity, which is related to magnetic induction B by $B = \mu H$, where μ is the magnetic permeability, D is the electric displacement vector, which is related to the electric field strength E by $D = \varepsilon E$, where ε is the permittivity, and P_e is the charge density.

Since plasma is a non-magnetic and weakly polarized medium, the magnetic permeability μ and permittivity ε can be taken as the vacuum magnetic permeability μ_0 and vacuum permittivity ε_0, respectively. For non-relativistic flow with a flow velocity V much lower than the light speed C, the ratio of the displacement current term to the modulus of the current density J in Eq. (6.15) is a small quantity of the order of V^2/C^2. Therefore, the displacement current can be neglected, simplifying Ampere's law to

$$\boldsymbol{J} = \frac{1}{\mu_0} \nabla \times \boldsymbol{B} \qquad (6.18)$$

Using the generalized Ohm's law, as shown in Eq. (6.4), in combination with Eqs. (6.14), (6.17) and (6.18), the following magnetic induction equation can be derived:

$$\frac{\partial \boldsymbol{B}}{\partial t} - \nabla \times (\boldsymbol{V} \times \boldsymbol{B}) = -\nabla \times [\eta \nabla \times \boldsymbol{B} + \nu(\nabla \times \boldsymbol{B}) \times \boldsymbol{B}] \qquad (6.19)$$

where $\eta = 1/\mu_0 \sigma_e$ and $V = 1/\mu_0 e n_e$. The terms in the equation from left to right represent the unsteady term, convection term, magnetic diffusion term, and Hall effect term, respectively.

Similar to ordinary fluids, the flow of plasma satisfying the continuity condition also follows the basic laws of classical mechanics, namely, the laws of conservation of mass, momentum and energy. Therefore, the flow equations of plasma also include the continuity equation, momentum equation and energy equation, which are essentially extensions of ordinary fluid dynamics equations with electromagnetic force.

The continuity equation reflects the conservation of mass in the fluid, and its differential form is

$$\frac{\partial \rho}{\partial t} + \nabla \cdot (\rho \boldsymbol{V}) = 0 \qquad (6.20)$$

where ρ and V represent the density and velocity of the plasma, respectively. This equation shows that the rate of change of mass within the volume enclosed by any closed surface in space is determined by the net mass flow through the closed surface.

In the framework of Newtonian mechanics, the rate of change in the momentum of a substance is equal to the sum of all the forces acting on the substance. For non-relativistic plasma flow, where plasma is subjected to thermal pressure, viscous force and electromagnetic force $J \times B$, the momentum equation can be derived using Newton's second law and Eqs. (6.17) and (6.18).

$$\frac{\partial(\rho V)}{\partial t} + \nabla \cdot \left[\rho\, VV + \left(p + \frac{B^2}{2\mu_0} \right)\overline{\overline{I}} - \frac{BB}{\mu_0} \right] = \nabla \cdot \overline{\overline{\tau}} \tag{6.21}$$

where P is the plasma pressure, $\overline{\overline{I}}$ is the unit tensor, and $\overline{\overline{\tau}}$ is the viscous stress tensor.

The energy equation is a mathematical expression of the law of energy conservation, which states that the rate of change of the total energy within any finite volume of the fluid is equal to the sum of the work done by the volume and surface forces, the heat transferred across the surface, and the heat generated by the internal heat source. For the flow of plasma, the Joule heat $E \cdot J$ caused by the electromagnetic field needs to be accounted for. Therefore, the energy equation becomes

$$\frac{\partial(\rho e_t)}{\partial t} + \nabla \cdot \left(\rho e_t + p + \frac{B^2}{2\mu_0} - \frac{BB}{\mu_0} \right) V = \nabla \cdot (\overline{\overline{\tau}} \cdot V) - \nabla \cdot q + \frac{J^2}{\sigma_e} \tag{6.22}$$

where q is the heat flux vector and e_t is the total specific energy, which can be expressed as

$$e_t = \frac{p}{\rho(\gamma - 1)} + \frac{V^2}{2} + \frac{B^2}{2\rho\mu_0} \tag{6.23}$$

where $\gamma = c_p/c_v$ is the specific heat ratio, in which c_p and c_v represent the specific heat of the fluid at constant pressure and constant volume, respectively.

It can be observed from the fluid dynamics equations that the distribution of the magnetic field affects the flow conditions and that the effect of the electric field is not directly reflected in the equations. Therefore, the system of MHD equations, which consists of the continuity equation, momentum equation, energy equation and magnetic induction equation, can reflect the main development and variation laws of the plasma flow and is written in vector form as follows:

$$\frac{\partial}{\partial t} \begin{bmatrix} \rho \\ \rho V \\ B \\ \rho e_t \end{bmatrix} + \nabla \cdot \begin{bmatrix} \rho V \\ \rho\, VV + \left(p + \frac{B^2}{2\mu_0} \right)\overline{\overline{I}} - \frac{BB}{\mu_0} \\ VB - BV \\ \left(\rho e_t + p + \frac{B^2}{2\mu_0} \right)V - \frac{B}{\mu_0}(V \cdot B) \end{bmatrix}$$
$$= \begin{bmatrix} 0 \\ \nabla \cdot \overline{\overline{\tau}} \\ -\nabla \times [\eta \nabla \times B + \nu(\nabla \times B) \times B] \\ \nabla \cdot (\overline{\overline{\tau}} \cdot V) - \nabla \cdot q + \frac{\eta}{\mu_0}|\nabla \times B|^2 \end{bmatrix} \tag{6.24}$$

2. System of GLM-MHD Equations

It is very difficult to directly solve the system of MHD equations Eq. (6.24). The first difficulty is the singularity of the system of equations. Equation (6.24) is a system of non-strictly hyperbolic equations, its inviscid Jacobian matrix is not full rank, and it has a zero eigenvalue, making the application of numerical schemes difficult [7]. Therefore, it is necessary to modify the form of this system of equations. The second difficulty in solving the system of MHD equations is spurious magnetic field divergence [8]. Equation (6.24) implies that the magnetic flux continuity equation $\nabla \cdot B = 0$. However, the divergence of the magnetic field is nonzero due to numerical discretization issues and the calculation accuracy in the solving process. Not effectively controlling error accumulation may lead to computational failure.

To address these two issues, Powell [9] proposed a system of eight-wave MHD equations. The non-full rank of the inviscid Jacobian matrix is avoided by adding a term proportional to ∇B to the original system of equations, and the accumulation of the pseudomagnetic field divergence is suppressed by convection. The Powell method is computationally straightforward and widely applicable. However, this eight-wave model is non-conservative in form, and the spurious magnetic field divergence persists in the convective propagation process, which may lead to divergence in calculations when dealing with flows with significant magnetic field variations, such as plasma motion in a PPT. Other methods for addressing the spurious magnetic field divergence include the projection method and the constrained transport method [10]. The projection method needs to solve a Poisson equation after each time step to correct the magnetic field; however, there is a considerable computational cost [11]. The constrained transport method involves discretization on staggered cells using a special form, thus imposing high requirements on the initial and boundary conditions [12]. For this reason, we adopt a hyperbolic divergence cleaning method [13] in this section. A generalized linear models (GLM) Ψ is introduced and solved simultaneously with the original system of MHD equations, thereby eliminating the singularity and attenuating the spurious magnetic field divergence in the convective dissipation process while preserving the conservative form of the system of MHD equations. The GLM form of the system of governing MHD equations (GLM-MHD equations) can be written as

$$
\frac{\partial}{\partial t}\begin{bmatrix} \rho \\ \rho V \\ B \\ \rho e_t \\ \psi \end{bmatrix} + \nabla \cdot \begin{bmatrix} \rho V \\ \rho\, VV + \left(p + \frac{B^2}{2\mu_0}\right)\bar{\bar{I}} - \frac{BB}{\mu_0} \\ VB - BV + \psi\bar{\bar{I}} \\ \left(\rho e_t + p + \frac{B^2}{2\mu_0}\right)V - \frac{B}{\mu_0}(V \cdot B) \\ c_h^2 B \end{bmatrix}
$$

$$
= \begin{bmatrix} 0 \\ \nabla \cdot \bar{\bar{\tau}} \\ -\nabla \times [\eta \nabla \times B + \nu(\nabla \times B) \times B] \\ \nabla \cdot (\bar{\bar{\tau}} \cdot V) - \nabla \cdot q + \frac{\eta}{\mu_0}|\nabla \times B|^2 \\ -\frac{c_h^2}{c_b^2}\psi \end{bmatrix} \qquad (6.25)
$$

where C_h and C_b are parameters that characterize the hyperbolic and parabolic properties, respectively.

$$c_h = \frac{c_{CFL}}{\Delta t} h_{min} \tag{6.26}$$

$$c_b = \sqrt{-\Delta t \frac{c_h^2}{\ln(c_d)}} \tag{6.27}$$

where $c_{CFL} \in (0,1)$, $C_d \in (0,1)$, Δt is the time step, and h_{min} represents the minimum cell scale.

3. Form of the System of Dimensionless MHD Equations in a 3D Cartesian Coordinate System

The physical quantities in Eq. (6.25) in the 3D Cartesian coordinate system are non-dimensionalized using selected characteristic reference quantities. The dimensionless quantities marked with the superscript "*" are as follows:

$$x^* = \frac{x}{L_{ref}}, y^* = \frac{y}{L_{ref}}, z^* = \frac{z}{L_{ref}}, u^* = \frac{u}{V_{ref}}, v^* = \frac{v}{V_{ref}}, w^* = \frac{w}{V_{ref}}, t^* = \frac{t}{L_{ref}/V_{ref}}$$

$$\rho^* = \frac{\rho}{\rho_{ref}}, p^* = \frac{p}{\rho_{ref}V_{ref}^2}, T^* = \frac{T}{T_{ref}}, B^* = \frac{B}{B_{ref}}, e_t^* = \frac{e_t}{V_{ref}^2}, \psi^* = \frac{\psi}{B_{ref}V_{ref}}$$

$$\mu_f^* = \frac{\mu_f}{\mu_{f\,ref}}, \kappa^* = \frac{\kappa}{\kappa_{ref}}, \sigma_e^* = \frac{\sigma_e}{\sigma_{e\,ref}}, n_e^* = \frac{n_e}{\sigma_{e\,ref}B_{ref}/e}, c_h^* = \frac{c_h}{V_{ref}}, c_b^* = \frac{c_b}{\sqrt{L_{ref}}\,V_{ref}}$$

In the process of non-dimensionalizing the system of equations, the dimensionless numbers of the magnetic Mach number, magnetic Reynolds number, Mach number, Reynolds number, and Prandtl number are as follows:

$$M_m = \frac{V_{ref}}{\sqrt{B_{ref}^2/\mu_0 \rho_{ref}}} \tag{6.28}$$

$$Re_m = \sigma_{e\,ref}\mu_0\,V_{ref}\,L_{ref} \tag{6.29}$$

$$M_a = \frac{V_{ref}}{\sqrt{\gamma p(\rho_{ref}, T_{ref})/\rho_{ref}}} \tag{6.30}$$

$$Re = \rho_{ref}V_{ref}\,L_{ref}\big/\mu_{f\,ref} \tag{6.31}$$

$$Pr = c_{pref}\,\mu_{fref}\big/\kappa_{ref} \tag{6.32}$$

Removing the superscripts of the dimensionless quantities, the dimensionless form of the system of MHD equations can be written in the following conservative form:

$$\frac{\partial \boldsymbol{Q}}{\partial t} + \frac{\partial \boldsymbol{E}}{\partial x} + \frac{\partial \boldsymbol{F}}{\partial y} + \frac{\partial \boldsymbol{G}}{\partial z} = \frac{\partial \boldsymbol{E}_v}{\partial x} + \frac{\partial \boldsymbol{F}_v}{\partial y} + \frac{\partial \boldsymbol{G}_v}{\partial z} + \boldsymbol{H} \tag{6.33}$$

where Q is a vector of conservative variables, E, F, G, E_V, F_V, and G_V are the inviscid flux and viscous flux vectors in the x, y, and z directions, respectively, in the Cartesian coordinate system, and H is the source term, with specific forms as follows:

$$\boldsymbol{Q} = \begin{bmatrix} \rho & \rho u & \rho v & \rho w & B_x & B_y & B_z & \rho e_t & \psi \end{bmatrix}^{\mathrm{T}} \tag{6.34}$$

$$\boldsymbol{E} = \begin{bmatrix} \rho u \\ \rho u^2 + p + \frac{-B_x^2 + B_y^2 + B_z^2}{2M_m^2} \\ \rho uv - \frac{B_x B_y}{M_m^2} \\ \rho uw - \frac{B_x B_z}{M_m^2} \\ \psi \\ uB_y - vB_x \\ uB_z - wB_x \\ \left(\rho e_t + p + \frac{B_x^2 + B_y^2 + B_z^2}{2M_m^2}\right)u - \frac{B_x(uB_x + vB_y + wB_z)}{M_m^2} \\ c_h^2 B_x \end{bmatrix} \tag{6.35}$$

$$\boldsymbol{F} = \begin{bmatrix} \rho v \\ \rho vu - \frac{B_y B_x}{M_m^2} \\ \rho v^2 + p + \frac{B_x^2 - B_y^2 + B_z^2}{2M_m^2} \\ \rho vw - \frac{B_y B_z}{M_m^2} \\ vB_x - uB_y \\ \psi \\ vB_z - wB_y \\ \left(\rho e_t + p + \frac{B_x^2 + B_y^2 + B_z^2}{2M_m^2}\right)v - \frac{B_y(uB_x + vB_y + wB_z)}{M_m^2} \\ c_h^2 B_y \end{bmatrix} \tag{6.36}$$

$$
G = \begin{bmatrix}
\rho w \\
\rho w u - \dfrac{B_z B_x}{M_m^2} \\
\rho w v - \dfrac{B_z B_y}{M_m^2} \\
\rho w^2 + p + \dfrac{B_x^2 + B_y^2 - B_z^2}{2 M_m^2} \\
w B_x - u B_z \\
w B_y - v B_z \\
\psi \\
\left(\rho e_t + p + \dfrac{B_x^2 + B_y^2 + B_z^2}{2 M_m^2} \right) w - \dfrac{B_z \left(u B_x + v B_y + w B_z \right)}{M_m^2} \\
c_h^2 B_z
\end{bmatrix}
\tag{6.37}
$$

$$
\overline{\overline{\tau}} = \begin{bmatrix}
\tau_{xx} & \tau_{xy} & \tau_{xz} \\
\tau_{yx} & \tau_{yy} & \tau_{yz} \\
\tau_{zx} & \tau_{zy} & \tau_{zz}
\end{bmatrix}
$$

$$
= \mu_f \begin{bmatrix}
\dfrac{4}{3} \dfrac{\partial u}{\partial x} - \dfrac{2}{3} \left(\dfrac{\partial v}{\partial y} + \dfrac{\partial w}{\partial z} \right) & \dfrac{\partial u}{\partial y} + \dfrac{\partial v}{\partial x} & \dfrac{\partial u}{\partial z} + \dfrac{\partial w}{\partial x} \\
\dfrac{\partial u}{\partial y} + \dfrac{\partial v}{\partial x} & \dfrac{4}{3} \dfrac{\partial v}{\partial y} - \dfrac{2}{3} \left(\dfrac{\partial w}{\partial z} + \dfrac{\partial u}{\partial x} \right) & \dfrac{\partial v}{\partial z} + \dfrac{\partial w}{\partial y} \\
\dfrac{\partial u}{\partial z} + \dfrac{\partial w}{\partial x} & \dfrac{\partial v}{\partial z} + \dfrac{\partial w}{\partial y} & \dfrac{4}{3} \dfrac{\partial w}{\partial z} - \dfrac{2}{3} \left(\dfrac{\partial u}{\partial x} + \dfrac{\partial v}{\partial y} \right)
\end{bmatrix}
\tag{6.38}
$$

$$
q_{\text{Joule}} = \begin{pmatrix} q_{Jx} \\ q_{Jy} \\ q_{Jz} \end{pmatrix} = \frac{1}{\sigma_e} \begin{pmatrix}
B_y \left(\dfrac{\partial B_y}{\partial x} - \dfrac{\partial B_x}{\partial y} \right) + B_z \left(\dfrac{\partial B_z}{\partial x} - \dfrac{\partial B_x}{\partial z} \right) \\
B_z \left(\dfrac{\partial B_z}{\partial y} - \dfrac{\partial B_y}{\partial z} \right) + B_x \left(\dfrac{\partial B_x}{\partial y} - \dfrac{\partial B_y}{\partial x} \right) \\
B_x \left(\dfrac{\partial B_x}{\partial z} - \dfrac{\partial B_z}{\partial x} \right) + B_y \left(\dfrac{\partial B_y}{\partial z} - \dfrac{\partial B_z}{\partial y} \right)
\end{pmatrix}
\tag{6.39}
$$

$$
E_v = \begin{bmatrix}
0 \\
\dfrac{1}{Re} \tau_{xx} \\
\dfrac{1}{Re} \tau_{xy} \\
\dfrac{1}{Re} \tau_{xz} \\
0 \\
-\dfrac{1}{Re_m \sigma_e} \left(\dfrac{\partial B_x}{\partial y} - \dfrac{\partial B_y}{\partial x} \right) - \dfrac{B_y}{Re_m n_e} \left(\dfrac{\partial B_y}{\partial z} - \dfrac{\partial B_z}{\partial y} \right) + \dfrac{B_x}{Re_m n_e} \left(\dfrac{\partial B_z}{\partial x} - \dfrac{\partial B_x}{\partial z} \right) \\
\dfrac{1}{Re_m \sigma_e} \left(\dfrac{\partial B_z}{\partial x} - \dfrac{\partial B_x}{\partial z} \right) + \dfrac{B_x}{Re_m n_e} \left(\dfrac{\partial B_x}{\partial y} - \dfrac{\partial B_y}{\partial x} \right) - \dfrac{B_z}{Re_m n_e} \left(\dfrac{\partial B_y}{\partial z} - \dfrac{\partial B_z}{\partial y} \right) \\
\dfrac{1}{Re} \left(u \tau_{xx} + v \tau_{xy} + w \tau_{xz} \right) - \dfrac{1}{Re Pr (\gamma - 1) M_a^2} q_x + \dfrac{1}{Re_m M_m^2} q_{Jx} \\
0
\end{bmatrix}
\tag{6.40}
$$

$$
F_v = \begin{bmatrix}
0 \\
\dfrac{1}{\mathrm{Re}}\tau_{yx} \\
\dfrac{1}{\mathrm{Re}}\tau_{yy} \\
\dfrac{1}{\mathrm{Re}}\tau_{yz} \\
\dfrac{1}{\mathrm{Re}_m\sigma_e}\left(\dfrac{\partial B_x}{\partial y}-\dfrac{\partial B_y}{\partial x}\right)-\dfrac{B_x}{\mathrm{Re}_m n_e}\left(\dfrac{\partial B_z}{\partial x}-\dfrac{\partial B_x}{\partial z}\right)+\dfrac{B_y}{\mathrm{Re}_m n_e}\left(\dfrac{\partial B_y}{\partial z}-\dfrac{\partial B_z}{\partial y}\right) \\
0 \\
-\dfrac{1}{\mathrm{Re}_m\sigma_e}\left(\dfrac{\partial B_y}{\partial z}-\dfrac{\partial B_z}{\partial y}\right)-\dfrac{B_z}{\mathrm{Re}_m n_e}\left(\dfrac{\partial B_z}{\partial x}-\dfrac{\partial B_x}{\partial z}\right)+\dfrac{B_y}{\mathrm{Re}_m n_e}\left(\dfrac{\partial B_x}{\partial y}-\dfrac{\partial B_y}{\partial x}\right) \\
\dfrac{1}{\mathrm{Re}}\left(u\tau_{yx}+v\tau_{yy}+w\tau_{yz}\right)-\dfrac{1}{\mathrm{RePr}(\gamma-1)M_a^2}q_y+\dfrac{1}{\mathrm{Re}_m M_m^2}q_{Jy} \\
0
\end{bmatrix}
\tag{6.41}
$$

$$
G_v = \begin{bmatrix}
0 \\
\dfrac{1}{\mathrm{Re}}\tau_{zx} \\
\dfrac{1}{\mathrm{Re}}\tau_{zy} \\
\dfrac{1}{\mathrm{Re}}\tau_{zz} \\
-\dfrac{1}{\mathrm{Re}_m\sigma_e}\left(\dfrac{\partial B_z}{\partial x}-\dfrac{\partial B_x}{\partial z}\right)-\dfrac{B_x}{\mathrm{Re}_m n_e}\left(\dfrac{\partial B_x}{\partial y}-\dfrac{\partial B_y}{\partial x}\right)+\dfrac{B_z}{\mathrm{Re}_m n_e}\left(\dfrac{\partial B_y}{\partial z}-\dfrac{\partial B_z}{\partial y}\right) \\
\dfrac{1}{\mathrm{Re}_m\sigma_e}\left(\dfrac{\partial B_y}{\partial z}-\dfrac{\partial B_z}{\partial y}\right)+\dfrac{B_z}{\mathrm{Re}_m n_e}\left(\dfrac{\partial B_z}{\partial x}-\dfrac{\partial B_x}{\partial z}\right)-\dfrac{B_y}{\mathrm{Re}_m n_e}\left(\dfrac{\partial B_x}{\partial y}-\dfrac{\partial B_y}{\partial x}\right) \\
0 \\
\dfrac{1}{\mathrm{Re}}\left(u\tau_{zx}+v\tau_{zy}+w\tau_{zz}\right)-\dfrac{1}{\mathrm{RePr}(\gamma-1)M_a^2}q_z+\dfrac{1}{\mathrm{Re}_m M_m^2}q_{Jz} \\
0
\end{bmatrix}
\tag{6.42}
$$

$$
H = \begin{bmatrix} 0\ 0\ 0\ 0\ 0\ 0\ 0\ 0\ -\dfrac{c_h^2}{c_b^2}\psi \end{bmatrix}^T
\tag{6.43}
$$

6.1.4 Thermochemical Model

To solve the governing system of MHD equations in a closed manner, the relevant plasma physical parameters must be calculated. This requires first determining the composition of the plasma. For the plasma generated by PTFE ablation, Kovitya, Schmahl, Cassibry, and Sonoda [14–17] used the minimum free energy method or the equilibrium constant method to calculate the state parameters of more than twenty components, including electrons, carbon and fluorine atoms, molecules, and ions. Since no characteristic lines of fluorocarbons are observed through emission spectroscopy diagnostics in this section and basic data on relevant molecules are lacking, it is assumed that the plasma contains only the most basic electrons, i.e., carbon and fluorine atoms and the ions generated by ionization, namely e^-, C, C^+, $\cdots$, C^{Nc+}, F, F^+, …, F^{NF+}, where N_C and N_F represent the highest valence of carbon and fluorine ions, respectively. Thus, a thermochemical model for plasma with a relatively small computational burden can be established based on the Saha equation.

When the thermal motion velocity of each component particle satisfies the Maxwell velocity distribution, each component in the plasma satisfies the ideal gas equation of state. Thus, according to Dalton's law of partial pressure,

$$p = \left(\sum_{j=0}^{N_C} n_{C^{j+}} + \sum_{j=0}^{N_F} n_{F^{j+}} \right) \kappa T_h + n_e \kappa T_e \tag{6.44}$$

Using m_c and m_F to represent the masses of carbon and fluorine atoms, respectively, the plasma density ρ can be expressed as

$$\rho = m_C \sum_{j=0}^{N_C} n_{C^{j+}} + m_F \sum_{j=0}^{N_F} n_{F^{j+}} + m_e n_e \approx m_C \sum_{j=0}^{N_C} n_{C^{j+}} + m_F \sum_{j=0}^{N_F} n_{F^{j+}} \tag{6.45}$$

Based on the molecular structure of PTFE and according to the condition of element conservation, we have

$$\sum_{j=0}^{N_C} n_{C^{j+}} / \sum_{j=0}^{N_F} n_{F^{j+}} = 1/2 \tag{6.46}$$

According to the assumption of the quasi-neutral plasma,

$$n_e = \sum_{j=1}^{N_C} j n_{C^{j+}} + \sum_{j=1}^{N_F} j n_{F^{j+}} \tag{6.47}$$

Under the assumption of local thermodynamic equilibrium, the number densities of the reacting components in the plasma satisfy the law of mass action. In the case of ionization reactions, this law is expressed as the Saha equation [18].

$$\frac{n_{M^{(j+1)+}} n_e}{n_{M^{j+}}} = \frac{2Z_{j+1}}{Z_j} \left(\frac{2\pi m_e \kappa T_e}{h^2} \right)^{3/2} \exp\left(-\frac{E_j}{\kappa T_e} \right) \tag{6.48}$$

where M represents the component element, Z_j and Z_{j+1} are the partition functions of the particles ionized j times and $j + 1$ times, respectively, h is Planck's constant, and E_j is the ionization energy of the ionization reaction. The partition function of heavy particles is equal to the product of the translational, rotational, vibrational, and electronic partition functions [19]. For the carbon and fluorine atoms or ions in this section, since there is no rotational or vibrational energy, Z_j and Z_{j+1} are calculated solely based on the translational partition function Q_{tr} and electronic partition function Q_{el}, which take the following forms:

$$Q_{\mathrm{tr}} = \left(\frac{2\pi m\kappa T}{h^2}\right)^{3/2} V_s \tag{6.49}$$

$$Q_{\mathrm{el}} = \sum_{l=0}^{\infty} g_l \exp\left(-\frac{\varepsilon_l}{\kappa T_e}\right) \tag{6.50}$$

where V_s is the system volume, g_l is the electron degeneracy, and ε_l is the energy level. By simultaneously solving the above equations, first, it is easy to obtain the number density of heavy particles n_h according to the conservation of elements. The ratio of the number density of each component to the number density of heavy particles is defined as $\alpha_e = n_e/n_h$ and $\alpha j\, M = n_{mj+}/n_h$. Setting the right-hand side term of the Saha equation to be f_M^{j+1}, we have

$$f(\alpha_e) = \frac{\sum_{j=1}^{N_C} \frac{j \prod_{j=1}^{N_C} f_C^j}{(\alpha_e n_h)^j}}{1 + \sum_{j=1}^{N_C} \frac{\prod_{j=1}^{N_C} f_C^j}{(\alpha_e n_h)^j}} + \frac{2\sum_{j=1}^{N_F} \frac{j \prod_{j=1}^{N_F} f_F^j}{(\alpha_e n_h)^j}}{1 + \sum_{j=1}^{N_F} \frac{\prod_{j=1}^{N_F} f_F^j}{(\alpha_e n_h)^j}} - \frac{(m_C + 2m_F)\alpha_e n_h}{\rho} = 0 \tag{6.51}$$

By repeatedly iterating to find α_e, the particle number density of each component in the plasma under the given state parameters can be calculated, and then, the thermodynamic properties of the plasma can be conveniently calculated using the partition function.

6.1.5 Transport Coefficient

In non-equilibrium plasma, due to the uneven distribution of parameters, factors such as the velocity gradient, temperature gradient, and potential gradient within the system can cause transport processes such as momentum transfer, heat transfer, and charge migration. The intensity of these processes is characterized by transport coefficients, including the viscosity coefficient, thermal conductivity and electrical conductivity [20]. The strict theoretical formulas of the transport coefficients are derived from the Boltzmann equation by using the Chapman–Enskog expansion, and their calculation requires data of the collision cross-sections between various particles. Given the large errors in the measurement data of collision cross-sections, the uncertainty introduced by theoretical calculations due to a lack of accurate knowledge about the interaction potential between particles, and the complexity of the calculations, simplified models are adopted for calculation in this section.

1. Viscosity Coefficient

Shear stress is generated when there is relative motion between fluid layers, and the shear stress caused by the unit velocity gradient is represented by the viscosity coefficient. Since the momentum of the plasma is mainly concentrated on the heavy particles, the role of electrons can be ignored. According to the Braginskii equation [21], with v_h representing the collision frequency of heavy particles, the viscosity coefficient is expressed as

$$\mu_f = 0.96 \frac{n\kappa T_h}{v_h} \tag{6.52}$$

2. Thermal Conductivity

Thermal conductivity represents the heat flux due to a unit temperature gradient. For heavy particles and electrons, with V_e and V_{ei} representing the electron collision frequency and electron–ion collision frequency, respectively, their thermal conductivities are calculated as [22]

$$k_h = \frac{\mu_f c_p}{Pr} \tag{6.53}$$

$$k_e = \frac{2.4}{1 + v_{ei}/\left(\sqrt{2}v_e\right)} \frac{\kappa^2 n_e T_e}{m_e v_e} \tag{6.54}$$

3. Electrical Conductivity

Under the action of an electric field, electrons and ions in the plasma undergo relative motion, resulting in a continuous increase in the current. At the same time, Coulomb collisions between electrons and ions hinder their acceleration. When the motion of electrons and ions tends to be balanced, the current density is proportional to the electric field strength, and the proportionality coefficient is defined as the electrical conductivity of the plasma. For weakly ionized plasma, the electrical conductivity can be expressed according to the Krook collision model [23].

$$\sigma_e = \frac{e^2 n_e}{m_e v_e} \tag{6.55}$$

6.2 Numerical Calculation Methods

6.2.1 Coordinate Transformation

1. Coordinate Transformation of the Ablation Heat Transfer Equation

During the ablation process of the propellant, the positions of the ablated surface and the phase interface are constantly changing, posing significant challenges with respect to the numerical calculation of the ablation process. To address this issue, a coordinate transformation is performed as follows:

$$\chi = \frac{x - s(y, z, t)}{\ell(y, z, t)} \tag{6.56}$$

$$\xi = \frac{x - s(y, z, t) - \ell(y, z, t)}{l(y, z, t)} \tag{6.57}$$

After the transformation, the new coordinate x in the amorphous region and the ζ in the crystal region are both in a value range of $[0, 1]$. In the new coordinate systems (ζ, y, z) and (x, y, z), the transformed forms of the ablation heat transfer equations (Eqs. (6.7) and (6.8), respectively) are

$$
\begin{aligned}
\rho_c c_c \frac{\partial T_c}{\partial t} = {} & \frac{k_c \left(A_y^2 + A_z^2 + 1\right)}{l^2} \frac{\partial^2 T_c}{\partial \xi^2} + k_c \frac{\partial^2 T_c}{\partial y^2} \\
& + k_c \frac{\partial^2 T_c}{\partial z^2} - \frac{2k_c A_y}{l} \frac{\partial^2 T_c}{\partial \xi \partial y} - \frac{2k_c A_z}{l} \frac{\partial^2 T_c}{\partial \xi \partial z} \\
& + \frac{1}{l} \frac{\partial T_c}{\partial \xi} \left(\rho_c c_c A_t + \frac{A_y^2 + A_z^2 + 1}{l} \frac{\partial k_c}{\partial \xi} - A_y \frac{\partial k_c}{\partial y} - A_z \frac{\partial k_c}{\partial z} \right. \\
& + \frac{2k_c A_y}{l} \frac{\partial l}{\partial y} + \frac{2k_c A_z}{l} \frac{\partial l}{\partial z} - k_c \frac{\partial A_y}{\partial y} - k_c \frac{\partial A_z}{\partial z} \bigg) \\
& + \frac{\partial T_c}{\partial y} \left(\frac{\partial k_c}{\partial y} - \frac{A_y}{l} \frac{\partial k_c}{\partial \xi} \right) + \frac{\partial T_c}{\partial z} \left(\frac{\partial k_c}{\partial z} - \frac{A_z}{l} \frac{\partial k_c}{\partial \xi} \right) \tag{6.58}
\end{aligned}
$$

$$
\begin{aligned}
\rho_a c_a \frac{\partial T_a}{\partial t} = {} & \frac{k_a \left(A_y^2 + A_z^2 + 1\right)}{\ell^2} \frac{\partial^2 T_a}{\partial \chi^2} \\
& + k_a \frac{\partial^2 T_a}{\partial y^2} + k_a \frac{\partial^2 T_a}{\partial z^2} - \frac{2k_a A_y}{\ell} \frac{\partial^2 T_a}{\partial \chi \partial y} - \frac{2k_a A_z}{\ell} \frac{\partial^2 T_a}{\partial \chi \partial z} \\
& + \frac{1}{\ell} \frac{\partial T_a}{\partial \chi} \left(\rho_a c_a A_t + \frac{A_y^2 + A_z^2 + 1}{\ell} \frac{\partial k_a}{\partial \chi} - A_y \frac{\partial k_a}{\partial y} - A_z \frac{\partial k_a}{\partial z} \right. \\
& + \frac{2k_a A_y}{\ell} \frac{\partial \ell}{\partial y} + \frac{2k_a A_z}{\ell} \frac{\partial \ell}{\partial z} - k_a \frac{\partial A_y}{\partial y} - k_a \frac{\partial A_z}{\partial z} \bigg) \\
& + \frac{\partial T_a}{\partial y} \left(\frac{\partial k_a}{\partial y} - \frac{A_y}{\ell} \frac{\partial k_a}{\partial \chi} \right) + \frac{\partial T_a}{\partial z} \left(\frac{\partial k_a}{\partial z} - \frac{A_z}{\ell} \frac{\partial k_a}{\partial \chi} \right) + Q_p(T_a) \tag{6.59}
\end{aligned}
$$

where

$$
A_{y,z,t} = \begin{cases} (1 - \xi) \left(\dfrac{\partial s(y, z, t)}{\partial y, z, t} + \dfrac{\partial \ell(y, z, t)}{\partial y, z, t} \right), & \text{Crystal region} \\[4mm] \dfrac{\partial s(y, z, t)}{\partial y, z, t} + \chi \dfrac{\partial \ell(y, z, t)}{\partial y, z, t}, & \text{Amorphous region} \end{cases} \tag{6.60}
$$

2. Coordinate Transformation of the Governing System of MHD Equations

To facilitate calculations, it is usually necessary to convert the physical domain into the computational domain via coordinate transformation. The dimensionless form of the system of GLM-MHD equations (Eq. 6.33) in a general curvilinear coordinate system (ζ, η, ζ) is given by

$$\frac{\partial \overline{Q}}{\partial t} + \frac{\partial \overline{E}}{\partial \xi} + \frac{\partial \overline{F}}{\partial \eta} + \frac{\partial \overline{G}}{\partial \zeta} = \frac{\partial \overline{E}_v}{\partial \xi} + \frac{\partial \overline{F}_v}{\partial \eta} + \frac{\partial \overline{G}_v}{\partial \zeta} + \overline{H} \tag{6.61}$$

Let J^{-1} represent the determinant of the coordinate transformation matrix

$$J^{-1} = \left| \frac{\partial(x, y, z)}{\partial(\xi, \eta, \zeta)} \right| = x_\xi \left(y_\eta z_\zeta - y_\zeta z_\eta \right) + x_\eta \left(y_\zeta z_\xi - y_\xi z_\zeta \right) + x_\zeta \left(y_\xi z_\eta - y_\eta z_\xi \right) \tag{6.62}$$

Then, the various fluxes in Eq. (6.61) can be expressed as

$$\begin{aligned}
\overline{Q} &= J^{-1} Q, \overline{H} = J^{-1} H \\
\overline{E} &= J^{-1} \left(\xi_x\, E + \xi_y\, F + \xi_z\, G \right), \overline{E}_v = J^{-1} \left(\xi_x\, E_v + \xi_y\, F_v + \xi_z\, G_v \right) \\
\overline{F} &= J^{-1} \left(\eta_x\, E + \eta_y\, F + \eta_z\, G \right), \overline{F}_v = J^{-1} \left(\eta_x\, E_v + \eta_y\, F_v + \eta_z\, G_v \right) \\
\overline{G} &= J^{-1} \left(\zeta_x\, E + \zeta_y\, F + \zeta_z\, G \right), \overline{G}_v = J^{-1} \left(\zeta_x\, E_v + \zeta_y\, F_v + \zeta_z\, G_v \right)
\end{aligned} \tag{6.63}$$

By defining the Jacobian matrices of the inviscid fluxes, the governing system of MHD equations (Eq. 6.61) can be rewritten as

$$\frac{\partial \overline{Q}}{\partial t} + \overline{A} \frac{\partial \overline{Q}}{\partial \xi} + \overline{B} \frac{\partial \overline{Q}}{\partial \eta} + \overline{C} \frac{\partial \overline{Q}}{\partial \zeta} = \frac{\partial \overline{E}_v}{\partial \xi} + \frac{\partial \overline{F}_v}{\partial \eta} + \frac{\partial \overline{G}_v}{\partial \zeta} + \overline{H} \tag{6.64}$$

where the Jacobian matrices are calculated as

$$\begin{aligned}
\overline{A} &= \frac{\partial \overline{E}}{\partial \overline{Q}} = \xi_x \frac{\partial E}{\partial Q} + \xi_y \frac{\partial F}{\partial Q} + \xi_z \frac{\partial G}{\partial Q} \\
\overline{B} &= \frac{\partial \overline{F}}{\partial \overline{Q}} = \eta_x \frac{\partial E}{\partial Q} + \eta_y \frac{\partial F}{\partial Q} + \eta_z \frac{\partial G}{\partial Q} \\
\overline{C} &= \frac{\partial \overline{G}}{\partial \overline{Q}} = \zeta_x \frac{\partial E}{\partial Q} + \zeta_y \frac{\partial F}{\partial Q} + \zeta_z \frac{\partial G}{\partial Q}
\end{aligned} \tag{6.65}$$

6.2.2 Jacobian Matrix and Eigenvalues

To solve the governing system of MHD equations, it is necessary to determine the specific forms and eigenvalues of the Jacobian matrices $\overline{A}$, $\overline{B}$, and $\overline{C}$. Due to the high complexity of the system of equations, its Jacobian matrices cannot be directly obtained. Therefore, the primitive variables are introduced

$$\Phi = \begin{bmatrix} \rho & u & v & w & B_x & B_y & B_z & p & \psi \end{bmatrix}^{\mathrm{T}} \tag{6.66}$$

Then, we can obtain

$$A_\Phi = \frac{\partial \Phi}{\partial Q}\frac{\partial E}{\partial \Phi} = \begin{bmatrix}
u & \rho & 0 & 0 & 0 & 0 & 0 & 0 & 0 \\
0 & u & 0 & 0 & -\frac{B_x}{\rho M_m^2} & \frac{B_y}{\rho M_m^2} & \frac{B_z}{\rho M_m^2} & \frac{1}{\rho} & 0 \\
0 & 0 & u & 0 & -\frac{B_y}{\rho M_m^2} & -\frac{B_x}{\rho M_m^2} & 0 & 0 & 0 \\
0 & 0 & 0 & u & -\frac{B_z}{\rho M_m^2} & 0 & -\frac{B_x}{\rho M_m^2} & 0 & 0 \\
0 & 0 & 0 & 0 & 0 & 0 & 0 & 0 & 1 \\
0 & B_y & -B_x & 0 & -v & u & 0 & 0 & 0 \\
0 & B_z & 0 & -B_x & -w & 0 & u & 0 & 0 \\
0 & \gamma p & 0 & 0 & \frac{\overline{\gamma}}{M_m^2}V \cdot B & 0 & 0 & u & -\frac{\overline{\gamma}B_x}{M_m^2} \\
0 & 0 & 0 & 0 & c_h^2 & 0 & 0 & 0 & 0
\end{bmatrix}, \tag{6.67}$$

where $\overline{\gamma} = \gamma - 1$; thus, the Jacobian matrix corresponding to the inviscid flux E can be obtained as follows:

$$A = \frac{\partial E}{\partial Q} = \left(\frac{\partial Q}{\partial \Phi}\right)A_\Phi\left(\frac{\partial Q}{\partial \Phi}\right)^{-1}$$

$$= \begin{bmatrix}
0 & 1 & 0 & 0 & 0 & 0 & 0 & 0 & 0 \\
\frac{\overline{\gamma}V^2}{2} - u^2 & 2u - \overline{\gamma}u & -\overline{\gamma}v & -\overline{\gamma}w & -\frac{\gamma B_x}{M_m^2} & \frac{(1-\overline{\gamma})B_y}{M_m^2} & \frac{(1-\overline{\gamma})B_z}{M_m^2} & \overline{\gamma} & 0 \\
-uv & v & u & 0 & -\frac{B_y}{M_m^2} & -\frac{B_x}{M_m^2} & 0 & 0 & 0 \\
-uw & w & 0 & u & -\frac{B_z}{M_m^2} & 0 & -\frac{B_x}{M_m^2} & 0 & 0 \\
0 & 0 & 0 & 0 & 0 & 0 & 0 & 0 & 1 \\
\frac{vB_x-uB_y}{\rho} & \frac{B_y}{\rho} & -\frac{B_x}{\rho} & 0 & -v & u & 0 & 0 & 0 \\
\frac{wB_x-uB_z}{\rho} & \frac{B_z}{\rho} & 0 & \frac{-B_x}{\rho} & -w & 0 & u & 0 & 0 \\
\Xi & \Omega & -uv\overline{\gamma} - \frac{B_xB_y}{\rho M_m^2} & -uw\overline{\gamma} - \frac{B_xB_z}{\rho M_m^2} & \Lambda & \Theta & \Upsilon & \gamma u & 0 \\
0 & 0 & 0 & 0 & c_h^2 & 0 & 0 & 0 & 0
\end{bmatrix} \tag{6.68}$$

where

$$\Xi = \frac{(\gamma - 2)V^2 u}{2} - \frac{\gamma p u}{\overline{\gamma} \rho} + \frac{B_x\left(vB_y + wB_z\right) - u\left(B_y^2 + B_z^2\right)}{\rho M_m^2}$$

$$\Omega = \frac{\gamma p}{\overline{\gamma} \rho} + \frac{V^2}{2} - \overline{\gamma} u^2 + \frac{B_y^2 + B_z^2}{\rho M_m^2}, \quad \Lambda = -\frac{\gamma u B_x + vB_y + wB_z}{M_m^2}$$

$$\Theta = \frac{(1 - \overline{\gamma})uB_y - vB_x}{M_m^2}, \quad \Upsilon = \frac{(1 - \overline{\gamma})uB_z - wB_x}{M_m^2} \tag{6.69}$$

Using the same method, the Jacobian matrices B and C corresponding to the inviscid fluxes F and G can be obtained. Then, the Jacobian matrices $\overline{A}$, $\overline{B}$, and $\overline{C}$ can be determined by Eq. (6.65). The specific form and derivation of the Jacobian matrices can be found in Appendix A. Since A and A_Φ have the same eigenvalues, based on A_Φ as well as B_Φ and C_Φ, which have similar forms, the eigenvalues of $\overline{A}$, $\overline{B}$, and $\overline{C}$ can be calculated. The eigenvalues of $\overline{A}$. $\overline{B}$, and $\overline{C}$ have similar expressions. Taking $\overline{A}$ as an example, its nine eigenvalues are

$$\lambda_1 = V_\varsigma, \lambda_{2,3} = V_\varsigma \pm v_{a\varsigma}, \lambda_{4,5}$$
$$= V_\varsigma \pm v_{p+}, \lambda_{6,7} = V_\varsigma \pm v_{p-}, \lambda_{8,9} = \pm \varsigma c_h \tag{6.70}$$

where

$$V_\varsigma = \xi_x u + \xi_y v + \xi_z w, B_\varsigma = \xi_x B_x + \xi_y B_y + \xi_z B_z, \varsigma = \sqrt{\xi_x^2 + \xi_y^2 + \xi_z^2} \tag{6.71}$$

$$c_s = \sqrt{\frac{\gamma p}{\rho}}, v_a = \frac{B}{\sqrt{\rho M_m^2}}, v_{a\varsigma} = \frac{B_\varsigma}{\sqrt{\rho M_m^2}}$$

$$v_{p+} = \sqrt{\frac{1}{2}\left(\varsigma^2\left(c_s^2 + v_a^2\right) + \sqrt{\varsigma^4\left(c_s^2 + v_a^2\right)^2 - 4\varsigma^2 c_s^2 v_{a\varsigma}^2}\right)} \tag{6.72}$$

$$v_{p-} = \sqrt{\frac{1}{2}\left(\varsigma^2\left(c_s^2 + v_a^2\right) - \sqrt{\varsigma^4\left(c_s^2 + v_a^2\right)^2 - 4\varsigma^2 c_s^2 v_{a\varsigma}^2}\right)}$$

where C_s is the thermodynamic speed of sound, V_a and $V_{a\varsigma}$ are the Alfven velocities, and V_{p+} and V_{p-} are the phase velocities of fast and slow magnetosonic waves, respectively, reflecting the combined effects of the plasma aerodynamic pressure and magnetic pressure. The eigenvalue $\lambda_{8,9}$ indicates that the physical significance of the parameter C_h is to propagate the magnetic field divergence to the surrounding area, thus avoiding the accumulation of the spurious magnetic field divergence at local positions. The above nine eigenvalues are all nonzero, showing that the nine-wave form of the system of GLM-MHD equations has a great advantage in not only reducing the spurious magnetic field divergence and maintaining the conservative form of the system of MHD equations but also solving the singularity problem.

6.2.3   Time Discretization Methods

1. WSSOR Algorithm

In this book, the finite difference method is applied to discretize and solve the ablation heat transfer equation. During the PTFE ablation process, the thickness of the amorphous region is very small, i.e., on the order of micrometers. To avoid the problem of a too small time steps and excessively long computation time due to the stability constraints in explicit schemes, implicit schemes are adopted here for calculation. The time derivative term of the ablation heat transfer equation is determined using the first-order forward difference, the spatial derivative term is determined using the second-order central difference, and the internal heat source term is expanded in a Taylor series. As a result, the ablation heat transfer equation after coordinate transformation can be discretized into the following fifteen-point scheme:

$$
\begin{aligned}
&a_0 T_{i,j,k}^{n+1} + a_1 T_{i+1,j,k}^{n+1} + a_2 T_{i-1,j,k}^{n+1} + a_3 T_{i,j+1,k}^{n+1} + a_4 T_{i,j-1,k}^{n+1} + a_5 T_{i,j,k+1}^{n+1} \\
&+ a_6 T_{i,j,k-1}^{n+1} + a_7 T_{i+1,j+1,k}^{n+1} + a_8 T_{i+1,j-1,k}^{n+1} + a_9 T_{i-1,j+1,k}^{n+1} + a_{10} T_{i-1,j-1,k}^{n+1} \\
&+ a_{11} T_{i+1,j,k+1}^{n+1} + a_{12} T_{i+1,j,k-1}^{n+1} + a_{13} T_{i-1,j,k+1}^{n+1} + a_{14} T_{i-1,j,k-1}^{n+1} = f\left(T_{i,j,k}^{n}\right)
\end{aligned}
$$

$$(6.73)$$

where $\alpha_0, \alpha_1, \ldots, \alpha_{14}$ are coefficients. Directly solving Eq. (6.73) requires applying the matrix inversion operation on a system of large sparse matrix algebraic equations, which is very difficult. Iterative methods are usually used for this type of system of linear algebraic equations. Currently, iterative methods mainly include the Jacobi iteration, Gauss–Siedel iteration, successive overrelaxation (SOR) iteration, and symmetric SOR (SSOR) iteration [24], all of which are derived by splitting the coefficient matrix of the linear equation system. In this section, the weighted symmetric successive overrelaxation (WSSOR) algorithm, obtained by improving the SOR and SSOR iterative algorithms, is used for the solution. The WSSOR algorithm performs a weighted average on the vectors generated by the SOR and SSOR iterations [25]. This algorithm has a faster convergence rate and higher computational accuracy. Its iterative scheme is given by

$$
T^{n+1} = \vartheta\left(H_{\text{SOR}} T^n + V_{\text{SOR}}\right) + (1 - \vartheta)\left(H_{\text{SSOR}} T^n + V_{\text{SSOR}}\right) \tag{6.74}
$$

where ζ is the weighting factor, H_{SOR} and H_{SSOR} are the iteration matrices of the SOR and SSOR iterative algorithms, respectively, and V_{SOR} and V_{SSOR} are vectors on the right-hand side of the two iterative schemes.

2. Dual Time-Stepping Method

In addition to the issues of singularity and spurious magnetic field divergence, the severe stiffness problem must be addressed when numerically solving the system of MHD equations. When the conductivity of the local plasma in the flow field is low, the magnetic Reynolds number in the flow field is small, causing the magnetic diffusion term in the magnetic induction equation to be significantly larger than the convection term. This leads to great difficulty in solving the system of MHD equations. When the Hall effect term is included in the magnetic induction equations, the stiffness problem of the system of MHD equations becomes more prominent, posing a greater challenge. Since the stiffness of the system of equations originates from the magnetic viscosity term, commonly used methods for handling stiffness problems such as the point implicit methods, relaxation methods, and decoupling methods are not suitable. As a result, in general, the only option is to reduce the time step, which is highly disadvantageous for MHD simulations. This is because the computational burden of these simulations is several times larger than that of computational fluid dynamics (CFD) simulations of the same cell size. To minimize the adverse effect of the stiffness problem, the implicit dual time-stepping method is used in this section to discretize and solve the system of MHD control equations, improving the computational stability and efficiency.

The basic idea of the dual time-stepping method is to introduce a virtual time iterative process at the frozen real-time points to improve the time accuracy [26]. By discretizing the time derivative term of the governing system of MHD equations (Eq. 6.61) using a three-point backward difference with second-order accuracy and implicitly treating the inviscid and source terms, we obtain

$$
J^{-1} \frac{3\boldsymbol{Q}^{n+1} - 4\boldsymbol{Q}^n + \boldsymbol{Q}^{n-1}}{2\Delta t} + (\delta_\xi \overline{\boldsymbol{E}} + \delta_\eta \overline{\boldsymbol{F}} + \delta_\zeta \overline{\boldsymbol{G}})^{n+1}
$$
$$
= (\overline{\delta}_\xi \overline{\boldsymbol{E}}_v + \overline{\delta}_\eta \overline{\boldsymbol{F}}_v + \overline{\delta}_\zeta \overline{\boldsymbol{G}}_v)^n + J^{-1} \boldsymbol{H}^{n+1} \tag{6.75}
$$

where Δt is the real-time step, δ represents the inviscid term difference operator, $\overline{\delta}$ represents the central difference operator, n is the number of real-time steps. By introducing a virtual time iteration process, with $\Delta \tau$ representing the virtual time step and P representing the number of virtual time iteration steps, we obtain

$$
J^{-1} \frac{\boldsymbol{Q}^{p+1} - \boldsymbol{Q}^p}{\Delta \tau} + J^{-1} \frac{3\boldsymbol{Q}^{p+1} - 4\boldsymbol{Q}^n + \boldsymbol{Q}^{n-1}}{2\Delta t} + (\delta_\xi \overline{\boldsymbol{E}} + \delta_\eta \overline{\boldsymbol{F}} + \delta_\zeta \overline{\boldsymbol{G}})^{p+1}
$$
$$
= (\overline{\delta}_\xi \overline{\boldsymbol{E}}_v + \overline{\delta}_\eta \overline{\boldsymbol{F}}_v + \overline{\delta}_\zeta \overline{\boldsymbol{G}}_v)^p + J^{-1} \boldsymbol{H}^{p+1} \tag{6.76}
$$

By using the Jacobian matrix of inviscid flux and taking a first-order Taylor expansion as the source term, we can obtain

$$
\left[\left(\frac{J^{-1}}{\Delta \tau} + \frac{3J^{-1}}{2\Delta t} \right) \boldsymbol{I} - J^{-1} \left(\frac{\partial \boldsymbol{H}}{\partial \boldsymbol{Q}} \right)^p + J^{-1} (\delta_\xi \overline{\boldsymbol{A}} + \delta_\eta \overline{\boldsymbol{B}} + \delta_\zeta \overline{\boldsymbol{C}})^p \right] \Delta \boldsymbol{Q}^p = \boldsymbol{RHS}^p
$$
$$
= - \left[J^{-1} \frac{3\boldsymbol{Q}^p - 4\boldsymbol{Q}^n + \boldsymbol{Q}^{n-1}}{2\Delta t} \right.
$$
$$
\left. + (\delta_\xi \overline{\boldsymbol{E}} + \delta_\eta \overline{\boldsymbol{F}} + \delta_\zeta \overline{\boldsymbol{G}} - \overline{\delta}_\xi \overline{\boldsymbol{E}}_v - \overline{\delta}_\eta \overline{\boldsymbol{F}}_v - \overline{\delta}_\zeta \overline{\boldsymbol{G}}_v - J^{-1} \boldsymbol{H})^p \right] \tag{6.77}
$$

where I is an identity matrix and $\Delta Q^\rho = Q^{P+1} - Q^P$. When $\Delta Q^P \to 0$, setting $Q^{n+1} = Q^{P+1}$ yields the unsteady solution at the corresponding time.

3.   LU-SGS Method

Among the current implicit methods, the lower–upper symmetric Gauss–Seidel (LU-SGS) iterative method proposed by Yoon and Jameson is widely used [26]. The LU-SGS method uses a spectral radius splitting method to construct the approximate Jacobian matrices $\widehat{A}^\pm$, $\widehat{B}^\pm$, and $\widehat{C}^\pm$ for $\overline{A}, \overline{B}, \overline{C}$, respectively.

$$
\begin{cases}
\widehat{A}^\pm = \frac{1}{2}\left[\overline{A} \pm \rho\left(\overline{A}\right)I\right] \\
\widehat{B}^\pm = \frac{1}{2}\left[\overline{B} \pm \rho\left(\overline{B}\right)I\right] , \\
\widehat{C}^\pm = \frac{1}{2}\left[\overline{C} \pm \rho\left(\overline{C}\right)I\right]
\end{cases}
\tag{6.78}
$$

where $P(\overline{A}) = bmax \left[|\lambda(\overline{A})|\right]$, in which b is a constant greater than or equal to 1 that is used to adjust the stability and $\lambda(\overline{A})$ is the eigenvalue of the matrix. The expressions for $P(\overline{B})$ and $P(\overline{C})$ are similar to that for $P(\overline{A})$. By performing the approximate LU decomposition on Eq. (6.77) rewritten using the above approximate Jacobian matrices, we obtain

$$
LD^{-1}U\Delta Q^p = RHS^p
\tag{6.79}
$$

where D is a diagonal matrix and L and U are triangular matrices.

$$
\begin{cases}
D = \left(\dfrac{1}{\Delta\tau} + \dfrac{3}{2\Delta t} + \rho\left(\overline{A}\right) + \rho\left(\overline{B}\right) + \rho\left(\overline{C}\right)\right)J^{-1}I - J^{-1}\left(\dfrac{\partial H}{\partial Q}\right)^p \\
L = D + J^{-1}(\delta_\xi^- \widehat{A}^+ + \delta_\eta^- \widehat{B}^+ + \delta_\zeta^- \widehat{C}^+ - \widehat{A}^+ - \widehat{B}^+ - \widehat{C}^+)^p \\
U = D + J^{-1}(\delta_\xi^+ \widehat{A}^- + \delta_\eta^+ \widehat{B}^- + \delta_\zeta^+ \widehat{C}^- + \widehat{A}^- + \widehat{B}^- + \widehat{C}^-)^p
\end{cases}
\tag{6.80}
$$

After the decomposition, ΔQ^P can be obtained by two scans and one scalar inversion. The calculation procedure is as follows:

$$
\begin{cases}
L\Delta Q^* = RHS^p \\
\Delta Q^{**} = D\Delta Q^* \\
U\Delta Q^p = \Delta Q^{**}
\end{cases}
\tag{6.81}
$$

The LU-SGS method has good computational stability and convergence, and it does not require complex matrix inversion, greatly simplifying the matrix operations and improving the computational efficiency.

6.2.4 Spatial Discretization Method

1. M-AUSMPW+ Scheme

The system of GLM-MHD equations is purely hyperbolic. Therefore, the inviscid term is differentiated using an upwind scheme. Upwind schemes can usually be divided into flux vector splitting (FVS) and flux difference splitting (FDS) schemes. These two schemes have their own advantages and disadvantages in terms of numerical dissipation and computational accuracy. Therefore, Liou and Steffen proposed the advection upstream splitting method (AUSM) scheme [27]. In terms of scheme construction, the AUSM scheme is an improvement of the van Leer scheme, but in terms of its dissipation term, it is a composite of the flux vector splitting (FVS) and flux difference splitting (FDS) schemes. After years of development, a series of AUSM-type schemes have been developed. Among them, the AUSM scheme with weighting functions based on pressure (M-AUSMPW+) has the advantages of high computational efficiency, high resolution, good robustness, small numerical oscillation, and strong mesh adaptability, making it particularly suitable for multidimensional flow calculations. Considering that the AUSM scheme does not requires the calculation of the overly complex eigenvectors of the system of nine-wave GLM-MHD equations, in this section, the M-AUSMPW+ scheme is selected for spatial discretization of the inviscid flux. In this scheme, the linear and nonlinear fields of flow convection characteristics are processed separately. For example, the numerical flux term $\overline{E}$ is constructed as follows:

$$\overline{E}_{1/2} = \overline{M}_L^+ c_{1/2} \Psi_{L,1/2} + \overline{M}_R^- c_{1/2} \Psi_{R,1/2} + P_L^+ P_L + P_R^- P_R + \frac{1}{2}\left(F_{B,L} + F_{B,R}\right) \quad (6.82)$$

where $\overline{M} \pm L$ and R are the Mach number splitting functions, $C_{1/2}$ is the unified sound speed at the element interface, $P \pm L$ and R are the pressure splitting functions, and the vectors ψ, P and F_B have the following forms:

$$\Psi = \begin{bmatrix} \rho \\ \rho u \\ \rho v \\ \rho w \\ B_x \\ B_y \\ B_z \\ \rho e_t + p_t \\ 0 \end{bmatrix}, P = \begin{bmatrix} 0 \\ J^{-1}\xi_x p_t \\ J^{-1}\xi_y p_t \\ J^{-1}\xi_z p_t \\ -\overline{B}_{n,1/2}u \\ -\overline{B}_{n,1/2}v \\ -\overline{B}_{n,1/2}w \\ -\overline{B}_{n,1/2}(V \cdot B) \\ 0 \end{bmatrix}, F_B = \begin{bmatrix} 0 \\ -B_x\overline{B}_{n,1/2}/M_m^2 \\ -B_y\overline{B}_{n,1/2}/M_m^2 \\ -B_z\overline{B}_{n,1/2}/M_m^2 \\ J^{-1}\xi_x\psi \\ J^{-1}\xi_y\psi \\ J^{-1}\xi_z\psi \\ 0 \\ c_h^2\overline{B}_n \end{bmatrix}$$

$$(6.83)$$

where $P_t = P + B^2/2M_m^2$ is the total pressure, $\overline{B} = J^{-1}B_\zeta$, $\overline{B}_{n,\,1/2} = 1/2(\overline{B}_{n,\,L} + \overline{B}_{n,R})$. By setting $\overline{V}_n = J^{-1}V_\zeta$, the unified sound speed at the element interface can be defined as

$$
c_{1/2} = \begin{cases} \dfrac{c_a^2}{\max(|\overline{V}_{n,L}|,c_a)}, & \overline{V}_{n,L} + \overline{V}_{n,R} \geq 0 \\[2ex] \dfrac{c_a^2}{\max(|\overline{V}_{n,R}|,c_a)}, & \overline{V}_{n,L} + \overline{V}_{n,R} < 0 \end{cases} \tag{6.84}
$$

where

$$
c_a = \sqrt{2(\gamma - 1)\,(\gamma + 1)H_{\text{normal}}}
$$

$$
H_{\text{normal}} = \min\left(\left(\frac{\gamma}{\gamma - 1}\frac{p}{\rho} + \frac{B^2}{\rho M_m^2}\right)_L, \left(\frac{\gamma}{\gamma - 1}\frac{p}{\rho} + \frac{B^2}{\rho M_m^2}\right)_R\right) \tag{6.85}
$$

The Mach number on both sides of the interface is defined as $M_{L,\,R} = \overline{V}_{n,\,L,\,R}/C_{1/2}$. Then, the pressure splitting function is

$$
P_{L,R}^\pm = \begin{cases} \frac{1}{4}(M_{L,R} \pm 1)^2(2 \mp M_{L,R}), & |M_{L,R}| \leq 1 \\[1ex] \frac{1}{2}(1 \pm \text{sign}(M_{L,R})), & |M_{L,R}| > 1 \end{cases} \tag{6.86}
$$

By setting $P_{ts} = P + LP_{t,\,L} + P\text{- }Rp_{t,\,R}$, a pressure correction weight function is introduced as follows

$$
f_{L,R} = \begin{cases} \frac{p_{t,L,R}}{p_{ts}} - 1, & p_{ts} \neq 0 \\[1ex] 0, & p_{ts} = 0 \end{cases} \tag{6.87}
$$

$$
w = 1 - \min\left(\frac{p_{t,L}}{p_{t,R}}, \frac{p_{t,R}}{p_{t,L}}\right)^3 \tag{6.88}
$$

The Mach number splitting function is calculated as

$$
\begin{cases} M_{L,R}^\pm = \begin{cases} \pm\frac{1}{4}(M_{L,R} \pm 1)^2, & |M_{L,R}| \leq 1 \\[1ex] \frac{1}{2}(M_{L,R} \pm |M_{L,R}|), & |M_{L,R}| > 1 \end{cases} \\[3ex] \overline{M}_L^+ = \begin{cases} M_L^+ + M_R^-[(1 - w)(1 + f_R) - f_L], & M_L^+ + M_R^- \geq 0 \\[1ex] M_L^+ w(1 + f_L), & M_L^+ + M_R^- < 0 \end{cases} \\[3ex] \overline{M}_R^- = \begin{cases} M_R^- w(1 + f_R), & M_L^+ + M_R^- \geq 0 \\[1ex] M_R^- + M_L^+[(1 - w)(1 + f_L) - f_R], & M_L^+ + M_R^- < 0 \end{cases} \end{cases} \tag{6.89}
$$

During the calculation of the M-AUSMPW+ scheme, the convection vector $\Psi_{L,\,R,\,1/2}$ at the element interface is calculated based on the primitive variable $\Phi_{L,\,R,\,1/2}$.

$$\Phi_{L,\frac{1}{2}} = \Phi_L + \frac{\max\left[0, (\Phi_R - \Phi_L)(\Phi_{L,\text{superbee}} - \Phi_L)\right]}{(\Phi_R - \Phi_L)\left|\Phi_{L,\text{superbee}} - \Phi_L\right|}$$

$$\min\left[a\frac{|\Phi_R - \Phi_L|}{2}, \left|\Phi_{L,\text{superbee}} - \Phi_L\right|\right]$$

$$\Phi_{R,\frac{1}{2}} = \Phi_R + \frac{\max\left[0, (\Phi_L - \Phi_R)(\Phi_{R,\text{superbee}} - \Phi_R)\right]}{(\Phi_L - \Phi_R)\left|\Phi_{R,\text{superbee}} - \Phi_R\right|}$$

$$\min\left[a\frac{|\Phi_L - \Phi_R|}{2}, \left|\Phi_{R,\text{superbee}} - \Phi_R\right|\right] \tag{6.90}$$

where $a = 1 - \min(1, \max(|M_L|, |M_R|))^2$, and the subscript superbee represents the value of the variable at the interface calculated using the superbee limiter.

$$\phi_{\text{superbee}}(r) = \max(0, \min(2r, 1), \min(r, 2)) \tag{6.91}$$

2. MLP Method

By constructing a limiter, the multidimensional limiting process (MLP) method extends the 1D monotonic condition that controls numerical oscillation to the multidimensional case [28]. Taking the ζ direction as an example, $\Phi_{L,R}$ at the element interface is calculated by the MLP method as follows:

$$\Phi_{L,i+1/2,j,k} = \Phi_{i,j,k} + \frac{1}{2}\max\left(0, \min\left(\alpha_L, \alpha_L r^{\xi}_{L,i,j,k}, \beta_L\right)\right)\Delta\Phi_{i-1/2,j,k}$$
$$\Phi_{R,i+1/2,j,k} = \Phi_{i+1,j,k} - \frac{1}{2}\max\left(0, \min\left(\alpha_R, \alpha_R r^{\xi}_{R,i+1,j,k}, \beta_R\right)\right)\Delta\Phi_{i+3/2,j,k} \tag{6.92}$$

where $r^{\xi}_{L,R}$ is the ratio of parameter changes in the ζ direction, $\alpha_{L,R}$ is the multidimensional limiting coefficient, and $\beta_{L,R}$ is the interpolation coefficient.

$$r^{\xi}_{L,i,j,k} = \frac{\Delta\Phi_{i+1/2,j,k}}{\Delta\Phi_{i-1/2,j,k}}, \quad r^{\xi}_{R,i+1,j,k} = \frac{\Delta\Phi_{i+1/2,j,k}}{\Delta\Phi_{i+3/2,j,k}}$$

$$\alpha_L = g\left[\frac{2\max\left(1, r^{\xi}_{L,i,j,k}\right)\left(\Phi^{\max}_{p,q,r} - \Phi_{i,j,k}\right)}{\left(1 + \frac{\Delta\Phi^q_\eta}{\Delta\Phi^p_\xi} + \frac{\Delta\Phi^r_\zeta}{\Delta\Phi^p_\xi}\right)_{i,j,k}\Delta\Phi_{i+1/2,j,k}}\right]$$

$$\alpha_R = g\left[\frac{2\max\left(1, 1/r^{\xi}_{R,i+1,j,k}\right)\left(\Phi^{\min}_{p,q,r} - \Phi_{i+1,j,k}\right)}{\left(1 + \frac{\Delta\Phi^q_\eta}{\Delta\Phi^p_\xi} + \frac{\Delta\Phi^r_\zeta}{\Delta\Phi^p_\xi}\right)_{i+1,j,k}\Delta\Phi_{i+3/2,j,k}}\right] \tag{6.93}$$

where $g(x) = \max(1, \min(2, x))$ and (p, q, r) represents the geometric center of the cells around cell points (i, j, k). To achieve higher-order computational accuracy, the MLP method is combined with third-order polynomial interpolation (MLP3), and the interpolation coefficients in Eq. (6.92) are taken as

$$\beta_L = \frac{1 + 2r^{\xi}_{L,i,j,k}}{3}, \beta_R = \frac{1 + 2r^{\xi}_{R,i+1,j,k}}{3} \tag{6.94}$$

6.2.5 Boundary Conditions

Boundary conditions are an important part of a complete mathematical description of a specific flow and heat transfer problem. For the propellant ablation process and plasma flow process in a PPT, the boundary conditions can be categorized into propellant heat transfer boundary conditions, inflow/outflow boundary conditions and electrode surface boundary conditions.

1. Propellant Heat Transfer Boundary Conditions

On the ablated surface of PTFE, the external incoming heat flux q_{in} needs to be determined. A large number of calculations show that when the discharge energy is low, heat conduction is the main way of transferring heat from the plasma to the propellant in the PPT [29]. Therefore, convection and radiation heat transfer are ignored. Denoting n as the unit outward normal vector to the surface, the heat flux transferred to the propellant surface is determined by the thermal conductivity coefficient of heavy particles and electrons near the surface, as well as the temperature gradient of the plasma, as expressed below:

$$q_{in} = -\boldsymbol{n} \cdot (k_h \nabla T_h + k_e \nabla T_e) \tag{6.95}$$

The adiabatic boundary condition is applied to the remaining surface of PTFE as follows:

$$\boldsymbol{n} \cdot \nabla T_c = 0, \boldsymbol{n} \cdot \nabla T_a = 0 \tag{6.96}$$

2. Inflow/Outflow Boundary Conditions

At the inflow boundary, the ablation products of PTFE leave the ablated surface and enter the plasma flow calculation region, with their temperature set as the propellant surface temperature T_S. The pressure at the inflow boundary is calculated according to the Clausius–Clapeyron equation as follows [30]:

$$p = p_c \exp\left(\frac{h_{\text{eff}}}{R_s}\left(\frac{1}{T_c} - \frac{1}{T_s}\right)\right) \tag{6.97}$$

where $P_C = 1333.2$ Pa, $T_c = 748$ K, $h_{\text{eff}} = 1.768 \times 10^6$ J/kg, and $R_s = 83.1$ J/(kg K). The density and velocity at the inflow boundary are determined by the ideal gas law and the ablation rate

$$\rho = p/R_s T_s, \quad V = \dot{m}(y, z, t)/\rho \tag{6.98}$$

Since PTFE is insulating, no current flows through it in the normal direction. Therefore, only the B_Z component the magnetic induction is considered at the inflow boundary. Denoting the electrode width as W_e, according to Ampere's law, we have

$$B_z = \mu_0 \frac{I(t)}{W_e} \tag{6.99}$$

For the outflow boundary, the flow is considered to be supersonic. Considering that the divergence of the actual magnetic field is zero, the following extrapolated boundary conditions are applied for the conservative variable vector Q:

$$\boldsymbol{n} \cdot \nabla Q = 0 \tag{6.100}$$

3. Electrode Surface Boundary Conditions

The electrode surface is non-penetrating, and the plasma velocity satisfies

$$\boldsymbol{n} \cdot \boldsymbol{V} = 0 \tag{6.101}$$

The Knudsen number is in the range of 10^{-3} to 10^{-1}. Within this range, velocity slip and temperature jump occur near the object surface [31]. The boundary conditions for velocity and temperature are

$$u - u_w = \frac{2 - \sigma_v}{\sigma_v} \frac{Kn}{1 - bKn} \frac{\partial u}{\partial n} \tag{6.102}$$

$$T - T_w = \frac{2 - \sigma_a}{\sigma_a} \frac{2\gamma}{\gamma + 1} \frac{Kn}{Pr} \frac{\partial T}{\partial n} \tag{6.103}$$

where u_w is the velocity of the object surface (which is zero for the electrode solid wall), T_w is the actual temperature of the object surface, σ_v and σ_a are the tangential momentum accommodation coefficient and thermal accommodation coefficient, respectively, and b is a set parameter.

The electrode is considered an ideal conductor. Since the skin depth of the magnetic field penetrating into the ideal conductor is close to zero and the ideal conductor is an equipotential body, the electromagnetic field on the electrode surface satisfies [32]

$$\boldsymbol{n} \cdot \boldsymbol{B} = 0 \tag{6.104}$$

$$\boldsymbol{n} \times \boldsymbol{E} = 0 \tag{6.105}$$

Using the generalized Ohm's law and Ampere's law, Eq. (6.105) can be rewritten as

$$n \times (\eta \nabla \times B - V \times B + v(\nabla \times B) \times B) = 0 \tag{6.106}$$

The velocity V at the electrode surface is a very small slip velocity. If an approximation of no-slip flow is adopted and the Hall effect term is ignored, the following equation is easily derived using the rules of vector operations

$$(n \cdot \nabla)B = 0 \tag{6.107}$$

Equations (6.104) and (6.107) together form the magnetic field boundary conditions of an ideal conducting wall. If the Hall effect is considered, we can obtain

$$\left(n \cdot \left(\nabla + \frac{v}{\eta}(\nabla \times B)\right)\right)B = 0 \tag{6.108}$$

In this case, the magnetic field boundary conditions are mixed boundary conditions

$$\begin{cases} B_x = B_z \dfrac{\partial B_x}{\partial y} \Big/ \dfrac{\partial B_z}{\partial y} \\[2mm] B_y = 0 \\[2mm] B_z = \dfrac{\eta}{v} \dfrac{\partial B_z}{\partial y} \Big/ \dfrac{\partial B_z}{\partial x} \end{cases} \tag{6.109}$$

6.3 Numerical Simulation Results and Analysis

6.3.1 Numerical Example Validation

1. Numerical Example of PTFE Ablation

Heat Conduction Process

Consider 1D unsteady heat conduction in a single-phase body. Assuming constant physical parameters and an initial temperature of T_0, when the heat flux q is held constant, there exists an analytical solution for the temperature distribution of the object [33]:

$$T(x, t) = \frac{q}{k}\sqrt{\frac{4\alpha t}{\pi}} \exp\left(-\frac{x^2}{4\alpha t}\right) - \frac{qx}{k}\operatorname{erfc}\left(\frac{x}{\sqrt{4\alpha t}}\right) + T_0 \tag{6.110}$$

where α is the thermal diffusivity. The 1D calculation is performed using the Fortran program, and the calculation results are shown in Fig. 6.2. The temperature distribution curves of numerical solutions and analytical solutions at the three time points in the figure are in high agreement, indicating that the program is capable of accurately simulating unsteady heat conduction processes.

Phase Transition Process

The influences of phase interface motion and phase transition latent heat on the temperature distribution are considered. It is assumed that PTFE is initially in a crystalline solid phase, with an initial temperature at the phase transition temperature T_m, and that at $t > 0$, an external heat source with a heat flux q is applied to the front end surface. If the endothermic decomposition of PTFE is ignored and the temperature distribution within the region after phase transition is assumed to be a quadratic function of coordinates, an approximate solution exists for the temperature distribution under the condition of constant physical parameters [34]

$$
\begin{aligned}
T(x, t) = &\frac{h_m}{2c_a x_m}\left(1 - \sqrt{1 + 4\mu_m}\right)(x - x_m) \\
&+ \frac{h_m}{8c_a x_m^2}\left(1 - \sqrt{1 + 4\mu_m}\right)^2 (x - x_m)^2 + T_m
\end{aligned}
\tag{6.111}
$$

where the coefficient μ_m and the position x_m of the phase interface are calculated by the following equation:

$$
\frac{\mu_m}{6}\left(\mu_m + 5 + \sqrt{1 + 4\mu_m}\right) = \frac{q^2 t}{\alpha\rho^2 h_m^2}, \quad x_m = \frac{\mu_m \alpha \rho h_m}{q}
\tag{6.112}
$$

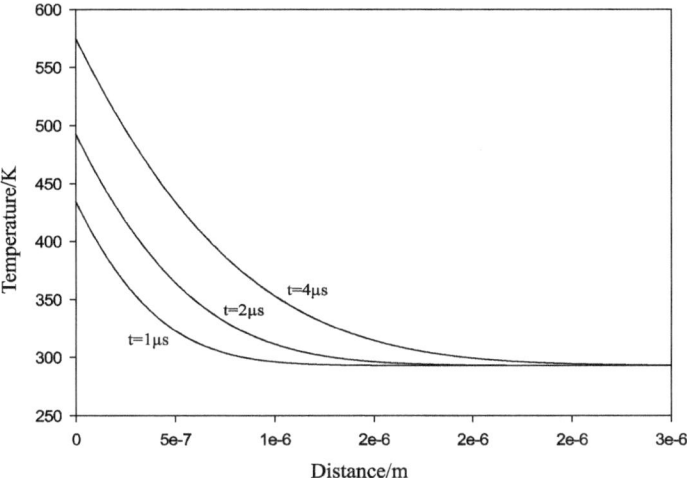

Fig. 6.2 Temperature distributions during the heat conduction process

Figure 6.3 presents the temperature distribution curve considering the phase transition process. The numerical solutions and the approximate solutions in the figure are similar, indicating that the program is able to simulate the phase transition process of PTFE well.

Ablation Process

Assuming a linear distribution of temperature near the ablated surface and a constant density of PTFE, Kemp [35] provided approximate calculation formulas for determining the ablated mass flux and the ablated surface recession velocity based on the ablated surface temperature T_s under the steady ablation condition.

$$\dot{m} = \sqrt{\frac{A_p \rho \kappa T_s^2}{B_p(h_s - h_{-\infty})}} \exp\left(-\frac{B_p}{T_s}\right) \tag{6.113}$$

$$v = \sqrt{\frac{A_p \kappa T_s^2}{B_p \rho(h_s - h_{-\infty})}} \exp\left(-\frac{B_p}{T_s}\right) \tag{6.114}$$

where B_p is the activation temperature and $h_s - h_{-\infty}$ is the difference in the enthalpy of the propellant before and after ablation.

Using the custom-developed computer program, the characteristic ablation parameters of PTFE when the ablation process reaches a steady state under the constant heat flux are calculated. The calculation results are compared with the approximate values obtained from Eqs. (6.113) and (6.114), as shown in.

In Fig. 6.4, the numerical solutions match well with the approximate solutions, indicating that the program can calculate the PTFE ablation characteristics accurately.

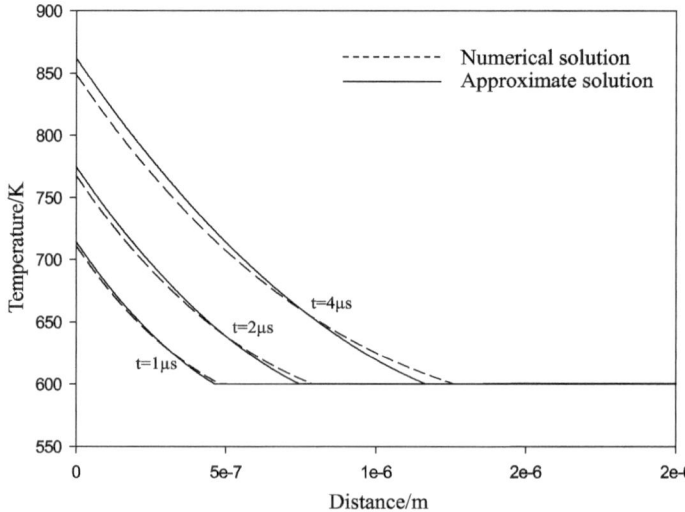

Fig. 6.3 Temperature distributions during the phase transition process

Fig. 6.4 Steady-state ablated mass flux and ablated surface recession velocity

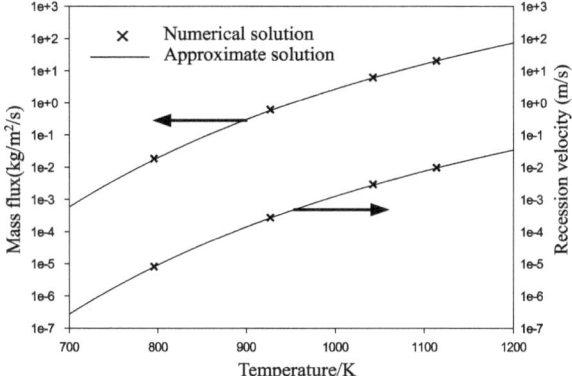

2. Numerical Example of MHD Flow

Orszag–Tang Vortex Problem

To examine the reliability of the numerical algorithm and calculation program for MHD simulations, the well-known Orszag–Tang vortex problem [36] is investigated as an example in this section. This problem considers the process in which the initially smooth flow field generates shock waves and evolves to MHD turbulent flow over time. By solving this problem, we can examine the program's ability to simulate complex interactions between various wave systems in magnetohydrodynamics.

In the Orszag–Tang vortex problem, the initial conditions of the flow field are given as

$$\rho(x, y, 0) = \gamma^2, u(x, y, 0) = -\sin y, v(x, y, 0) = \sin x$$
$$p(x, y, 0) = \gamma, B_x(x, y, 0) = -\sin y, B_y(x, y, 0) = \sin 2x$$

(6.115)

where $\gamma = 5/3$. The calculation region is taken as yx, where $y \in [0, 2\pi]$, and periodic boundary conditions are adopted. The fluid viscosity term is not considered. The density contours at each time point are obtained from 2D calculations using the custom-developed 3D program, as shown in Fig. 6.5. The figure clearly reflects the formation and mutual interference of shock waves in the flow field. The flow field structure is consistent with the calculation results obtained by Li [37] using the discontinuous Galerkin method, indicating that the numerical algorithm used in this section is accurate and that the computer program can effectively calculate complex MHD flows.

Rayleigh Problem

The MHD Rayleigh problem is an important basic problem of unsteady MHD flow [38] that accounts for the molecular viscosity, magnetic field diffusion and wave propagation and can be used to assess the simulation accuracy of MHD boundary layer flow. In the MHD Rayleigh problem, a uniform magnetic field B_0 in the y direction and perpendicular to the plate is applied to an infinite plate. At time $t = 0$,

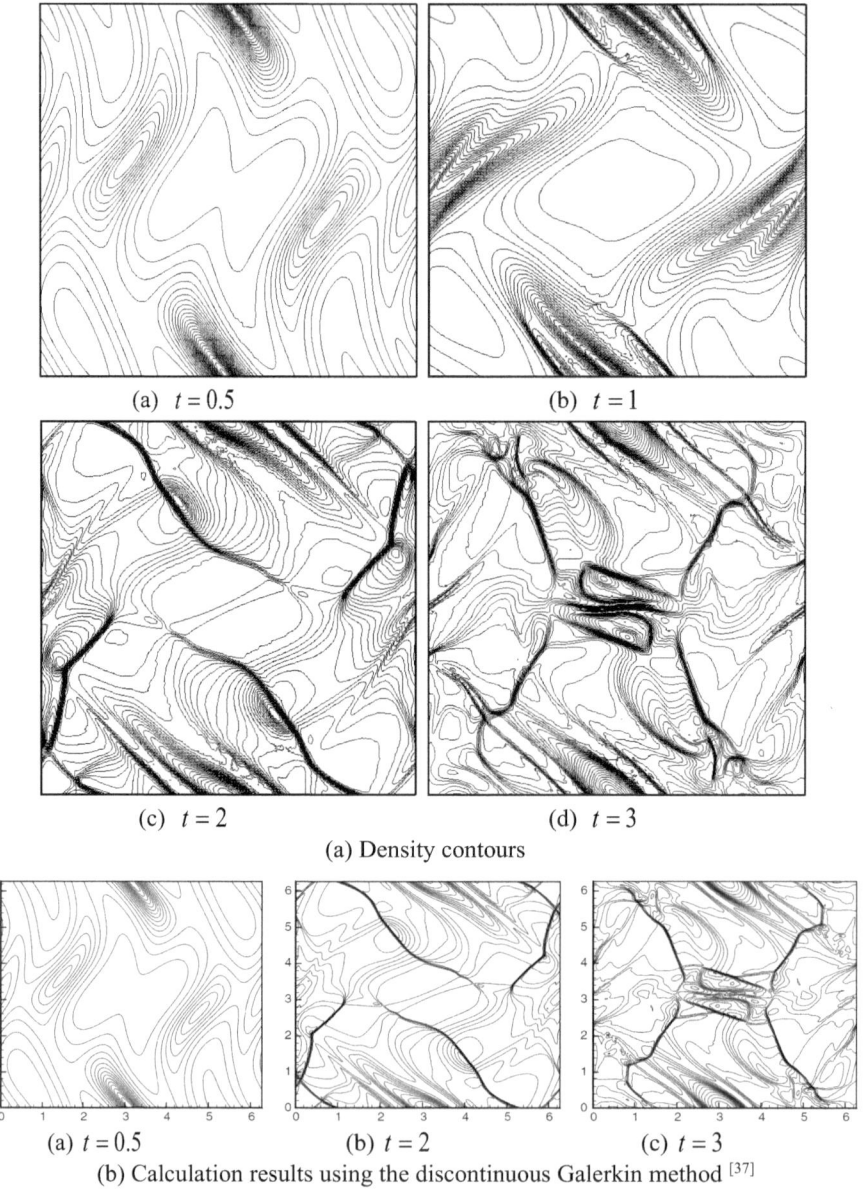

(a) $t = 0.5$ (b) $t = 1$

(c) $t = 2$ (d) $t = 3$

(a) Density contours

(a) $t = 0.5$ (b) $t = 2$ (c) $t = 3$

(b) Calculation results using the discontinuous Galerkin method [37]

Fig. 6.5 Calculation results for the Orszag–Tang vortex problem

Fig. 6.6 Velocity distributions for the ayleigh problem

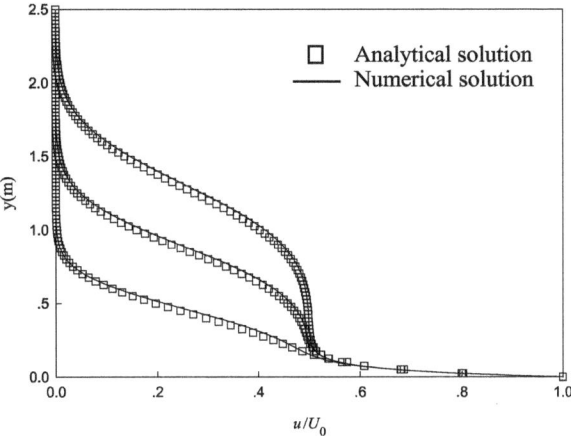

the plate suddenly moves at a uniform speed of U_0 along the x direction, causing an accelerated flow of the nearby conducting fluid with a viscosity coefficient of μ_f and a conductivity of σ_e. When the magnetic Prandtl number is $Pr_m = \mu_f / \mu_0 \sigma_e \rho = 1$, the MHD Rayleigh problem has an analytical solution. For an insulated plate, the fluid velocity and the induced magnetic field intensity are expressed as follows:

$$\frac{u}{U_0} = \frac{1}{4}\left[2.0 - (\mathrm{erf}(\lambda_+) + \mathrm{erf}(\lambda_-)) + e^{\frac{-A_0 y}{v_f}} \mathrm{erfc}(\lambda_-) + e^{\frac{A_0 y}{v_f}} \mathrm{erfc}(\lambda_+) \right]$$

$$\frac{B_x}{B_{\mathrm{ref}}} = \frac{1}{4}\left[erf(\lambda_-) - \mathrm{erf}(\lambda_+) + e^{\frac{-A_0 y}{v_f}} \mathrm{erfc}(\lambda_-) - e^{\frac{A_0 y}{v_f}} \mathrm{erfc}(\lambda_+) \right] \qquad (6.116)$$

where erf and erfc are the error function and the complementary error function, respectively, A_0 is the Alfvén velocity, v_f is the kinematic viscosity coefficient, B_{ref} is the reference magnetic induction, and $\lambda_\pm$ is a parameter determined by position and time.

Given $B_0 = 1.5 \times 10^{-4}$ T, $\rho = 4 \times 10^{-5}$ kg/m^3, and $\sigma_e = 1/\mu_0$, the calculation length in the y direction is set to 2.5 m, and the region is evenly meshed into 100 cells. Figures 6.6 and 6.7 show comparisons of the analytical solutions and numerical solutions of the velocity and induced magnetic field intensity distributions, respectively, at different time points. It is observed that the numerical results are close to the analytical solutions, indicating that the computer program is capable of simulating viscous flow and magnetic field diffusion well.

6.3.2 Analysis of Plasma Flow in the Discharge Channel

The PPT relies on the high-speed ejection of the plasma generated by discharge ablation out of the discharge channel to generate a reactive thrust. The plasma flow

Fig. 6.7 Distributios of the induced magnetic field for the Rayleigh problem

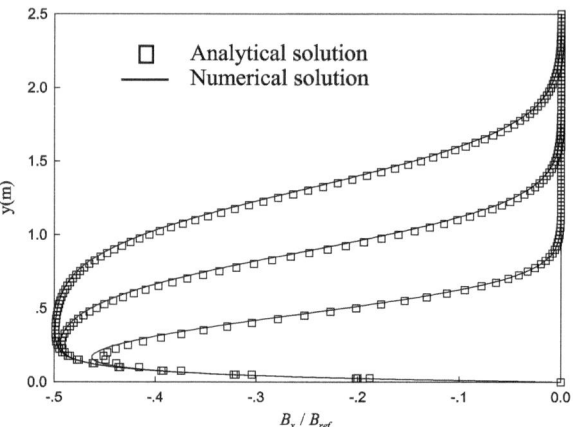

process directly affects the thrust performance. At present, most studies on plasma flow in PPTs are based on simple electromechanical models or 1D MHD models, which cannot reflect the multidimensional characteristics observed during the PPT operation. Even higher-dimensional MHD research, represented by the MACH2 2D simulation program, is still preliminary and insufficient to reveal the complex flow conditions in PPTs. To gain a deeper understanding of the plasma flow process in the PPT discharge channel, the 3D flow characteristics and acceleration process of the plasma in the PPT are analyzed, while the Hall effect is ignored; then, the influence of the Hall effect on the plasma motion in the PPT is investigated. For physical quantities with minimal change in the z direction, only the analysis results within the spanwise center plane ($z = 0.5w_e$) are presented.

The calculation regions for numerical simulation are divided into two parts, namely, the PTFE ablation heat transfer region A and the plasma flow calculation region consisting of the discharge channel B1 and the downstream extension region B2 of the thruster exit, as shown in Fig. 6.8. For the calculation of the PTFE ablation process in region A, a uniform structured mesh is used, the heat transfer distance l_0 is set to 20 μm, the initial temperature is set to 300 K, and the density, specific heat capacity, thermal conductivity, and pyrolysis heat of PTFE are all considered as functions of temperature. Their physical parameters are listed in Appendix B. For the calculation of the plasma flow region, a structured mesh with appropriate refinement near the ablated surface and the electrode surface is used. The ablative PPT (APPT) typically operates continuously for multiple pulses. Therefore, it can be considered that there is a certain mass of plasma in the flow field at the beginning. The initial mass of the plasma is set to 0.1 μg, the temperature is set to 0.2 eV, and both the initial velocity and magnetic induction intensity are set to zero. Considering that the density of components with a particularly high valence is very low, in the actual numerical simulation, the carbon and fluorine components are only calculated up to the highest tetravalent and pentavalent ions, respectively, that is, there are a total of twelve components, namely, e^-, C, C^+, C^{2+}, C^{3+}, C^{4+}, F, F^+, F^{2+}, F^{3+}, F^{4+}, F^{5+}.

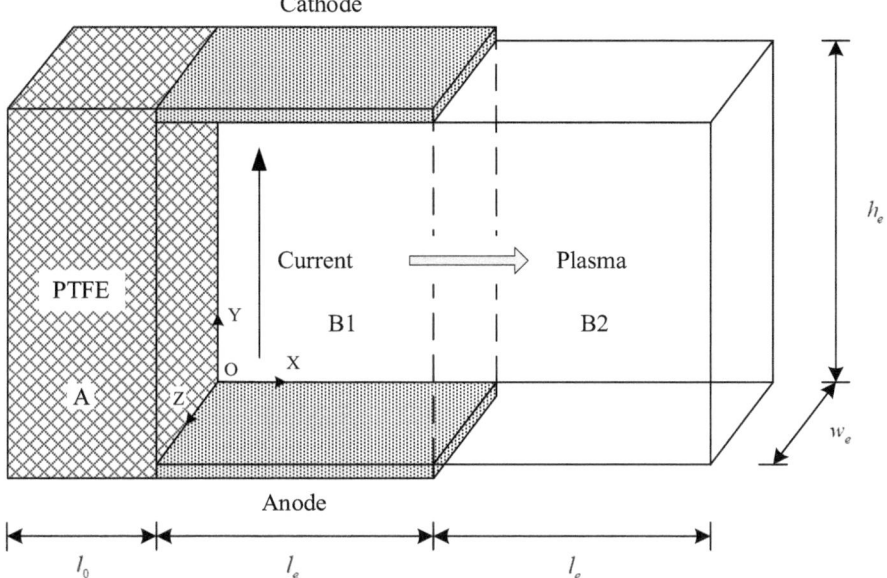

Fig. 6.8 Schematic diagram of the numerical simulation regions

1. Flow Field Structure

a. Density Distribution

The density distribution of the plasma in the PPT discharge channel is shown in Fig. 6.9. It is observed that as a large amount of propellant is ablated in the first 2 μs of discharge, a high-density plasma gathers near the propellant surface, and the plasma density is higher at locations where the surface temperature of the propellant is near the electrode. Under the effect of strong electromagnetic and aerodynamic acceleration in the initial stage of discharge, the high-density plasma near the propellant surface starts to expand and move downstream, leading to contour surfaces protruding downstream, as shown in Fig. 6.9a and b. In the middle and late stages of discharge, as the propellant ablation rate decreases and the plasma continues to move toward the thruster nozzle downstream of the channel, the density of plasma in the discharge channel decreases continuously. The plasma in the central region of the PPT discharge channel rapidly expands outwards to accelerate and be ejected out of the thruster, while the plasma in the vicinity of the electrode moves more slowly due to viscous effects; as a result, the density distribution contour surfaces in the later stage of discharge is concave upstream. In the later stage of PPT discharge, the current is small, and the electromagnetic acceleration effect of the thruster on the plasma is very weak, causing many plasma particles to stay in the discharge channel for a long time. Therefore, the variation in the plasma density, as shown in the figure, appears very slow during this period.

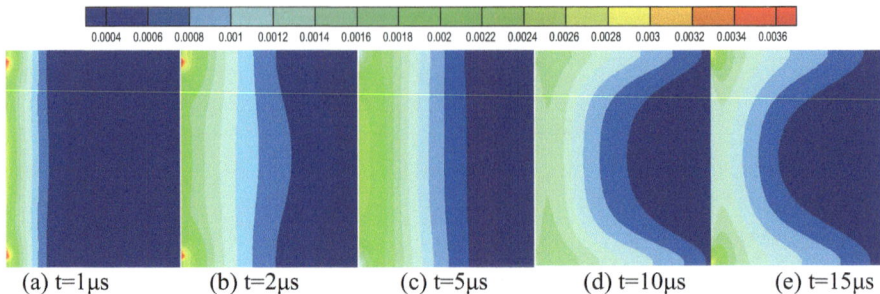

(a) t=1μs   (b) t=2μs   (c) t=5μs   (d) t=10μs   (e) t=15μs

Fig. 6.9 Density distribution

Figure 6.10 shows the plasma flow velocity distribution. Due to the presence of strong pulse discharge in the first 2 μs, the plasma is subjected to a large electromagnetic force, rapidly accelerating to a high speed of about 20 km/s and being quickly ejected out of the discharge channel between the electrodes. After the first half of the oscillation period of the PPT discharge, the current starts to reverse. Before $t = 5\mu s$, the reverse current reaches its peak, and the plasma is again subjected to relatively strong electromagnetic acceleration, forming a region with a high velocity in the middle of the discharge channel downstream. Since most of the initial energy stored in the capacitor is released in the first half of the oscillation period of the discharge, the plasma velocity at $t = 5\mu s$ is significantly lower than that at the initial stage of the discharge, reaching a maximum of only 5.5 km/s. As the discharge approaches its end, the plasma velocity further decreases to below 1–2 km/s in the later stage of discharge. According to the analysis of the plasma density in the previous section, there are still a large number of plasma particles in the discharge channel at this time. Most of these particles are located in regions with a velocity of only a few hundred meters per second. Therefore, the impulse and thrust generated by this part of the plasma are very small, greatly reducing the thrust efficiency of the PPT.

From the distribution of the plasma velocity component in the y direction (Fig. 6.11), it is observed that the plasma flow between the upper and lower electrodes is symmetric, and the velocity of the plasma in the direction perpendicular to

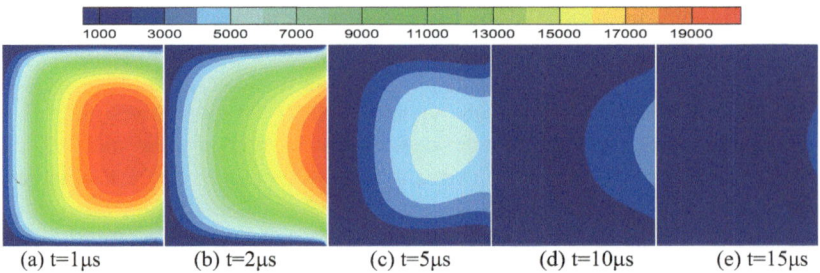

(a) t=1μs   (b) t=2μs   (c) t=5μs   (d) t=10μs   (e) t=15μs

Fig. 6.10 Velocity distribution in the x direction

the electrodes is significantly lower than the velocity in the flow direction. At the beginning of the PPT discharge, the plasma moves from a high-density region in the vicinity of the ablated surface of the propellant near the electrode to the central region of the discharge channel, which has a very high flow velocity. The plasma in the vicinity of the thruster nozzle expands from the central position toward the free-flow region outside the ends of the electrodes on both sides. In the middle and late stages of the discharge, the ablation rate of the propellant decreases rapidly, and the flow process near the ablated surface becomes less noticeable. The plasma in the entire discharge channel flows slowly downstream against the side of the closer electrode.

Figure 6.12 shows the velocity distribution of the plasma in the spanwise direction. It is observed that the plasma expands slowly from the discharge channel between the electrodes to both sides in the spanwise direction, with a velocity much slower than that in the x and y directions, indicating that the force acting on the plasma in the z direction can be ignored.

c. Temperature Distribution

The plasma temperature distribution calculated under the assumption of local thermodynamic equilibrium is shown in Fig. 6.13. The calculations show that during the initial stage of the pulse discharge, the plasma is subjected to strong Joule heating due to the large discharge current of the PPT, causing the temperature to rise rapidly, followed by a gradual decrease with the oscillatory decay of the discharge current. Comparing Figs. 6.10 and 6.13 reveals that the plasma temperature distribution and the flow velocity distribution have similar variation patterns. This is because, on the one hand, both the plasma temperature and the $J \times B$ electromagnetic force are directly related to the discharge current, and, on the other hand, a higher temperature of plasma leads to a higher degree of ionization, making the plasma more susceptible to acceleration to high velocity under the electromagnetic force.

d. Magnetic Field Distribution

During the PPT operation, a strong discharge occurs between the two electrodes, and the current mainly flows in the y direction perpendicular to the electrode surface,

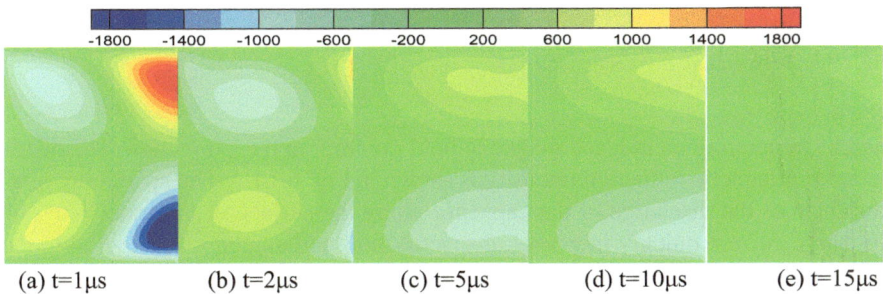

(a) t=1μs   (b) t=2μs   (c) t=5μs   (d) t=10μs   (e) t=15μs

Fig. 6.11 Velocity distribution in the y direction

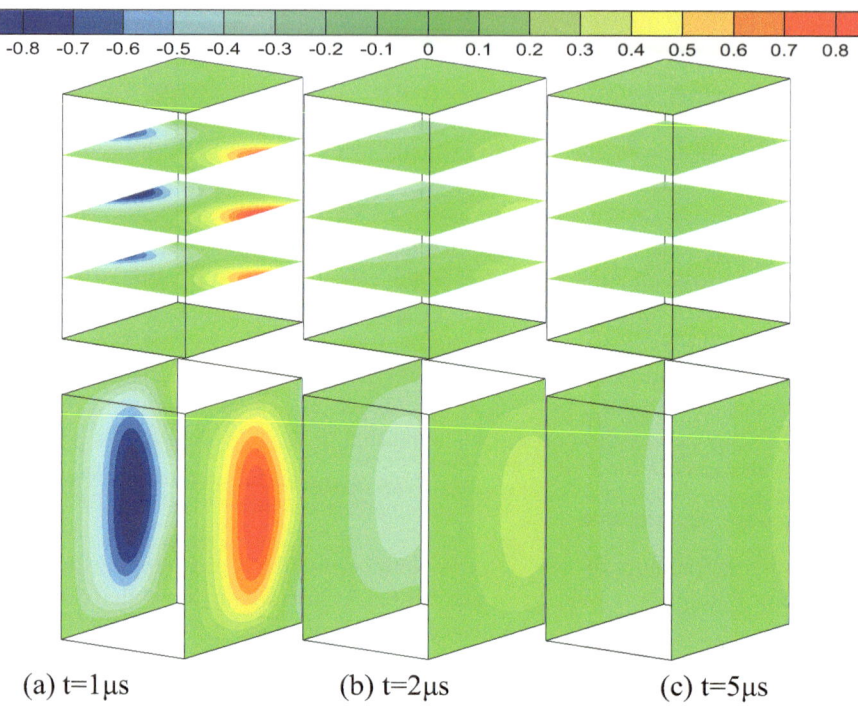

Fig. 6.12 Velocity distribution in the *z* direction

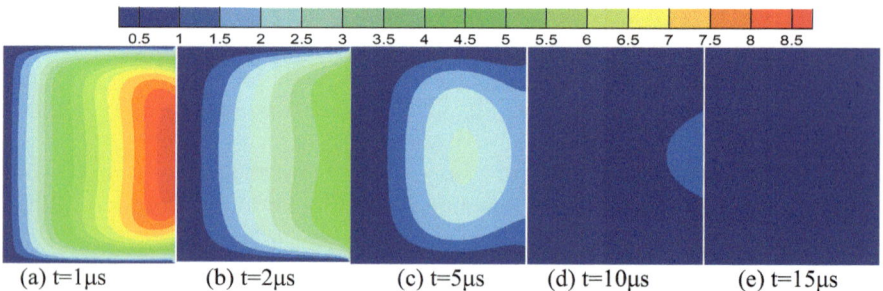

Fig. 6.13 Temperature distribution

thereby generating an induced magnetic field pointing in the *z* direction. The distribution of magnetic induction intensity is shown in Fig. 6.14. When spark plug ignition induces the discharge of the capacitor along the surface of the propellant, the discharge current quickly increases to several thousand amperes. According to the boundary conditions, the magnetic induction intensity Bz near the ablated surface of the propellant is proportional to the discharge current. Therefore, the value of Bz also increases rapidly, forming a large magnetic field gradient near the surface of the propellant. Due to magnetic diffusion and magnetic freezing effects, the magnetic field diffuses from regions of high intensity to regions of low intensity, and at the same time, magnetic induction lines move downstream with the plasma, forming the distribution shown in Fig. 6.14b. It is clear that the magnetic induction intensity near the electrodes is greater than the field strength at the center of the discharge channel, consistent with the experimental measurements in Reference [39]. At $t = 5\,\mu s$, due to the reversal of the current, the magnetic field near the propellant surface also reverses. In the later stage of discharge, as the oscillation of the discharge current decays, the magnetic induction intensity in the entire discharge channel tends to zero.

Figures 6.15 and 6.16 show the slice contours of the magnetic induction intensity in the x and y directions, respectively. The distributions of the magnetic induction intensity components B_x and B_y exhibit obvious three-dimensional characteristics. B_x has large values on the ablated surface of the propellant and B_y has large values near the thruster nozzle. B_x and B_y have comparable magnitudes but they are much smaller than the magnitude of B_z.

e. Pressure Distribution

According to Ampere's law and the magnetic flux continuity law and using vector calculation formulas, the thermal pressure and $J \times B$ electromagnetic force acting on a unit volume of the plasma can be expressed as

$$f = -\nabla p + J \times B = \nabla \cdot \left[-\left(p + \frac{B^2}{2\mu_0}\right)\bar{\bar{I}} + \frac{BB}{\mu_0} \right] = \nabla \cdot \bar{\bar{P}}_{\text{MHD}} \qquad (6.117)$$

where $\bar{\bar{P}}_{\text{MHD}}$ is the MHD pressure tensor. Since B_z is much larger than B_x and B_y, the magnetic induction lines are basically along the z direction. Let the unit vectors in the directions of the three coordinate axes of the Cartesian coordinate system be e_x, e_y, and e_z. Then, $\bar{\bar{P}}_{\text{MHD}}$ can be written as

$$\bar{\bar{P}}_{\text{MHD}} = \left(p + \frac{B^2}{2\mu_0}\right)e_x e_x + \left(p + \frac{B^2}{2\mu_0}\right)e_y e_y + \left(p - \frac{B^2}{2\mu_0}\right)e_z e_z \qquad (6.118)$$

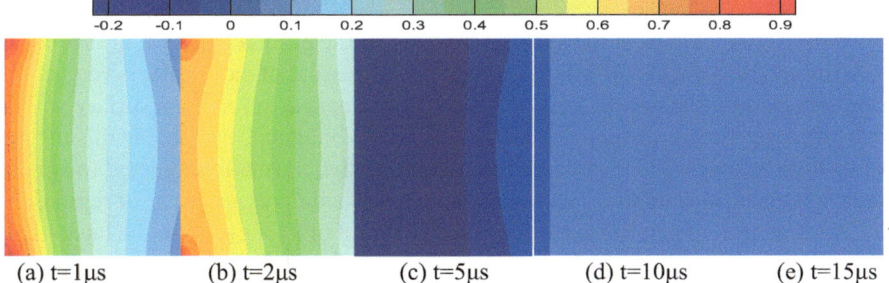

(a) t=1μs   (b) t=2μs   (c) t=5μs   (d) t=10μs   (e) t=15μs

Fig. 6.14 Distribution of magnetic induction intensity B_z

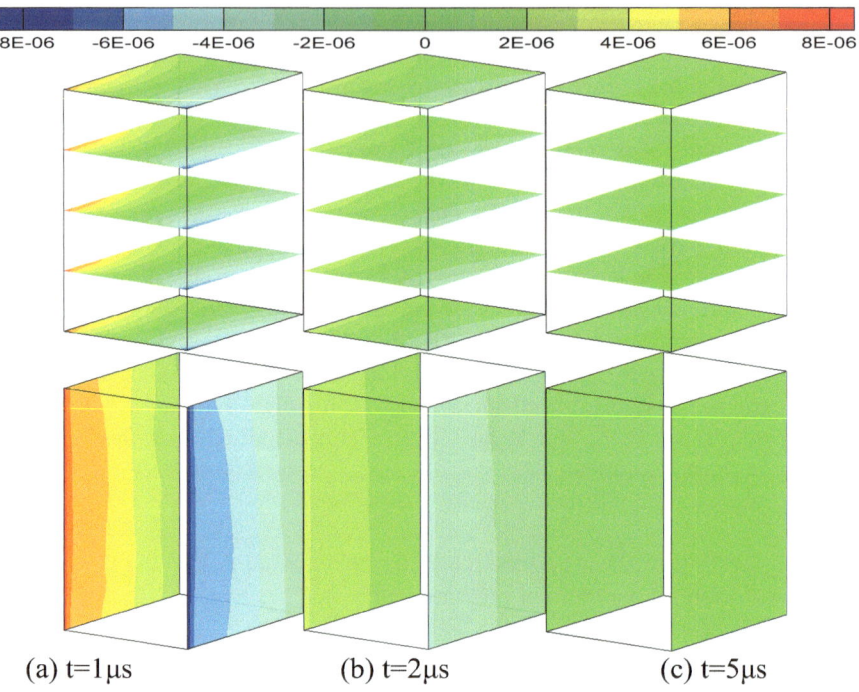

| -8E-06 | -6E-06 | -4E-06 | -2E-06 | 0 | 2E-06 | 4E-06 | 6E-06 | 8E-06 |

(a) t=1μs   (b) t=2μs   (c) t=5μs

Fig. 6.15 Distribution of magnetic induction intensity B_x

Thus, the force exerted on the plasma accelerating in the flow direction is equal to the gradient of the total pressure $p + B^2/2\mu_0$, in which the distribution of the magnetic pressure $B^2/2\mu_0$ can be obtained from the distribution of the magnetic induction intensity, as shown in Fig. 6.14. To obtain the relative changes in the thermal pressure and magnetic pressure on the plasma, the specific pressure of the plasma is defined as

$$\beta_p = \frac{p}{p + B^2/2\mu_0} \tag{6.119}$$

The variation pattern of the specific pressure of the plasma is shown in Fig. 6.17. In the first 2 μs of the PPT discharge, the specific pressure of the plasma is very small in most regions of the flow field, especially $\beta_p < 0.1$ near the propellant surface with a high magnetic induction intensity, indicating that the magnetic pressure in this region is much greater than the plasma pressure and that the plasma is subjected to strong electromagnetic acceleration. At $t = 5$ μs, the magnetic pressure in most regions of the flow field is comparable to the plasma pressure. For a long time in the later stage of discharge, the magnetic induction intensity is very small, and the plasma specific pressure is close to 1. At this time, the aerodynamic force acting on the plasma plays

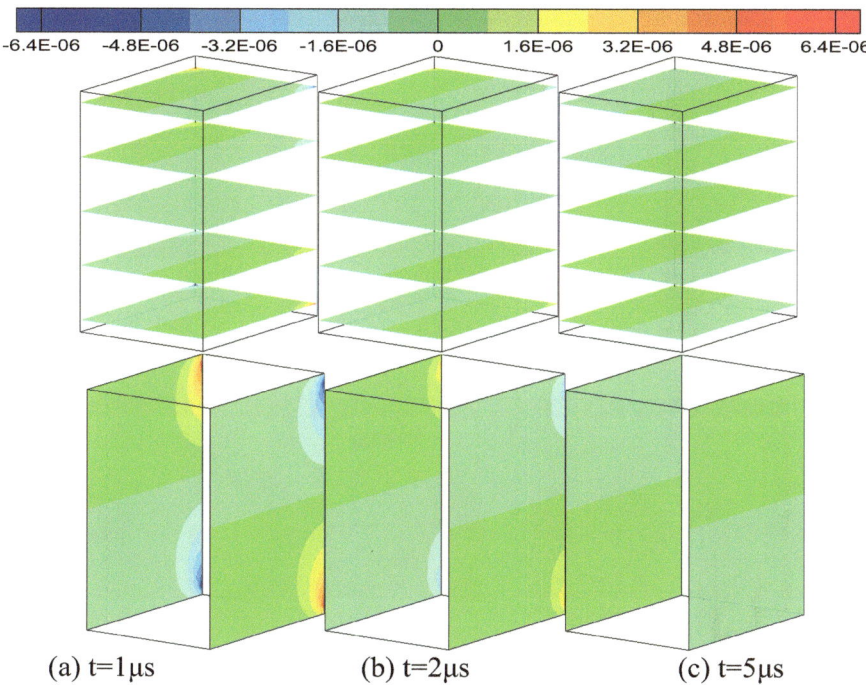

Fig. 6.16 Distribution of magnetic induction intensity B_y

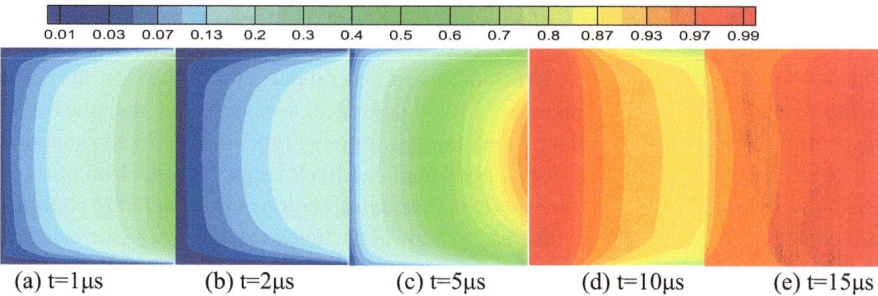

Fig. 6.17 Plasma specific pressure distribution

a predominant role in the acceleration process. Due to the small aerodynamic force, the motion of the plasma becomes very slow.

2. Changes in Component Properties

Figures 6.18 and 6.19 show the variations in the number density of neutral particles and electrons in the spanwise center plane of the discharge channel. A comparison reveals that in the early stage of PPT discharge, the neutral particles are mainly

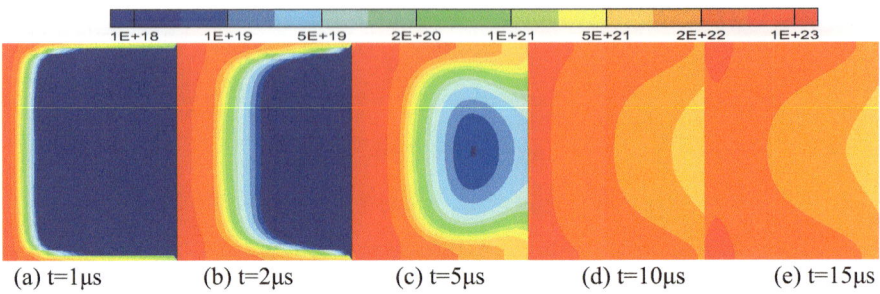

Fig. 6.18 Changes in the neutral particle number density

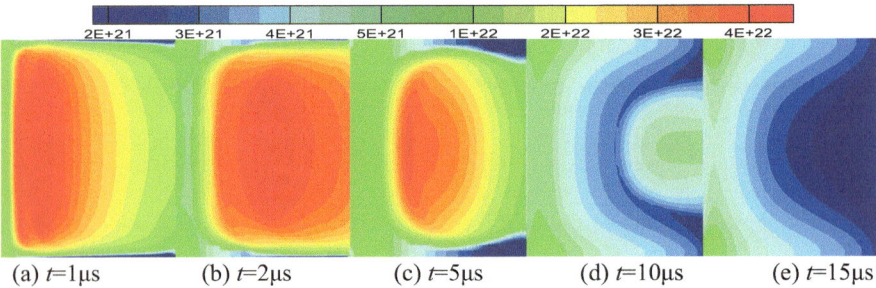

Fig. 6.19 Changes in the electron number density

distributed near the ablated surface of the propellant and the electrode surface, exactly corresponds to the locations with low plasma temperature in Fig. 6.13. The number density of electrons is much greater than that of neutral particles in the central region of the discharge channel, indicating a high degree of ionization of the plasma in the central region. Although the plasma temperature is higher at the electrode end in Fig. 6.13a and b, due to the lower plasma density at this location, the electron number density is greater upstream of the nozzle in the central region of the discharge channel. In the later stage of PPT pulse discharge, neutral particles slowly move downstream from the high-density region upstream of the discharge channel. Because a plasmoid with a high degree of ionization is ejected from the thruster at high speed and both the density and temperature of the plasma in the flow field in the later stage of discharge decrease simultaneously, the electron number density decreases rapidly. At this time, the plasma particles in the discharge channel are mostly neutral particles. In Fig. 6.19d, there exists a region with a high electron number density near the nozzle. Based on the electron number density distribution at multiple previous time points, it can be determined that the electrons in this region are mainly generated around the time when the discharge current reaches its reverse peak. According to the analysis of the velocity distribution of the flow field, the velocity of the plasma in this region is approximately 2 km/s, which allows it to move out of the nozzle before the end of the discharge pulse, resulting in a low-density electron distribution only near the propellant and electrode surfaces, as shown in Fig. 6.19e.

Twelve components are considered in the simulation of the plasma flow process in the PPT. To facilitate the understanding of the variation in the number density of each component, the density during the discharge process is set to a typical value of $\rho = 0.001$ kg/m^3 to obtain the variation in the mole fraction of each component with the plasma temperature, as shown in Fig. 6.20. It can be seen from the figure that when the temperature is less than 1 eV, neutral carbon and fluorine atoms are the main components, indicating a low ionization degree of the plasma. As the temperature increases, the fluorine and carbon atoms ionize to become monovalent ions, and the low-valent ions continue to ionize at a higher temperature to become higher-valent ions. At the same valence, the ionization energy of fluorine is higher than that of carbon, thus requiring a higher temperature for significant ionization. Due to the high plasma temperature in the early stage of discharge, the plasma in most regions of the flow field is highly ionized in this period. Many studies, including the emission spectroscopy diagnostic studies mentioned in this section, have measured the presence of multivalent ions, further confirming this point.

3. Thrust Action Process and Operation Performance Analysis

PPTs use pulse discharge to eject plasma from the thruster to generate an equivalent reactive thrust. The plasma acceleration process is the result of the combined action of electromagnetic force and aerodynamic force. To facilitate the evaluation of the thrust performance of PPTs, many researchers have proposed empirical estimation formulas for electromagnetic impulse, aerodynamic impulse, or impulse bit. Vondra

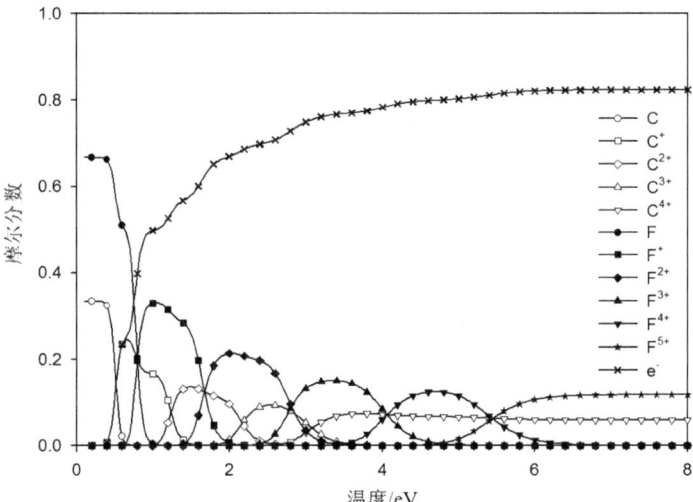

Fig. 6.20 Variations in the component mole fractions with temperature

[40] was the first to establish a theoretical formula for calculating the electromagnetic impulse I_{em}, which is expressed as

$$I_{em} = \frac{\mu_0}{2} \frac{h_e}{w_e} \int_0^{t_f} I^2 dt \qquad (6.120)$$

where the electromagnetic impulse is proportional to the aspect ratio of the ablated end face of the propellant. Using a similar form, Burton [41] proposed another formula for estimating the electromagnetic impulse by introducing the concept of inductance gradient.

$$I_{em} = \frac{1}{2} L' \int_0^{t_f} I^2 dt \qquad (6.121)$$

For the parallel-electrode breech-fed PPT, the inductance gradient L' is calculated as

$$L' = 0.6 + 0.4 \ln \frac{h_e}{w_e + d_e} \qquad (6.122)$$

where the unit of L' is $\mu H/m$, and d_e is the thickness of the electrode. For the aerodynamic impulse, Guman [42] derived the expression under the assumption of quasi-steady isentropic flow as follows:

$$I_{gas} = \sqrt{\frac{8(\gamma - 1)}{\gamma^2(\gamma + 1)} m_p E_0} \qquad (6.123)$$

where m_p is the pulse ablated mass of the propellant, which is empirically calculated as follows:

$$m_p = 1.32 \times 10^{-6} A_p^{0.65} E_0^{0.35} \qquad (6.124)$$

where A_P is the area of the ablated end surface of the propellant. By applying the exponential burning rate law of propellant combustion in solid rocket motors to the ablation process of PTFE, Henrikson [43] derived the following equation for calculating the aerodynamic impulse under the assumption of 1D quasi-steady isentropic flow with a high magnetic Reynolds number:

$$I_{gas} = 1.255 A_p \left(\frac{\mu_0}{4.404 a_0 w_e^2 \rho_c V_{crit}} \right)^{1/n} \int_0^{t_f} I^{2/n} dt \qquad (6.125)$$

where V_{crit} is the Alfven critical velocity and α_0 and n are the burning rate coefficient and burning rate pressure exponent fitted from the experimental results, respectively. Based on the above expressions for electromagnetic impulse and aerodynamic impulse, the impulse bit during PPT operation can be estimated as

$$I_{\text{bit}} = I_{\text{em}} + I_{\text{gas}} \tag{6.126}$$

Based on extensive experimental data, Guman [44] proposed an empirical formula for the specific impulse of breech-fed PPTs as follows:

$$I_{\text{sp}} = 317.5 \left(\frac{E_0}{A_p} \right)^{0.585} \tag{6.127}$$

Combining the above formula with the empirical formula for the ablated mass, another empirical formula for the impulse bit can be derived.

$$I_{\text{bit}} = 317.5 \left(\frac{E_0}{A_p} \right)^{0.585} m_p g \tag{6.128}$$

Reference [39] pointed out that the impulses calculated from the inductance gradient formula (Eq. 6.121) and the empirical formula (Eq. 6.128) are close to each other. However, both impulses are much smaller than the theoretical value of the electromagnetic impulse and the estimated value of Guman aerodynamic impulse. Further verification shows that the estimation results using the theoretical calculation formula (Eq. 6.120) for the electromagnetic impulse and the Guman aerodynamic impulse formula (Eq. 6.125) are inaccurate, which may cause the calculated thrust efficiency to be greater than 1. The electromagnetic impulses calculated by the inductance gradient formula (Eq. 6.121) and the theoretical formula (Eq. 6.120) are 73.2 μN s and 203.1 μN s, respectively. The aerodynamic impulses calculated by the Guman formula (Eq. 6.123) and the Henrikson formula (Eq. 6.125) are 356.2 μN s and 161.8 μN s, respectively. The impulse bit calculated by the empirical formula (Eq. 6.128) is 108.1 μN s. These results show that the estimated values of the aerodynamic impulse is significantly larger. Therefore, there may be significant errors in calculating aerodynamic impulse using estimation formulas derived under a large number of simplifying assumptions and the burning rate parameters fitted from the experimental measurements under specific conditions. The theoretical value of the electromagnetic impulse in the above data is also relatively large. In fact, the derivation process of Eq. (6.120) reflects the time integration of the magnetic pressure on the ablated surface of the propellant. The impulse generated by this pressure in the numerical simulation is 206.8 μN s, which is consistent with the theoretical value calculated by Eq. (6.120). In the numerical calculation process using the MACH2 program, Thomas [45] calculated the aerodynamic impulse and electromagnetic impulse by integrating the aerodynamic forces on the propellant and electrode surfaces as well as the $J \times B$ volume force in the entire discharge channel over time.

If the discharge channel is selected as the control volume, the electromagnetic force inside it can be investigated. According to the momentum equation (Eq. 6.21) and the divergence theorem in the system of MHD equations, we have

$$F_L = \iiint_\Omega (J \times B) \mathrm{d}\Omega = \iiint_\Omega \nabla \cdot \left(\frac{BB}{\mu_0} - \frac{B^2}{2\mu_0}\bar{\bar{I}} \right) \mathrm{d}\Omega = \oiint_\Sigma \left(\frac{BB}{\mu_0} - \frac{B^2}{2\mu_0}\bar{\bar{I}} \right) \cdot e_n \mathrm{d}S$$

(6.129)

where Σ is the boundary of the control volume Ω, $\mathrm{d}S$ is the surface element on the boundary, and e_n is the unit outward normal vector of the surface element. Since the magnetic field lines are basically along the z direction, the electromagnetic force in the flow direction is

$$F_{Lx} = \iint_{A_p} \frac{B^2}{2\mu_0} \mathrm{d}S - \iint_{A_e} \frac{B^2}{2\mu_0} \mathrm{d}S$$

(6.130)

where A_e represents the nozzle surface of the thruster. Equation (6.130) shows that the electromagnetic force in the flow direction exerted on the plasma in the discharge channel is equal to the difference between the magnetic pressure on the ablated surface of the propellant and the magnetic pressure at the nozzle of the thruster. However, the calculations reveal that the magnetic pressure at the nozzle is approximately an order of magnitude smaller than that on the ablated surface of the propellant, indicating that the theoretical formula for calculating the electromagnetic impulse is actually the integration of the force exerted on the plasma in the entire discharge channel over time. Since the plasma in the discharge channel is not an adiabatic rigid body with a mass equal to m_p that exists at the beginning, the effect of electromagnetic force doing work not only accelerates the plasma but also transfers energy to the components of the thruster such as the plasma, the propellant, and the wires through Joule heating. Therefore, the electromagnetic impulse predicted by Eq. (6.130) is bound to be significantly greater than the actual value, which also indicates that the Thomas method for calculating the impulse is inappropriate.

To accurately calculate the thrust impulse generated during the PPT operation, Newton's third law and the momentum theorem are applied to the control volume Ω. Ignoring the very small momentum input when the ablation products of the propellant enter the discharge channel, the instantaneous thrust generated by the thruster is given as follows:

$$F_T = \iint_{A_e} \left(\rho u^2 + p + \frac{B^2}{2\mu_0} \right) \mathrm{d}S$$

(6.131)

where the instantaneous thrust F_T includes three terms, namely, the momentum thrust, aerodynamic force at the nozzle and magnetic pressure. Integrating these terms over time yields the total impulse generated by the PPT during pulsed operation

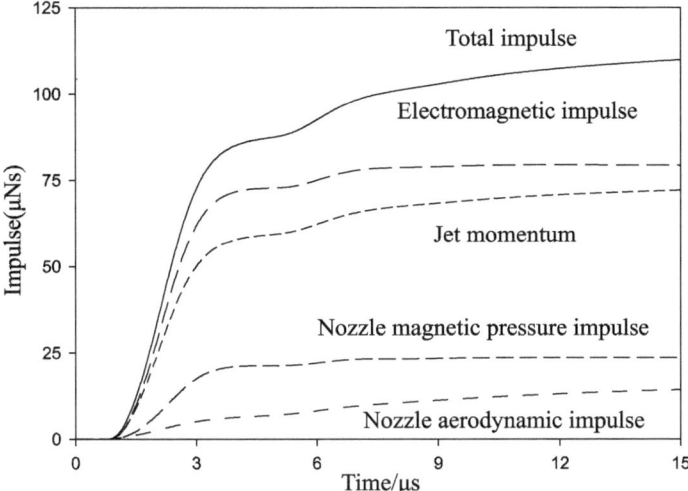

Fig. 6.21 Changes in impulse

and the corresponding impulses of the above three terms, as shown in Fig. 6.21. It is observed that the impulse generated by the thruster increases rapidly shortly after the start of PPT discharge, accounting for most of the impulse bits within the discharge pulse. The impulse increases slightly after the discharge current reaches its reverse peak, followed by a slower growth of the total impulse. The impulse generated by the magnetic pressure at the nozzle remains nearly constant, while the impulse generated by the aerodynamic pressure continues to increase. The impulse bit calculated for the entire discharge pulse is 109.9 μN s, which is 1.6% different from that estimated by the empirical formula for specific impulse, indicating that the impulses calculated by Eq. (6.131) and the empirical impulse formula given by Guman (Eq. 6.128) are reasonable and credible.

Since the momentum term of the plasma jet in Eq. (6.131) includes the combined action of electromagnetic force and aerodynamic force, it is not possible to accurately calculate the proportions of electromagnetic impulses and aerodynamic impulses in the impule bit. However, they can be estimated. Based on the analysis of the magnetic field and the plasma specific pressure, the magnetic induction intensity in the flow field in the later stage of discharge is close to zero. Thus, the effect of electromagnetic acceleration is very weak. The impulse generated in this time period can be considered to be the result of the aerodynamic force alone. In view of this, the electromagnetic impulse during PPT operation is estimated by adding the momentum of the plasma jet and the impulse of the magnetic pressure at the nozzle and then subtracting the aerodynamic impulse fitted based on its increase rate in the late stage of discharge. According to this calculation, the electromagnetic impulse in the discharge pulse is 79.2 μN s, which accounts for 72.1% of the impulse bit. If the electromagnetic impulse is estimated using the inductance gradient formula, this proportion is 66.6%. It is shown that, for typical operating conditions, the impulse generated by the electromagnetic force accounts for approximately 70%, while the aerodynamic impulse only accounts for about 30%, indicating that the parallel-plate PPT is an electric thruster dominated by electromagnetic acceleration.

Using the calculated impulse bit, the specific impulse of the thruster under the studied operating conditions is obtained as 614.1 s, and the thrust efficiency is 2.94%. This result confirms the very low performance of the current thrusters and also indicates that there is still much room for improvement in their performance. According to calculations, the mass of the ions that generate most of the thrust and impulse through electromagnetic acceleration accounts for only 10% of the pulse ablated mass of the propellant, while most of the remaining ablated mass is not effectively accelerated, indicating that it is absolutely possible to increase the specific impulse and thrust efficiency of the thruster by several times. According to the basic performance relationships, the specific impulse and thrust efficiency are directly proportional to the first and second powers of the impulse bit, respectively, and inversely proportional to the pulse ablated mass of the propellant. Therefore, to improve the thruster performance, it is necessary to increase the electromagnetic impulse and aerodynamic impulse, especially to enhance electromagnetic acceleration to further increase the proportion of electromagnetic impulse. On the other hand, it is essential to improve the propellant utilization efficiency and reduce the ablation lag and particle emission effects.

4. Influence of the Hall Effect on Plasma Motion

In the current literature, the Hall effect is essentially ignored in numerical studies of the plasma flow process in PPTs. To fully understand the operating characteristics of PPTs, numerical simulation are carried out using the generalized Ohm's law (Eq. 6.4) and the magnetic induction equation (Eq. 6.19) that include the Hall effect term to investigate its influence. When the conductivity of the local plasma in the flow field is low, the rigidity of the magnetic diffusion term in the governing set of equations is very high, requiring a very small time step for calculation. To reduce the computational load, considering that the plasma flow in the PPT discharge channel mainly exhibits 2D characteristics, a 2D simulation is performed in this section on the operating process in the first half of the oscillation period in the most important stage of PPT discharge.

Figure 6.22 shows the distribution contours of the velocity of the flow field in the y direction. It is observed that when the Hall effect is not considered, the plasma flow between the upper and lower electrodes is symmetric, and the plasma in the vicinity of the propellant surface near the electrodes flows to the central region of the discharge channel. Moreover, the plasma near the electrode exits expands outward from the center toward the outer sides of the two electrode exits. When the Hall effect is considered, the plasma near the electrode exits still expands toward the electrodes on both sides, but the plasma in most regions of the discharge channel has a positive y direction velocity. This indicates that the plasma in the discharge channel is ejected with a bias toward the cathode side, thereby resulting in a thrust that is biased toward the anode side and causing thrust loss. The discharge current at the time corresponding to Fig. 6.22d approaches its maximum value, and the y

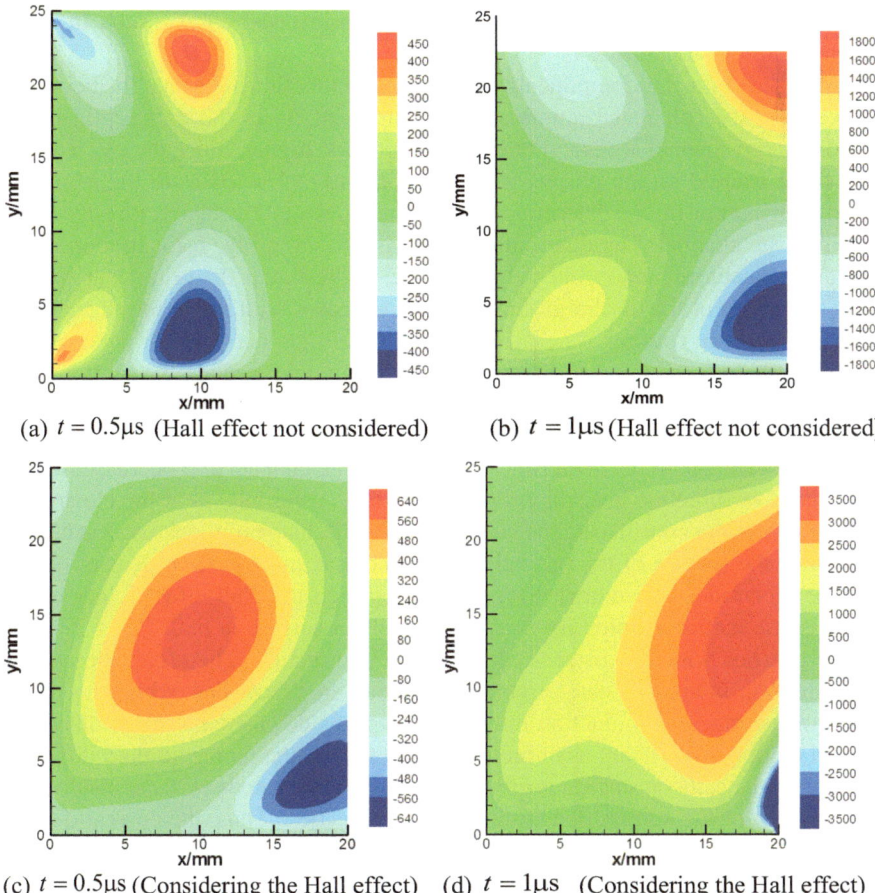

(a) $t = 0.5\mu s$ (Hall effect not considered)   (b) $t = 1\mu s$ (Hall effect not considered)

(c) $t = 0.5\mu s$ (Considering the Hall effect)   (d) $t = 1\mu s$ (Considering the Hall effect)

Fig. 6.22 Comparison of the velocity distribution in the y direction without and with considering the Hall effect, respectively

direction velocity of the plasma in most regions in the figure is greater than 2 km/ s, which is approximately 1/10 of the flow velocity. From this, it can be estimated that the angle of deviation of the thrust generated by the PPT from the x direction is approximately 5.7°. Correspondingly, in validating the performance of the PPT used on the Earth Observing-1 (EO-1) satellite, Arrington et al. [46] measured an obvious thrust component directed toward the anode, and the measured maximum angle of deviation of the thrust vector from the centerline of the discharge channel was 5.3°, which is close to the estimated angle. The plasma ejected toward the cathode side not only reduces the propulsion performance but also causes more severe plume contamination in the space on the cathode side than that on the anode side, which is highly detrimental to solar cell arrays and spaceborne optical instruments within this range. These non-uniform distribution characteristics of the PPT plume were mentioned in the literature [47, 48] and were confirmed by the results of triple Langmuir probe diagnostics.

Compared with the components of magnetic induction intensity in the x and y directions, the calculated amplitude of component B_Z in the z direction in the PPT discharge channel is several orders of magnitude higher. Since the electromagnetic force exerted on the plasma is proportional to $(\nabla \times B) \times B$, the evolution of the distribution of B_Z determines the accelerated motion of the plasma along the discharge channel. The distribution of B_Z is shown in Fig. 6.23. It is observed that the magnetic induction intensity B_Z is the highest near the propellant surface at the beginning of the discharge channel and decreases with increasing distance in the flow direction, forming a negative magnetic field gradient, which generates an electromagnetic force in the x direction, accelerating and ejecting the plasma from the discharge channel. The gradient change of the magnetic field corresponds to the current density. The distributions of magnetic induction intensity in Fig. 6.23a and b are symmetric about the centerline of the discharge channel and decrease rapidly along the flow direction near the propellant surface, especially near the electrodes. Therefore, without considering the Hall effect, the discharge of the thruster is mainly present on the surface of the propellant, and the maximum current density occurs near the electrodes. In contrast to the case in which the Hall effect is neglected, the magnetic induction intensity near the propellant surface in both Fig. 6.23c and d mainly varies with the distance along the flow direction. Therefore, the current flow direction is approximately perpendicular to the electrodes, while the distribution of the magnetic induction intensity in regions downstream of the discharge channel exhibits significant asymmetry. At $t = 0.5 \mu s$, the magnetic field diffusion distance on the cathode surface is greater than that on the anode surface, resulting in a component of the current density in the flow direction. Thus, the current density vector is no longer perpendicular to the electrodes. Instead, it points from the anode to the downstream of the cathode. As the discharge proceeds, the plasma continuously expands and accelerates downstream of the discharge channel, and the magnetic field also continuously diffuses accordingly. By $t = 1 \mu s$, the magnetic field has diffused beyond the end of the electrode. As shown from the magnetic induction intensity distribution in Fig. 6.23d, the current flow direction at this time becomes distorted downstream compared to that of the previous skewedness. Kumagai et al. [49] used a high-speed camera to monitor the PPT discharge process and observed that there were arc columns perpendicular to the electrode and a discharge path extending from the anode downstream to the cathode on the propellant surface. Taro et al. [50] observed similar ion emission images using emission spectroscopy and high-speed photography. In addition, Palumbo and Begun measured two discharge paths, i.e., one near the propellant surface that is approximately perpendicular to the electrodes and one that is twisted downstream and turns to the middle of the discharge channel. Therefore, the calculation results with the Hall effect term included are consistent with the experimental measurements reported in the literature.

In fact, denoting $E^* = E + V \times B$ as the equivalent electric field and choosing the magnetic field direction as the z-axis direction, the generalized Ohm's law can be written as $J = \overline{\overline{\sigma}}_e \cdot E^*$, where $\overline{\overline{\sigma}}_e$ is the conductivity tensor.

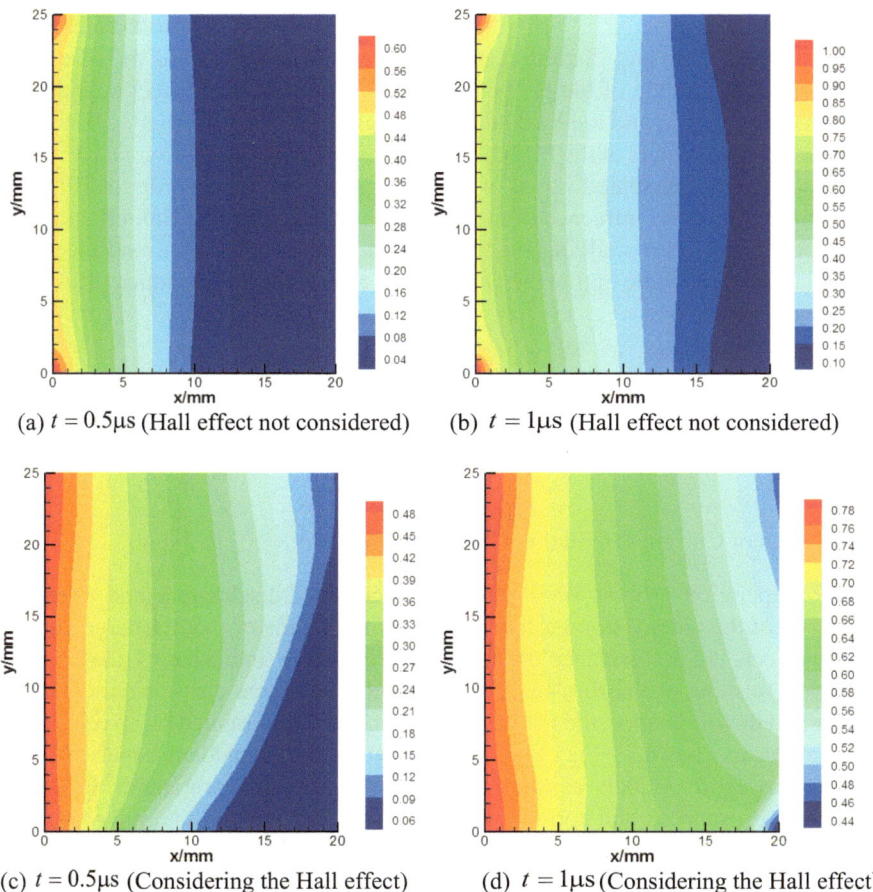

(a) $t = 0.5\mu s$ (Hall effect not considered) (b) $t = 1\mu s$ (Hall effect not considered)

(c) $t = 0.5\mu s$ (Considering the Hall effect) (d) $t = 1\mu s$ (Considering the Hall effect)

Fig. 6.23 Comparison of the distribution of the magnetic induction intensity B_z without and with considering the Hall effect

$$\overline{\overline{\sigma}}_e = \frac{\sigma_e}{1 + \omega_{ce}^2/\nu_{ei}^2} \begin{bmatrix} 1 & -\frac{\omega_{ce}}{\nu_{ei}} & 0 \\ \frac{\omega_{ce}}{\nu_{ei}} & 1 & 0 \\ 0 & 0 & 1 + \frac{\omega_{ce}^2}{\nu_{ei}^2} \end{bmatrix} \tag{6.132}$$

where ω_{ce} is the electron cyclotron frequency. The conductivity in the form of a tensor represents the anisotropy of the plasma. It is observed from the above equation that the conductivity in the direction parallel to the magnetic field is equal to σ_e, but the conductivity changes in the direction perpendicular to the magnetic field. Moreover, the electric field in the plane perpendicular to the magnetic field will result in a current component perpendicular to the electric field direction, namely the Hall current.

Therefore, the equivalent electric field perpendicular to the electrode direction will generate a current in the flow direction, and the equivalent electric field in the flow direction will generate a current in the direction perpendicular to the electrode, thus leading to skew and distortion of the current direction within the discharge channel. The magnitude of the Hall current induced by the Hall effect depends on the ratio of ω_{ce} to V_{ei}, and the Hall current is negligible when $\omega_{ce} \lesssim V_{ei}$. When the characteristic parameters of the PPT discharge process are $B = 0.1$ T, $n_e = 10^{22}$ m^{-3}, and $T_e = 3$ eV, the calculated values of ω_{ce} and V_{ei} are both on the order of 10^{10} s^{-1}, indicating a significant influence of the Hall effect. Therefore, the Hall effect needs to be considered when studying the accelerated motion of plasma, and electrode configurations need to be optimized to minimize their impact and reduce thrust loss when designing thrusters.

6.3.3 Influence of the Discharge Current on the Thruster Performance

1. Analysis of Current Oscillation Characteristics

The PPT operation process is essentially a process of discharging and releasing the energy of the energy storage capacitor. The magnitude of the discharge current directly affects the performance of the thruster, as clearly indicated by the impulse estimation formula of the thruster. To analyze the discharge characteristics of the thruster and their influence, the PPT discharge circuit is represented by an equivalent RLC circuit, as shown in Fig. 6.24. In this figure, the equivalent capacitance of the circuit is approximately equal to the capacitance C of the capacitor, while the equivalent resistance R and equivalent inductance L include the internal resistance and inductance of the energy storage capacitor and the resistance and inductance of the transmission wires, electrodes, and plasma. For a given thruster, the resistance and inductance of the capacitor, wires, and electrodes are constant, and only the resistance and inductance of the plasma change with time during the discharge process. To simplify the analysis, the equivalent resistance R and the equivalent inductance L are assumed to be constants. Based on circuit theory, the following second-order homogeneous differential equation with constant coefficients for the current can be obtained:

$$LC\frac{d^2 I}{dt^2} + RC\frac{dI}{dt} + I = 0 \qquad (6.133)$$

The two characteristic roots of this equation are

$$p_{1,2} = \frac{-R}{2L} \pm \sqrt{\frac{R^2}{4L^2} - \frac{1}{LC}} = -\delta \pm \sqrt{\delta^2 - \omega_0^2} \qquad (6.134)$$

where $\delta = R/2L$ and $\omega_0 = 1/\sqrt{LC}$. According to the values of the characteristic roots, there are three cases for the discharge circuit:

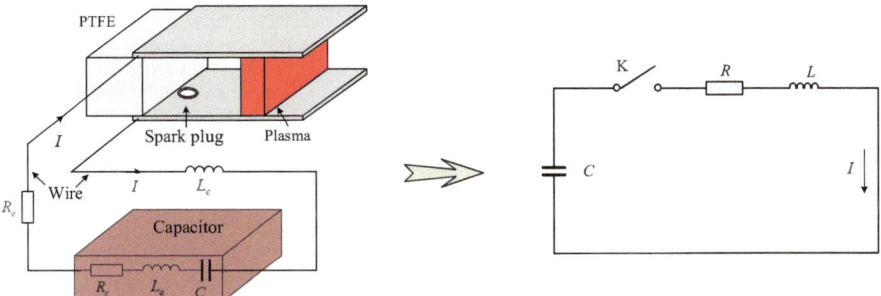

Fig. 6.24 Equivalent circuit of the discharge circuit

(1) Overdamped Case

When $\delta > \omega_0$, i.e., $L < CR^2/4$, the characteristic roots P_1 and P_2 are unequal real roots, and the discharge current is

$$I(t) = \frac{V_0}{2L\sqrt{\delta^2 - \omega_0^2}}\left(e^{p_1t} - e^{p_2t}\right) \tag{6.135}$$

(2) Critically Damped Case

When $\delta = \omega_0$, i.e., $L = CR^2/4$, the characteristic roots P_1 and P_2 are equal real roots, and the discharge current is

$$I(t) = \frac{V_0}{L}te^{-\delta t} \tag{6.136}$$

(3) Underdamped Case

When $\delta < \omega_0$, i.e., $L > CR^2/4$, the characteristic roots P_1 and P_2 are complex conjugate roots. Let $\omega = \sqrt{\omega20 - \delta2}$. Then, the discharge current is expressed as

$$I(t) = \frac{V_0}{\omega L}e^{-\delta t}\sin \omega t \tag{6.137}$$

Figure 6.25 shows schematic diagrams of the current waveforms under the three damping conditions. The discharge current exhibits decaying oscillations under the underdamped condition. According to Eq. (6.137), the attenuation coefficient of the current amplitude is δ, and the oscillation period is $2\pi/\omega$. Under overdamped and critically damped conditions, the current waveform is a non-oscillatory and the discharge is aperiodic.

The periodic oscillation of the discharge current, on the one hand, implies that the energy storage capacitor is repeatedly charged and discharged, which reduces its operational lifespan. On the other hand, this means that there are multiple conversions

Fig. 6.25 Current
waveforms under the three
damping conditions

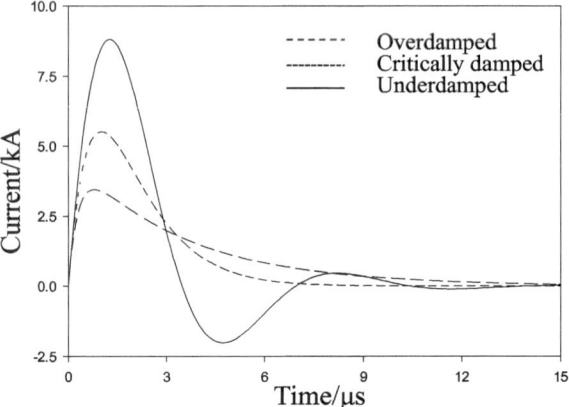

of energy between the electrical and magnetic forms, inevitably increasing the energy
loss in the circuit resistance and decreasing the energy transfer efficiency. Based on
the analysis of the propellant ablation process and the plasma flow process, the energy
utilization rate after the reverse current oscillation is low, with limited impulse generation. Therefore, the PPT performance is expected to improve by reducing current
oscillations or even transitioning discharge into an aperiodic form. The aperiodic
discharge requires satisfying $L \leq CR^2/4$. Since the capacitance is fixed and the resistive loss in the circuit needs to be minimized during the thruster design, the circuit
inductance must be minimized to reduce current oscillations. In the case of under-
damping, if $R < < 2\sqrt{LC}$, the maximum rising slope and the maximum amplitude
of the current are

$$\left(\frac{dI}{dt}\right)_{max} = \frac{V_0}{L}, \ I_{max} = V_0\sqrt{C/L} = \sqrt{\frac{2E_0}{L}} \tag{6.138}$$

Evidently, the rising slope and amplitude of the discharge current increase as the
inductance decreases. To enhance the $J \times B$ acceleration effect of the thruster, it is also
necessary to minimize the circuit inductance. Measures to reduce inductance mainly
include selecting capacitors with lower inductance, operating multiple capacitors in
parallel, keeping connections as short as possible, keeping wires carrying currents
in the same direction as far apart as possible to reduce mutual inductance, placing
wires carrying currents in opposite directions as close together as possible to increase
mutual inductance. However, due to factors such as the internal inductance of the
capacitors, the circuit inductance can be as high as several tens of nanohenries,
even for carefully designed thrusters, making it difficult to satisfy the condition of
aperiodic discharge. Therefore, the discharge current is usually in the form of damped
oscillation, as shown in Fig. 6.26.

Fig. 6.26 PPT discharge
current waveforms

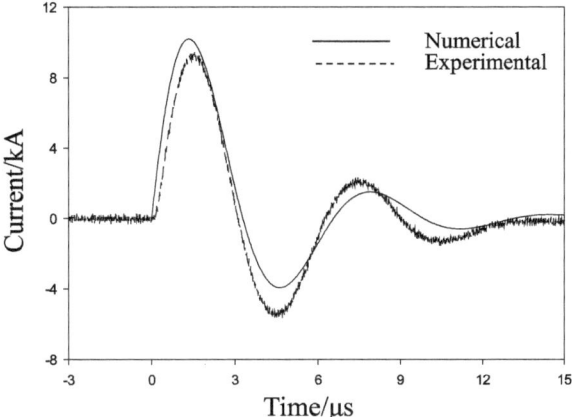

2. Evaluation of the Aperiodic Discharge Waveform

In cases where simply changing circuit parameters such as the inductance and resistance still cannot meet the requirements for aperiodic discharge, it is necessary to modify the circuit design to make the thruster generate aperiodic discharge currents. There are three main circuit designs for achieving aperiodic discharge: incorporating a rectifier circuit with a high-voltage silicon stack into the circuit, using an inductively driven circuit with inductor coils for capacitor energy storage, and employing a lumped parameter chain design circuit that generates square wave pulse currents using a pulse forming network (PFN), as shown in Figs. 6.27, 6.28 and 6.29. Among these three types of circuits, since the current waveform generated by the inductive drive circuit is similar to that rectified by the silicon stack, and it was noted in the Introduction of this book that the thrust and thrust efficiency of the inductively driven PPT decrease instead. Therefore, only the output currents of the silicon stack rectifier circuit and the PFN discharge circuit are analyzed in this section. For ease of comparison, a simulation analysis is performed on the waveforms of the underdamped oscillating current and the overdamped current with low inductance of the RLC circuit. Figure 6.30 presents the waveforms of the underdamped current and the three types of aperiodic currents under the same capacitor energy storage and the same load resistance.

The operation process of the thruster under four different current waveforms is simulated, and the curves of variations in the thruster impulse over time are obtained, as shown in Fig. 6.31. It is observed that at the beginning of discharge, the current increases rapidly, and the plasma quickly expands and accelerates to be ejected out of the PPT electrode exit. In the later stage of discharge, as the discharge current decays to near zero, the thruster impulse increases slowly. Among these four current waveforms, the initial rising slope and peak of the overdamped current are the largest, and the increase rate of the impulse generated by is the highest. However, the increase in the impulse quickly levels off due to the short peak duration. In sharp contrast, the pulse forming network (PFN) square wave current has the longest peak duration and

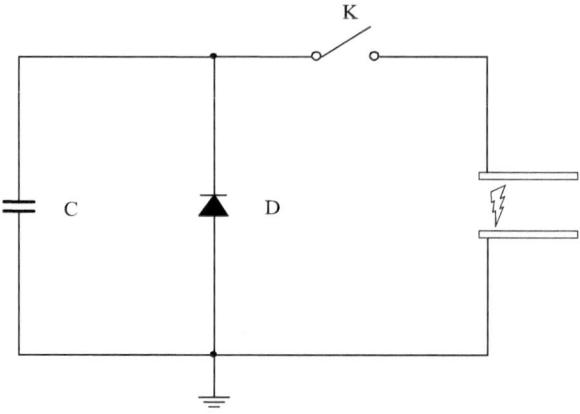

Fig. 6.27 Silicon stack rectifier circuit

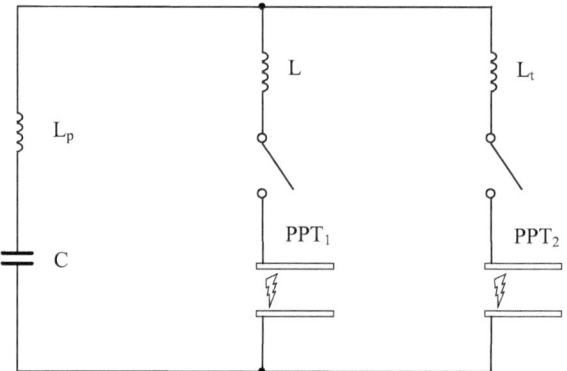

Fig. 6.28 Inductively driven PPT

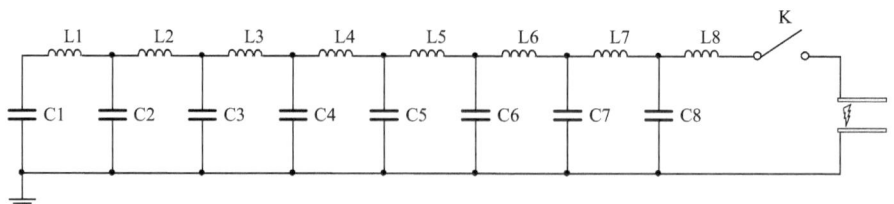

Fig. 6.29 PFN discharge circuit

also the longest sustained large increase in impulse. Although the initial rising slope of the PFN square wave current is not the smallest, its current amplitude is the smallest, causing the rate of impulse increase to be quickly lower than that of other current waveforms after the discharge begins. The small current amplitude also reduces the

Fig. 6.30 Waveforms of underdamped current and aperiodic current

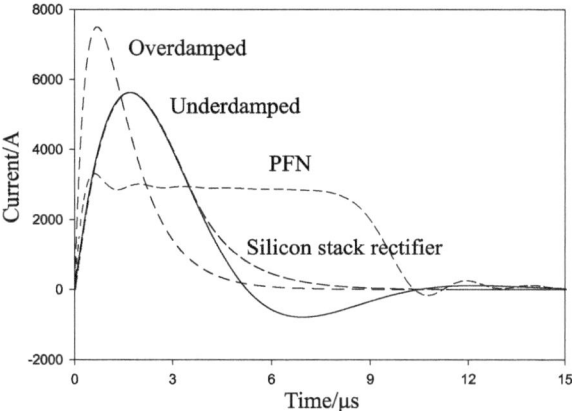

Fig. 6.31 Changes in impulse corresponding to four current waveforms

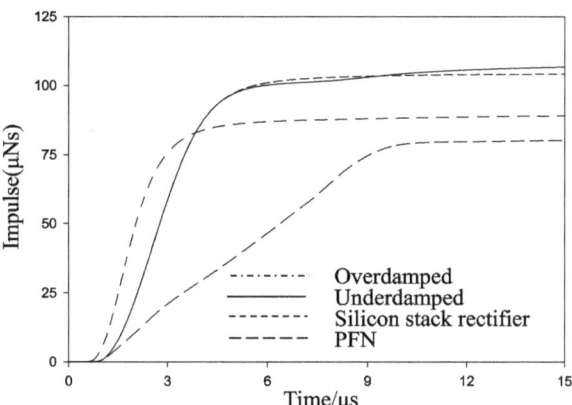

consumption of the propellant, resulting in a relatively small impulse bit generated by the PFN square wave current. Compared with those of the overdamped current and PFN square wave current, the amplitude and peak duration of the underdamped current and silicon stack rectified current are moderate, resulting in a significant and prolonged increase in thruster impulse.

Because the plasma impedance under the action of different current waveforms is not equal, it is not yet possible to simply compare the performance of the thruster under various current waveforms. Therefore, in this book, the proportion of the electromagnetic impulse to impulse bit under different discharge currents is estimated. The results are shown in Table 6.1. It is observed that the electromagnetic impulse corresponding to the underdamped oscillating current accounts for the smallest proportion, indicating that the aperiodic discharge current can enhance the electromagnetic acceleration effect and increase the proportion of the electromagnetic impulse to the impulse bit. The simulation results of the four current waveforms show that the pulse ablated mass of the propellant is the largest and the specific impulse is

Table 6.1 Calculation results of performance parameters under different currents

Current waveform	Underdamped	Overdamped	Silicon stack rectifier	PFN square wave
Impulse bit (μN s)	106.7	89.1	104.1	80.2
Proportion of electromagnetic impulse (%)	80.5	82.1	87.2	83.5
Specific impulse (s)	596.2	608.0	650.2	1077.3
Thrust efficiency (%)	2.77	2.36	2.95	3.76

the smallest under the underdamped current, confirming to a certain extent that the use of aperiodic discharge can improve the thruster performance.

Among the three aperiodic current waveforms analyzed in the simulation, although the overdamped current increases the proportion of the electromagnetic impulse, due to its short peak duration, both the impulse bit and the pulse ablated mass of the propellant are small, and the calculated thrust efficiency decreases. The PPT performance is improved to some extent after adding a high-voltage silicon stack rectifier. However, it is necessary to consider the resistive loss of the silicon stack itself and the adverse effects of the large mass and volume of the silicon stack in practical applications. The PFN square wave current has very small oscillations at the wave tail, and the current rapidly decreases as the discharge nears its end. This is precisely what is expected to reduce the propellant ablation and particle emission losses after the discharge ends. The calculated specific impulse and thrust efficiency in the table are also the highest among the four waveforms, indicating a significant improvement in the thruster performance. Notably, the significant improvement in the PPT operating performance with use of the PFN discharge circuit is calculated under the condition that the load resistance matches the impedance of the chain network. Since the plasma impedance varies with time in the PPT operation process, the plasma impedance and the chain network impedance only match within a certain range. This may cause multiple reflections of the square wave current with the same polarity or opposite polarity. Therefore, it is necessary to optimize the design of the number of network chains as well as the capacitance and inductance of each chain to increase the matching range, obtain the appropriate current amplitude and peak duration, suppress overshooting the wave front, reduce the flat-top drop during the duration of the square wave, and increase the rate of descent of the current wave tail, thereby optimizing the thruster performance. These issues await further study.

References

1. Qin Z. Electrical engineering. Beijing: Higher Education Press; 1999.
2. Keidar M, Boyed ID. Electrical discharge in the Teflon cavity of a coaxial pulsed plasma thruster. Trans Plasma Sci. 2000;28(2):376–85.
3. Kubota K, Funaki I, Okuno Y. Hall Effect on the magnetoplasmadynamic thruster flowfields. In: 43rd AIAA/ASME/SAE/ASEE joint propulsion conference, Cincinnati, OH. AIAA 2007-4385; 2007.
4. Arai N. Transient ablation of Teflon in intense radiative and convective environments. AIAA J. 1979;17(6):634–40.
5. Wu Q, Li H. Magnetohydrodynamics. Changsha: National University of Defence Technology Press; 2007.
6. Chen B, Shu Y, Hu W. Topical study of electromagnetism. Beijing: Higher Education Press; 2001.
7. Tian Z. Numerical simulation study of magnetohydrodynamic control of hypersonic flow. Changsha: National University of Defence Technology; 2008.
8. Li H, Tian Z. Numerical simulation study of magnetohydrodynamic control of hypersonic flow. Changsha: National University of Defence Technology Press; 2010.
9. Powell KG. An approximate Riemann solver for magnetohydrodynamics. Langley: ICASE Report; 1994.
10. Toth G. The $\nabla \cdot B = 0$ constraint in shock-capturing magnetohydrodynamics codes. J Comput Phys. 2000;161:605–52.
11. Evans CR, Hawley JF. Simulation of magnetohydrodynamic flows: a constrained transport method. J Astrophys. 1988;332:659–67.
12. Brackbill JU, Barnes DC. The effect of nonzero $\nabla \cdot b$ on the numerical solution of the magnetohydrodynamic equations. J Comput Phys. 1980;35:426–30.
13. Dedner A, Kemm F, Kroner D, et al. Hyperbolic divergence cleaning for the MHD equations. J Comput Phys. 2002;175:645–73.
14. Kovitya P. Thermodynamics and transport properties of ablated vapours of PTFE, alumina, perspex and PVC in the temperature range 5000–30000K. IEEE Trans Plasma Sci. 1984;12(1):38–42.
15. Schmahl CS. Thermochemical and transport processes in pulsed plasma microthrusters: a two-temperature analysis. The Ohio State University; 2002.
16. Cassibry JT. Numerical modelling studies of a coaxial plasma accelerator as a standoff driver for magnetized target fussion. The University of Alabama; 2004.
17. Sonoda S. A polytetrafluoroethylene thermochemical model for the study of pulsed plasma thrusters. Arizona State University; 2009.
18. Corpening JH. Computational analysis of a pulsed inductive plasma accelerator. Purdue University; 2008.
19. Bian Y, Xu L. Aerothermodynamics. Hefei: University of Science and Technology of China Press; 2011.
20. Huang Z, Ding E. Transport theory. Beijing: Science Press; 2008.
21. Braginskii SI. Transport processes in a plasma. Rev Plasma Phys. 1965;1:205–311.
22. Kubota K, Funaki I, Okuno Y. Comparison of simulated plasma flow field in a two-dimensional magnetoplasmadynamic thruster with experimental data. IEEE Trans Plasma Sci. 2009;37(12):2390–8.
23. Bittencourt JA. Fundamentals of plasma physics. Springer; 2004.
24. Li A. Iterative methods for solving the linear systems. Sci Technol Eng. 2007;7(14):3357–64.
25. Zeng M. Weighting-relaxation relative method. J Putian Univer. 2008;15(2):29–31.
26. Yan C. Methods and applications of computational fluid dynamics. Beijing: Beijing University of Aeronautics and Astronautics Press; 2006.
27. Kim KH, Kim C. Accurate, efficient and monotonic numerical methods for multidimensional compressible flows. J Comput Phys. 2005;208:527–69.

28. Yoon SH, Kim C, Kim KH. Multi dimensional limiting process for three dimensional flow physics analyses. J Comput Phys. 2008;227:6001–43.
29. Mikellides YG. Theoretical modelling and optimization of ablation-fed pulsed plasma thrusters. U.S.A: The Ohio State University; 1999.
30. Cassibry JT, Francis Thio YC, Markusic TE. Numerical modelling of a pulsed electromagnetic plasma thruster experiment. J Propul Power. 2006;22(3):628–36.
31. Chen X. Thermal plasma heat transfer and flow. Beijing: Science Press; 2009.
32. Peterson KJ. Computational magnetohydrodynamic investigation of flux compression and implosion dynamics in a Z-pinch plasma with an azimuthally opposed magnetic field configuration. The University of Tennessee; 2003.
33. Gatsonis NA, Juric D, Stechmann DP, et al. Numerical analysis of Teflon ablation in pulsed plasma thrusters. In: 43rd AIAA/ASME/SAE/ASEE joint propulsion conference. Cincinnati, OH. AIAA 2007-5227; 2007.
34. Stechmann DP. Numerical analysis of transient Teflon ablation in pulsed plasma thrusters. Worcester Polytechnic Institute; 2007.
35. Kemp NH. Surface recession rate of an ablating polymer. AIAA J. 1968;6(9):1790–1.
36. Orszag A, Tang CM. Small-scale structure of two-dimensional magnetohydrodynamic turbulence. J Fluid Dyn. 1979;90:129–45.
37. Li F, Shu C. Locally divergence-free discontinuous galerkin methods for MHD equations. J Sci Comput. 2005;22:413–42.
38. Hoffmann KA. An integrated computational tool for hypersonic flow simulation. AFRL-SR-ARTR-04-0232; 2004.
39. Li Z. Theoretical and experimental study on the design and performance of pulsed plasma thruster. Changsha: National University of Defense Technology; 2008.
40. Vondra RJ, Thomassen K, Solbes A. Analysis of solid Teflon pulsed plasma thruster. J Propul Power. 1970;7(12):1402–6.
41. Burton RL, Wilson MJ, Bushman SS. Energy balance and efficiency of the pulsed plasma thruster. Urbana, IL: University of Illinois. AIAA 98-3808; 1998.
42. Guman WJ. Pulsed plasma technology in microthrusters. NY: Fairchil Hiller Corp, AFAPL-TR-68-132; 1968.
43. Henrikson EM. An experimental and theoretical study towards performance improvements of the ablation fed pulsed plasma thruster. Arizona State University; 2010.
44. Guman WJ. Designing Solid propellant pulsed plasma thrusters. Farmingdale, New York: Fairchild Republic Company. AIAA 75-0410; 1975.
45. Thomas HD. Numerical simulation of pulsed plasma thrusters. U.S.A: The University of Tennessee; 2000.
46. Arrington L, Haag T. Multi-axis thrust measurements of the EO-1 pulsed plasma thruster. In: 35th AIAA/ASME/SAE/ASEE joint propulsion conference, Los Angeles, CA. AIAA 99-2290; 1999.
47. Gatsonis NA, Byrne L, Eckman R, et al. Pulsed plasma thruster plumes: experimental investigations and numerical modelling. In: 38th AIAA aerospace sciences meeting and exhibit, Reno, NV. AIAA 2000-0464; 2000.
48. Zhang R, Zhang D, Li Y, et al. Distribution and optical properties of the pulsed plasma thruster plume deposition. In: 2012 2nd international conference on electronic and mechanical engineering and information technology, Shenyang, Liaoning, China.
49. Kumagai N, Igarashi M, Sato K, et al. Plume diagnostics in pulsed plasma thruster. In: 38th AIAA/ASME/SAE/ASEE joint propulsion conference, Indianapolis, IN. AIAA 2002-4124; 2002.
50. Taro H, Atsushi N, Toshinori I, et al. Development of highly durable pulsed plasma thruster for active flare satellite constellation. In: 63rd international astronautical congress, Naples, Italy; 2012.

Part III
Plume

Chapter 7
Numerical Simulation of the PPT Plume Process Based on Hybrid Particle–Fluid Models

Pulsed plasma thrusters (PPTs) have advantages in terms of their structural mass and size, operational performance, and power supply requirements, making them widely applicable to tasks such as attitude control, drag compensation, orbit raising, and constellation phase control of microsatellites. With the development of space exploration technology, modern microsatellites are required to have higher functional density, lower manufacturing cost, and longer operational lifespans. However, the PPT plume may cause severe adverse effects (such as sputter erosion, sediment contamination, chemical contamination, thermal loads, and electromagnetic interference) on satellites. While integrating more highly sophisticated payloads and reducing design margins to lower costs, it is necessary to carry out in-depth studies on the flow characteristics of thruster plume to accurately assess the interaction effects between the plume and spacecraft and accordingly implement protective measures to ensure the normal operation of the spacecraft during its lifetime.

The PPT plume is a rarefied fluid containing various plasma components. Using the magnetohydrodynamic (MHD) simulation method based on the continuum assumption is no longer appropriate for numerical simulation of this plasma plume. Typically, the use of kinetics-based particle simulation methods are required. The direct simulation Monte Carlo (DSMC) method is employed to handle the motion of neutral particles and collisions between heavy particles, and the particle-in-cell (PIC) method to simulate the motion of charged particles in external and self-consistent electromagnetic fields. By combining the advantages of the DSMC and particle-in-cell (PIC) methods [1, 2], particle simulation methods can accurately describe the flow field characteristics and the variation process of the plasma plume, and have been widely used in numerical plume studies.

Using the PIC method to track the motion of charged particles in an electromagnetic field requires solving the Maxwell equations at scales below the Debye length with a time step smaller than the characteristic time corresponding to the plasma oscillation frequency and satisfying the stability conditions of the system of equations. As a result, the computational resources required for simulating a PPT plume with a high

© The Author(s) 2025
J. Wu et al., *Numerical Simulation of Pulsed Plasma Thruster*,
https://doi.org/10.1007/978-981-97-7958-1_7

plasma density are very large. To avoid these limitations, electrons are usually treated as a fluid. To obtain accurate information about the plume flow field and to deeply reveal plume flow patterns, in this chapter, the electromagnetic acceleration effect of the plume field is comprehensively considered, and a hybrid particle–fluid algorithm combining DSMC and PIC simulations with a fluid electron model is employed to conduct a three-dimensional (3D) numerical investigation of the PPT plume. This chapter mainly elaborates the basic ideas, calculation procedures and key techniques of the DSMC and PIC methods, presents a PPT flow model considering the influence of the magnetic field, and conducts research on the numerical algorithms for hybrid particle–fluid simulation.

7.1 Basic Theory of the Hybrid DSMC/PIC Fluid Algorithms

The motion of the PPT plume is very complex and includes the continuum flow in the core region, the transitional flow in the periphery, and the free flow of molecules. In addition, the collision processes of charged components in the plume differ from those of ordinary gases. PPT plume problems are primarily solved using kinetic methods, among which the DSMC, PIC, and their hybrid methods represent the main directions of development. The DSMC method [1, 3] is a physically based probabilistic simulation method and originates from the molecular dynamics method, which uses a probabilistic approach to determine whether intermolecular collisions occur. The PIC method, as a plasma particle simulation method, does not consider collisions between plasma particles. Instead, it uses computers to simulate particles and track the motion of a large number of charged particles in their self-consistent field and applied electromagnetic field to simulate the dynamics of the plasma. The hybrid DSMC/PIC fluid algorithm proposed by Gatsonis [4] can simulate an electric propulsion plume. Specifically, the motion of neutral particles and ions is simulated by the DSMC and hybrid PIC methods. Electrons are approximated as a massless fluid and are assumed to be in an equilibrium state, and the electric field distribution is obtained from the charge conservation equation.

7.1.1 Direct Simulation Monte Carlo Method

The direct simulation Monte Carlo (DSMC) method [1] originates from the molecular dynamics method. Moreover, the DSMC method does not directly solve the Boltzmann equation. Instead, it uses the physical process described in the equation. The DSMC method employs probabilistic rather than deterministic approaches to calculate and simulate intermolecular collisions. This physical simulation method is one of the most effective methods for numerically solving the problems in rarefied gas dynamics.

1. General Procedure of the DSMC Method

The DSMC simulation program can be roughly described by the following six steps, as shown in the flowchart in Fig. 7.1.

(1) Under the assumption of no collisions, the distance that each simulated molecule moves under its own velocity within Δt_m is obtained according to the uniform linear motion, and the new position coordinates of the simulated molecules are determined.

(2) Since the simulation region is always finite, the simulated molecules may interact with the boundary after undergoing migration, which must be addressed accordingly. If the boundary is a symmetry line (or plane), the simulated molecules undergo specular reflection at the boundary. If the boundary is a solid wall surface, commonly used methods include the specular reflection and diffuse reflection methods and a combination of these two methods. If the region outside the boundary is a vacuum, the simulated molecules are considered to have escaped. For the inlet boundary, it is necessary to determine the number and motion state of the simulated molecules entering the computational domain within Δt_m.

(3) The cell numbering of the simulated molecules is adjusted based on the new spatial position coordinates, and the simulated molecules are sorted.

(4) The collisions between the simulated molecules within Δt_m are calculated. Collision calculation is crucial in the DSMC method and will be discussed in detail in the next subsection. Here, Bird's no-time-counter method is used as an example to illustrate the calculation steps in the simulation. (1) The number of collisions N_t of simulated molecules within Δt_m is calculated. (2) The simulated molecules are randomly sampled to select possible collision pairs. (3) For each selected pair of simulated molecules, the ratio of $\sigma_T g$ to $(\sigma_T g)_{max}$ is calculated and compared with a random number R. If $\sigma_T g/(\sigma_T g)_{max} > R$, the pair of simulated molecules is retained, and a pair of colliding molecules forms. Otherwise, step (2) is repeated. (4) A determination of whether the actual number of collisions $N_{col} \leq N_t$ is made. If this condition holds, steps (2)–(4) are repeated to continue the calculation of collisions within this cell; otherwise, the calculation of collisions within this cell is considered to be completed, and the calculation of collisions in the next cell can be carried out. (5) The collision calculation within Δt_m is implemented for all cells according to steps (1)–(4).

(5) Following steps (1)–(4), after the program has run repeatedly for N time steps, whether the cumulative simulation time interval $\sum_{i=1}^{N} \Delta t_{mi}$ reaches the sampling time Δt_s, that is, whether the condition is satisfied, is determined. If the condition is met, a statistical calculation is performed on the simulated molecules within each cell to obtain the values of various macroscopic physical quantities of the flow field. If the simulated flow is steady macroscopically, it is necessary to confirm that the flow is in a steady state before performing statistical calculations on various physical quantities of the flow field.

(6) Due to computer memory limitations, the number of simulated molecules that can be arranged in each cell is limited. Therefore, there are significant fluctuations in the physical quantities of the flow field obtained by a single statistical calculation. To improve the computational accuracy, the common approach is to increase the sample size by repeated calculations, reducing the statistical errors in the physical quantities of the flow field.

In real gas flow, the motion and collision of gas molecules always occur simultaneously, with molecules coupling and influencing each other. It is difficult to accurately reflect this physical phenomenon in computer simulations due to the current level of computer development. Therefore, in the DSMC method, the motion and collision of gas molecules are actually decoupled. It is assumed that intermolecular collisions are instantaneous and do not alter the motion trajectory of the gas and that molecules move in straight line at constant speed between two successive collisions, greatly improving the speed and efficiency of DSMC simulations. The assumption that molecular motions and collisions of gas are decoupled is the theoretical foundation of the DSMC method. On the one hand, this assumption expands the application range of the method, making it possible to simulate complex flow fields using the DSMC method. On the other hand, it also imposes certain limitations on the application of DSMC simulations.

2. Representation of the Macroscopic State of the Gas Mixture

The velocity distribution function $f\left(t, \vec{X}, \vec{\zeta}\right)$ is the basis of molecular kinetic theory. This function not only provides an exact description of the motion state of gas molecules but can also be used to obtain the desired macroscopic physical quantities by averaging with the velocity distribution function as the weighting function.

The average value of the function $\varphi(\zeta)$ for any gas molecule velocity is expressed as follows:

$$\langle \varphi \rangle = \frac{1}{n} \int \varphi(\vec{\xi}) f(\vec{\xi}) d\vec{\xi} \tag{7.1}$$

From this, we have

$$\langle \varphi \rangle = \frac{1}{n} \int \varphi\left(\vec{\xi}\right) f\left(\vec{\xi}\right) d\vec{\xi}$$

$$n(t, \vec{x}) = \int f(t, \vec{X}, \vec{\xi}) d\vec{\xi}$$

$$P_{ij} = m \int \vec{c}_i \vec{c}_j f(t, \vec{X}, \vec{\xi}) d\vec{\xi}$$

$$q_i = \frac{1}{2} m \int c^2 \vec{c}_j f(t, \vec{X}, \vec{\xi}) d\vec{\xi} \tag{7.2}$$

where n, u, P_{ij}, and q_i are the number of gas molecules per unit volume or molecular number density, gas flow velocity, stress tensor, and energy flux vector, respectively.

Fig. 7.1 Flowchart of the
DSMC method

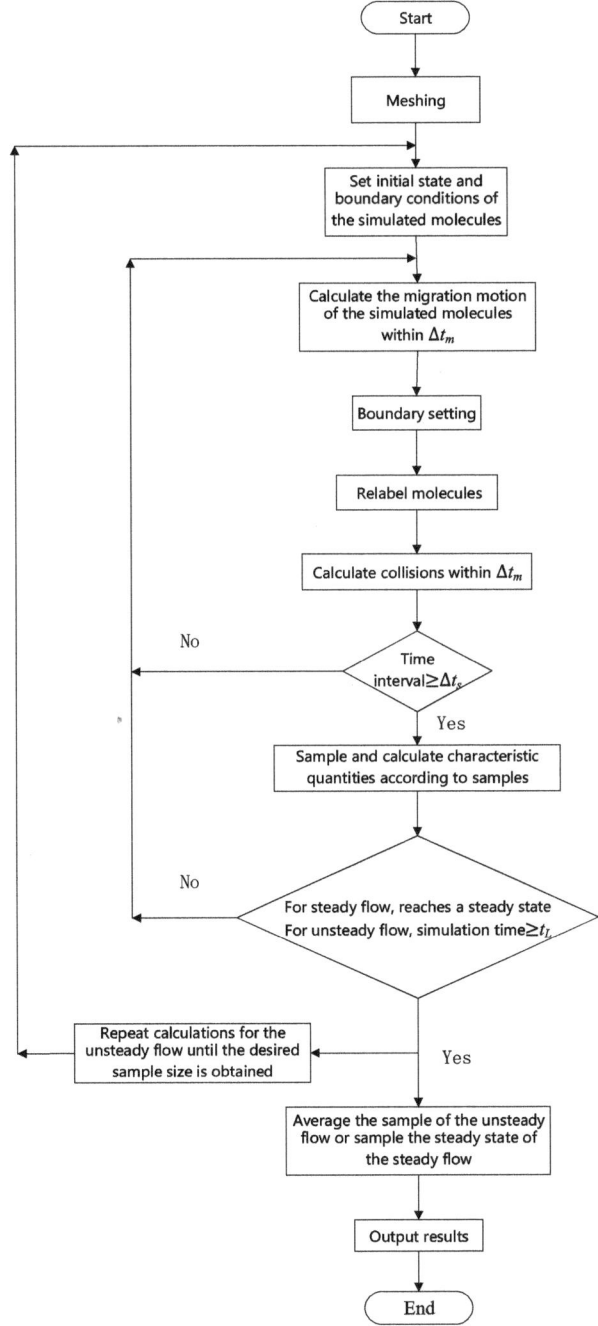

Let us define $\vec{c} = \vec{\xi} - \vec{u}$ as the thermal velocity or intrinsic velocity of the molecules. Then, the temperature T in molecular kinetics is defined as

$$\frac{3}{2}kT = \frac{1}{n}\int \frac{1}{2}mc^2 f(t, \vec{X}, \vec{\xi})d\vec{\xi} \tag{7.3}$$

where k is the Boltzmann constant. This definition indicates that energy is distributed equally among degrees of freedom (DOFs), that is, the energy fraction along the direction of each average DOF is equal to 1/2 KT. Here, only the temperature of the average DOFs but not the other DOFs of molecules is considered. The following quantity is also introduced in molecular kinetics:

$$P = \frac{1}{3}(P_{11} + P_{22} + P_{33}) \tag{7.4}$$

P is defined as a hydrostatic pressure or simply pressure, which is consistent with the concepts of pressure in classical fluid mechanics and thermodynamics. Evidently, ΣP_{ij} is a tensor invariant. Therefore, P is a scalar, and $P = P(\vec{X}, t)$. Consequently, the relationship between the pressure P and the temperature T is immediately obtained as follows:

$$P = nkT = \rho RT \tag{7.5}$$

which is the equation of state for an ideal gas $R = k/m$ is the gas constant.

A gas mixture is usually considered to be composed of S types of simple gases. For any of these simple gases, a velocity distribution function $f_i(t,)$ $i = 1, 2, \ldots S$ can be defined to describe the motion state of simple gas molecules. Then, each macroscopic hydrodynamic quantity of the gas mixture is expressed as the algebraic sum of the weighted average of the component gases according to their own velocity distribution functions. The specific expressions are as follows.

$$n = \sum_{i=1}^{S} n_i$$

$$\rho = \sum_{i=1}^{S} m_i n_i$$

$$\vec{u} = \frac{\sum_{i=1}^{S} m_i \int \vec{\xi} f_i d\vec{\xi}}{\sum_{i=1}^{S} m_i n_i}$$

$$\frac{3}{2}KT = \frac{1}{n}\sum_{i=1}^{S} \frac{1}{2}m_i \int c^2 f_i d\vec{\xi}$$

$$P_{ij} = \sum_{i=1}^{S} m_i \int \vec{c}_i \vec{c}_j f_i d\vec{\xi}$$

$$q_i = \sum_{i=1}^{S} \frac{1}{2} m_i \int c^2 \vec{c}_i f_i d\vec{\xi} \tag{7.6}$$

where n, p, μ, T, P_{ij}, and q_i are the molecular number density, mass density, macroscopic velocity, temperature, stress tensor, and energy flux of the gas mixture, respectively. The diffusion rate of the gas mixture components is defined as

$$\vec{W}_i = \vec{u}_i - \vec{u} = \frac{1}{n_i} \int (\vec{\xi}_i - \vec{u}) f_i d\vec{\xi} \tag{7.7}$$

3. Calculation of Collision in the DSMC Method

The collision calculation in the DSMC method includes the determination of the sampling function of collision pairs and the velocity of molecules after collision. Here, we introduce several relevant methods for determining the sampling function of collision pairs.

(1) Bird's Time Counter Scheme

According to the kinetic theory of gas molecules, the average collision frequency v of gas molecules of the same component in the equilibrium state is

$$v = n\overline{\sigma}_T \overline{g} \tag{7.8}$$

where the symbol "—" represents the average value. The above equation shows that the probability of collision between two specific gas molecules P_{col} is proportional to the product of the collision cross section σ_T and the relative velocity g of the molecule pair.

$$P_{col} \propto g\sigma_T \tag{7.9}$$

Therefore, the collision probability function P_{col} of the simulated molecule pairs in a cell element can be expressed as

$$P_{col}(g) = \frac{\sigma_T g}{(\sigma_T g)_{max}} \tag{7.10}$$

The relationship between the collision cross section σ_T and the relative velocity g of the colliding molecules varies depending on different molecular potential models. Therefore, the expression of the collision probability function P_{col} obtained from Eq. (7.10) differs for different molecular models. For the hard sphere molecular model, the collision cross section σ_T of the simulated molecule pairs is constant, and the sampling probability function of the hard sphere molecular collision pairs is $P_{col}(g) = g/(g)_{max}$. For the inverse power law molecular model, the collision sampling probability function is $P_{col}(g) = g^{1-4a}/(g^{1-4a})_{max}$.

On the other hand, from Eq. (4.8), the total number of collisions N_t of all simulated molecules in the cell within a time step Δt_m is obtained as follows:

$$N_t = \frac{1}{2} N_m n \overline{\sigma}_T \overline{g} \Delta t_m \tag{7.11}$$

where N_m is the total number of molecules in the cell, n is the number density of gas molecules in the cell, and the factor 1/2 is a weighting factor obtained from the fact that two molecules are involved in each collision. For the hard sphere molecular model, we have

$$N_t = \frac{1}{2} N_m n \sigma_T \overline{g} \Delta t_m \tag{7.12}$$

For the inverse power law molecule model, we have

$$N_t = \frac{\sigma_T}{2} N_m n g^{-1-4/\alpha} \Delta t_m \tag{7.13}$$

When N_t is calculated using any of the above equations, it is necessary to calculate the average value of a function with the relative velocity g of gas molecules as an independent variable. This not only causes difficulty in programming but also requires a large amount of computing time. To address this problem, Bird proposed a "time counter scheme". When using this scheme, a timer is set in each cell, and when a collision of simulated molecules occurs in a cell, a time interval Δt_{ci} corresponding to the colliding molecules is cumulatively added to the timer. For hard sphere molecules, inverse power law molecules, and Lennard–Jones molecules, Δt_{ci} is calculated as follows:

$$\Delta t_{ci} = \frac{2}{N_m n \sigma_T g} \tag{7.14}$$

$$\Delta t_{ci} = \frac{2}{N_m n \sigma_T g^{1-4/\alpha}} \tag{7.15}$$

Once the cumulative time displayed by the cell timer is greater than Δt_m, i.e., $\sum_i \Delta t_{ci} > \Delta t_m$, the collision calculation of the simulated molecules in this cell is stopped, and the next operation is performed.

Bird proved that by repeating the above operation for a certain number of simulated molecules, the average collision frequency of an individual molecule expressed in Eq. (7.8) can be accurately simulated. Therefore, the DSMC method can simulate the flow correctly when the number of simulated molecules in the cell is relatively small (generally, 20–30 simulated molecules are arranged in the cell), and its computational workload is proportional to the number of simulated molecules. On the other hand, the above analysis shows that even though this sampling method can be extended to different types of molecular collisions with slight modifications, it is not suitable for vectorized computation.

(2) Baganoff/McDonald Sampling Scheme

The application of supercomputers for vectorized computation has become the development trend in computational physics and computational mechanics. Since the DSMC method essentially tracks the motion trajectories of a large number of simulated molecules simultaneously, as long as a suitable collision sampling model for simulated molecules is established, vectorized computation can be realized, greatly improving the computational efficiency and achieving a high speedup ratio. Baganoff and McDonald proposed a collision sampling model that enables the vectorized computation of the DSMC method. They noted that the quantity $n_a n_b/(1 + \delta_{ab})$ represents the number of collision pairs that may occur between n_α type-a molecules and n_b type-b molecules within a unit volume of the physical space. $R(g)$ is the distribution function of these colliding molecules with respect to the relative velocity g. Therefore, the quantity $n_a n_b/(1 + \delta_{ab})R(g)dg$ is the number of collision pairs that may form between molecules with a relative velocity modulus between g and $g + dg$ within a unit volume of the physical space. The quantity $g\sigma_T \Delta t_m$ represents the volume of physical space swept by a molecule with a collision cross section of σ_T at a velocity g within the time interval Δt_m. If this volume is relatively small compared to the volume of the physical space under investigation and is used as a dimensionless parameter, the value of $g\sigma_T \Delta t_m$ for the simulated molecules sweeping across this volume can be interpreted as the probability of the collision of gas molecules within the unit volume. According to this interpretation, the total number of collisions between type-a and type-b molecules with a relative velocity in the range of g to $g + dg$ within the time step Δt_m can be expressed as

$$Z_{ab}dg\,\Delta t_m = S_{ab}R(g)dgP_{col} \tag{7.16}$$

$$S_{ab} = \frac{n_a n_b}{1 + \delta_{ab}} \tag{7.17}$$

$$P_{col} = \sigma_T g \Delta t_m \tag{7.18}$$

where S_{ab} is the number of colliding molecules sampled per unit volume, P_{col} is the sampling probability function of gas molecule collision pairs, and $R(g)$ is the distribution function of the relative velocity g of colliding molecules. Thus, Eq. (7.17) provides a sampling method for collision molecules that enables the implementation of vectorized computation. First, S_{ab} colliding molecules are selected within a unit volume of the physical space. Then, a determination of whether collisions of these molecule pairs indeed occur based on the collision probability is made.

When the gas is in equilibrium, following the above procedure yields

$$Z_{ab}d\vec{g}\,\Delta t_m = S_{ab}H\left(\vec{g}\right)d\vec{g}\,P_{col} \tag{7.19}$$

$$S_{ab} = \frac{n_a n_b}{1 + \delta_{ab}} \tag{7.20}$$

$$P_{\text{col}} = \sigma_T \, \vec{g} \, \Delta t_m \tag{7.21}$$

According to the analysis, $H\left(\vec{g}\right)$ is highly approximate to a normal function. Therefore, Eqs. (7.16)–(7.18) can be used instead of Eqs. (7.19)–(7.21) for collision molecule sampling. The aim of this approach is to achieve the same accuracy but greatly reduce the number of samples needed. From a practical point of view, when implementing a DSMC simulation, the number of collisions in a cell within a time step Δt_m is always limited within an acceptable range. For a small number of samples, only sampling from one-dimensional (1D) smooth functions is feasible.

To achieve the statistical calculation of macroscopic quantities, a sufficient number of simulated molecules must be set in each cell; as a result, the number of colliding molecules sampled S_{ab} calculated by Eq. (7.20) becomes very large, greatly reducing the computational efficiency of the DSMC method. To decrease S_{ab} while proportionally increasing the sampling probability P_{col} so that the product of S_{ab} and P_{col} remains unchanged, a proper selection method involves making P_{col} exactly equal to the collision probability. According to the experience in simulation using the DSMC method, it is appropriate to determine S_{ab} in the cell using use the following relationship:

$$S_{ab} = \frac{K}{2}\sqrt{n_a n_b} \tag{7.22}$$

where K is a constant determined by the capacity of the vector computer, and n is the number density of gas molecules. Therefore, S_{ab} is proportional to the so-called "natural sampling dimension" n, avoiding dependence on other data in the calculation process. Using Eq. (7.22) to determine S_{ab} not only reduces the total number of samples in the cell by one to two orders of magnitude but also includes the time-averaged value $\sqrt{nanb}$ in the molecule pair sampling probability P_{col}, thereby lowering fluctuations in P_{col}, reducing statistical errors, and improving the computational molecule efficiency.

(3) Bird's No-Time-Counter Scheme

Bird transformed the time counter scheme into a no-time-counter scheme, making the constructed DSMC method suitable for vectorized computation. In this scheme, the collision sampling probability function P_{col} of the simulated molecule pairs in the cell remains expressed as Eq. (7.10), but the number of simulated molecular collisions N_t in the cell within the time step Δt_m is given in advance. Bird provided the calculation formula for N_t of a gas with the same composition as

$$N_t = \frac{1}{2}N_m n(\sigma_T g)_{\text{max}} \Delta t_m \tag{7.23}$$

which serves as S_{ab}, the number of colliding molecules sampled in the cell. Its advantage lies in normalizing the collision sampling probability function P_{col} by the maximum value of $\sigma_T g$ in the cell, which meets the definition of the sampling probability. However, in the Baganoff/McDonald sampling method, P_{col} does not match the definition of the sampling probability. Therefore, the molecule pairs with P_{col} greater than 1 should be excluded from the simulation using the DSMC method.

4. Key Techniques in Simulation

(1) Meshing the Computational Region of the Flow

Similar to computational fluid dynamics, an important part of initializing a DSMC method program is to generate the mesh of the flow field to perform collision calculations and carry out statistical analysis of macroscopic physical quantities within the cells. To generate the mesh, the boundary of the computational flow field must first be described. For an internal flow field, the computational boundary consists of object surfaces, exits, and inlets; for an external flow field, a reasonable truncation of the infinite boundary must be determined according to certain principles. Once the boundary of the computational flow field is determined, the flow field is meshed into computational cells according to certain rules. In general, the meshing techniques in computational fluid dynamics can be applied to DSMC method simulation programs. Because the DSMC method does not suffer from discretization or computational instability issues, it has more flexibility in meshing and does not require a regularly shaped mesh. Based on Bird's experience, the cell dimension $\Delta x \backsim 1/3\ \overline{\lambda}$, where $\overline{\lambda}$ is the average free path of the gas molecules in the cell.

(2) Selection of the Time Step

There are two requirements for selecting the time step Δt_m in the DSMC algorithm. First, Δt_m should be smaller than the time required for simulated molecules to move one cell length. Second, Δt_m should be much smaller than the time required for simulated molecules to undergo a collision to enable the decoupling of the migration and collision of simulated molecules. Different Δt_m values are used for different computational domains. In each computational domain, Δt_m should satisfy the following conditions:

$$\Delta t_m < \min\left[\min\left(\frac{\Delta x}{u}\right), \min\left(\frac{\Delta y}{v}\right)\right] \quad \text{and} \quad \Delta t_m << \frac{1}{\nu} \tag{7.24}$$

where ν is the collision frequency of simulated molecules.

(3) Weight Function

Some flows are not meshed in a completely uniform manner. Thus, if the element number densities are similar, the number of simulated molecules may vary greatly within different cells. Cells with a large number of simulated molecules will waste computer time, while cells meshed with a small number of simulated molecules will cause large statistical fluctuations, leading to computational distortion. For regions with large density variations in the flow field, such as backflow regions, the number of simulated molecules may also differ significantly across cells. When calculating the flow of a gas mixture, if the number density of a certain component is much smaller than that of the other components, the number of simulated molecules corresponding to that component will be greatly reduced, posing difficulties for simulation.

This type of problem can be solved by using the weighting factor technique, which is an important technique in Monte Carlo methods. In the calculation, different weight factors W_i are assigned to different cells, where W_i represents the number of real gas molecules contained in the simulated molecules in each cell. If a cell has a volume of V_i, then the number of simulated molecules in the cell is $N_i = n_i \times V_i/W_i$, where n_i is the number density of real molecules in the gas. The basic principle of the weight factor configuration is to minimize the difference in the number of simulated molecules contained in each computational cell in the flow field. Therefore, the configuration should be implemented in combination with meshing to save computer time and memory while ensuring computational accuracy.

After the weighting factor technique is applied, the difference between the actual and weighted numbers of molecules must be carefully identified when the simulated molecules move from one cell to another. Then, the following is processed accordingly.

Delete–copy method. To ensure that the flux of the simulated molecules passing through the mesh boundary is conserved in a statistical sense, when the simulated molecules move from a cell with a weight factor W_1 to a cell with a weight factor W_2, the following deletions must be performed:

If $W_2 > W_1$, then the probability of this simulated molecule being removed is

$$P_{\text{removal}} = 1 - \frac{W_1}{W_2} \tag{7.25}$$

If $W_2 < W_1$, then the number of simulated molecules that need to be duplicated is

$$N_{\text{duplicate}} = \text{int}\left[\frac{W_1}{W_2} - 1\right] \tag{7.26}$$

where the square brackets represent rounding to the nearest integer. The probability that an additional simulated molecule needs to be duplicated is

$$P_{\text{duplicate}} = \frac{W_1}{W_2} - N_{\text{duplicate}} - 1 \tag{7.27}$$

The ideal situation is that $N_{\text{duplocate}}$ is zero. Then, the maximum number of simulated molecules that may be duplicated in each time step is 1. However, when W_2 differs significantly from W_1, a large number of simulated molecules will be duplicated, which will cause the following issues. The probability of sampling two duplicated molecules with zero collision probability increases with the increase in the number of duplicated molecules, resulting in a waste of sampling time; on the other hand, the duplication of simulated molecules will cause additional statistical fluctuations, thus affecting the computational efficiency. Therefore, when configuring weight factors, it is advisable to ensure the smoothness of the changes in weight factors and avoid large gradients.

7.1.2 Hybrid DSMC/PIC Fluid Algorithm

A hybrid algorithm can perform axisymmetric simulations on a PPT plume. This algorithm includes the interactions of neutral particles with ions and electrons. Information about the electric field and temperature field in the plume is obtained by solving the Poisson equation and the electron energy equation.

1. General Procedure of the Hybrid DSMC/PIC Fluid Algorithm

The hybrid algorithm, whose flowchart is shown in Fig. 7.2, includes the following steps:

(1) Under the assumption of no collision, determine the distances traveled by various types of simulated molecules in each time step. Neutral particles move at a uniform velocity in a straight line as follows:

$$\Delta \mathbf{r} = \mathbf{v} \Delta t \tag{7.28}$$

Charged particles are accelerated under the action of an electric field.

$$m_i d\mathbf{v}_i / dt = q_i \mathbf{E} + \mathbf{F}_{ie} \tag{7.29}$$

$$\mathbf{F}_{ie} = \upsilon_{ie} m_{ie} (\langle \mathbf{v}_e \rangle - \langle \mathbf{v}_i \rangle) - \upsilon_{ie} \frac{m_{ie}^2}{m_i} \frac{(\langle \mathbf{v}_e \rangle - \langle \mathbf{v}_i \rangle)^2}{\langle \mathbf{v}_i^2 \rangle - \langle \mathbf{v}_i \rangle^2} (\langle \mathbf{v}_i \rangle - \mathbf{v}_i)$$

$$+ \upsilon_{ie}^* \frac{k(T_i - T_e)}{\langle \mathbf{v}_i^2 \rangle - \langle \mathbf{v}_i \rangle^2} (\langle \mathbf{v}_i \rangle - \mathbf{v}_i) \tag{7.30}$$

$$\upsilon_{ie}^* = 16\sqrt{\pi} \left(\frac{Z_1 Z_2 e^2}{4\pi \varepsilon_0} \right)^2 \frac{n_e \ln \Lambda_{ie}}{m_i m_e \upsilon_{th}^3} \exp\left(-\frac{\Delta \upsilon^2}{\upsilon_{th}^2} \right) \tag{7.31}$$

The possible interactions with the boundary are determined and processed accordingly to establish the new position coordinates of the simulated molecules.

(2) Determine the number and motion state of each type of simulated molecule entering the computational domain at the current time.

(3) Adjust the cell number of the simulated molecules based on their new spatial position coordinates and sort the simulated molecules.

(4) Calculate the collisions between the simulated molecules within a time step, including the elastic collisions between neutral particles and between neutral particles and charged particles as well as the charge-exchange (CEX) collisions between neutral particles and charged particles. The sampling of collision pairs is performed using a no-time-counter (NTC).

Fig. 7.2 Flowchart of the hybrid DSMC/PIC fluid algorithm

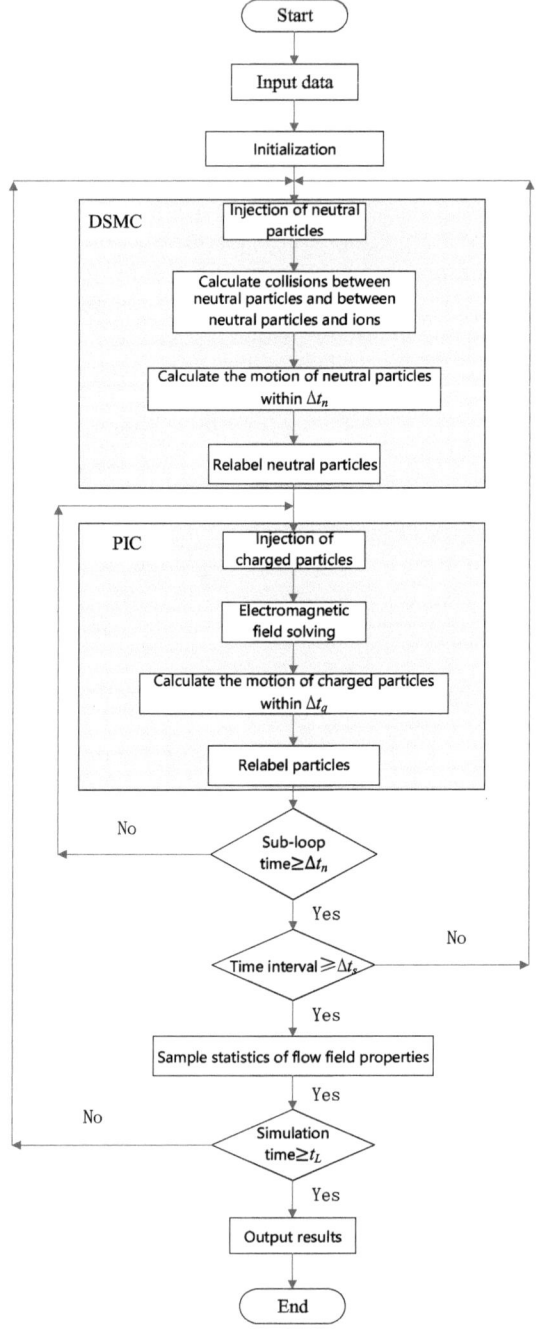

Table 7.1 VHS parameters
in the PPT plume

Component	d_{ref} (10^{-10} m)
F	3.0
C	2.5

(5) Carry out statistical calculations of various physical quantities in the flow field to obtain macroscopic physical quantities, with statistical calculations of charges based on cell points.

(6) Use the difference method to update the electric field and temperature field.

2. Collision Calculation in the Hybrid DSMC/PIC Fluid Algorithm

The PPT plume includes various intermolecular collisions, including elastic collisions and inelastic collisions (including CEX collisions). Since Coulomb collisions between ions can be ignored for the ion energy distribution, this type of collision can be not considered in the simulation of plasma plumes in most cases. Currently, CEX collisions between ions are not considered in the simulation due to a lack of experimental and theoretical collision cross sections. In addition, the interaction between electrons and various molecules is reflected by the electron fluid model.

(1) Elastic Collisions Between Neutral Particles

Bird proposed a variable hard sphere (VHS) model in 1994. In this model, the particle diameter d is a function of the relative motion rate g.

$$\frac{d}{d_{ref}} = \left(\frac{g_{ref}}{g}\right)^{\omega} \tag{7.32}$$

where ω is the temperature exponent and d_{ref} is the reference diameter. Table 7.1 lists some reference diameters of the components in the PPT plume.

For a gas mixture, the average collision frequency of components p and q is expressed as

$$v_{pq} = 2\sqrt{\pi}(d_{ref})_{pq}^2 n_q \sqrt{\left[\frac{T}{(T_{ref})_{pq}}\right]^{1-\omega} \frac{2k(T_{ref})_{pq}}{m_r}} \tag{7.33}$$

where $(T_{ref})_{pq}$ is the reference temperature, $(d_{ref})_{pq}$ is the reference diameter, and m_r is the reduced mass.

(2) Neutral-Ion Collisions

The main collision process in a weakly ionized plasma occurs between charged particles and neutral particles. Elastic collisions mainly occur at low energy levels, while CEX collisions mainly occur at high energy levels.

Lieberman and Lichtenberg [5] provided an expression for the average collision frequency of elastic collisions between neutral particles and ions.

Table 7.2 Relative polarizability

α_R (C)	α_R (F)
12	3.8

$$v_{in} = \frac{4}{3} n_n \overline{g}_{in} \overline{\sigma}_{in}^M \qquad (7.34)$$

where the average relative motion rate is equal to

$$\overline{g}_{in} = \sqrt{\frac{8k}{\pi}\left(\frac{T_i}{m_i} + \frac{T_e}{m_e}\right)} \qquad (7.35)$$

and the average momentum exchange cross section is

$$\overline{\sigma}_{in}^M = \frac{1}{c^6} \int g^5 \sigma_{in}^M \exp\left(-\frac{g^2}{c^2}\right) dg \qquad (7.36)$$

in which c is the thermal velocity is

$$\sigma_{in}^M = \sqrt{\frac{\pi \alpha e^2}{\varepsilon_0 m_r} \frac{1}{g}} \qquad (7.37)$$

where α is the polarizability of the atom. Typically, the relative polarizability is

$$\alpha_R = \frac{\alpha}{a_0^3} \qquad (7.38)$$

where $a_0 = 5.2918 \times 10^{-11}$ m is the Bohr radius. The relative polarizability of some neutral particles in the PPT plume is shown in Table 7.2.

During collisions, charge-exchange reactions occur between the neutral particles and ions.

$$X_{\text{fast(slow)}}^+ + X_{\text{slow(fast)}}^N = X_{\text{fast(slow)}}^N + X_{\text{slow(fast)}}^+ \qquad (7.39)$$

Sakabe and Izawa [6] used experimental data of the PPT plume to obtain its momentum exchange cross section as follows:

$$\sigma_{in}^{M,\text{CEX}} = A + B \lg(g) \qquad (7.40)$$

where the coefficients A and B are given in Table 7.3.

Then, the average collision frequency is equal to

Table 7.3 Coefficients of the momentum exchange cross section

Type of reaction	A	B
$F^+ - F$	8.3343×10^{-19}	-1.2522×10^{-19}
$C^+ - C$	1.7771×10^{-18}	-2.6797×10^{-19}

$$\bar{v}_{in} = \frac{4}{3}\bar{\sigma}_{in}^M n_n \sqrt{\frac{8k}{\pi m_{in}}(T_i + T_n)} \tag{7.41}$$

(3) Neutral-Electron Collisions

Mithner and Kruger [7] presented the average momentum exchange cross section between a neutral particle and an electron as

$$\bar{\sigma}_{en}^M = \left(\frac{m_e}{2kT_e}\right)^3 \int_0^\infty v^5 \sigma_{en}^M(v) \exp\left(-\frac{m_e v^3}{2kT_e}\right) dv \tag{7.42}$$

For most cases,

$$\frac{T_e}{m_e} \gg \frac{T_n}{m_n} \tag{7.43}$$

The average collision frequency is

$$\bar{v}_{en} = \frac{3}{4}\bar{\sigma}_{en}^M n_n \sqrt{\frac{8kT_e}{\pi m_e}} \tag{7.44}$$

Bittencourt provided the following expression for the electron-neutral particle collision frequency.

$$v_{en} = 2.60 \times 10^4 \sigma_{en}^2 n_n \sqrt{T_e} \tag{7.45}$$

The elastic collision cross section between an electron and a neutral particle is usually equal to 10^1 to $10^3 \pi a_0^2$.

(4) Ion–Electron Collisions

Without considering the relative motion, for the Maxwellian velocity distribution, the average momentum exchange collision frequency of an ion and an electron is expressed as

$$\bar{v}_{ie} = \frac{16\sqrt{\pi}}{3}\left(\frac{Z_i e^2}{4\pi\varepsilon_0 m_{ie}}\right)^2 \frac{n_e \ln \Lambda_{ie}}{v_{th}^3} = \frac{4\sqrt{2\pi}}{3}\left(\frac{Z_i e^2}{4\pi\varepsilon_0 m_{ie}}\right)^2 \left(\frac{m_e}{kT_e}\right)^{\frac{3}{2}} n_e \ln \Lambda_{ie} \tag{7.46}$$

where

$$v_{th} = \sqrt{2k\left(\frac{T_i}{m_i} + \frac{T_e}{m_e}\right)} \tag{7.47}$$

When considering the relative motion, the above collision frequency becomes

$$\bar{v}_{ie} = 8\sqrt{\pi}\left(\frac{Z_i e^2}{4\pi \varepsilon_0 m_{ie}}\right)^2 \frac{n_e \ln \Lambda_{ie}}{(\Delta v)^3}\left[\frac{\sqrt{\pi}}{2}\text{erf}\left(\frac{\Delta v}{v_{th}}\right) - \left(\frac{\Delta v}{v_{th}}\right)\exp\left(-\frac{\Delta v^2}{v_{th}^2}\right)\right] \tag{7.48}$$

where

$$\Delta v = |\langle \mathbf{v}_i \rangle - \langle \mathbf{v}_e \rangle| \tag{7.49}$$

When the relative speed is very small, $\Delta v/v_{th} < < 1$, and Eq. (4.48) can be solved by a series of function expansions.

$$\text{erf}(x) = \frac{2}{\sqrt{\pi}}\sum_{n=0}^{\infty}\frac{(-1)^n x^{2n+1}}{n!(2n+1)} \tag{7.50}$$

$$\exp(x) = \frac{2}{\sqrt{\pi}}\sum_{n=0}^{\infty}\frac{(-1)^n x^n}{n!} \tag{7.51}$$

Equation (4.48) can be approximately equal to

$$\bar{v}_{ie} = 8\sqrt{\pi}\left(\frac{Z_1 Z_2 e^2}{4\pi \varepsilon_0 m_{ie}}\right)^2 n_e \ln \Lambda_{ie}\frac{1}{v_{th}^3}\left[\sum_{n=0}^{\infty}\frac{(-1)^n y^{2n-2}}{n!}\left(\frac{1}{2n+1} - 1\right)\right]$$
$$= 8\sqrt{\pi}\left(\frac{Z_1 Z_2 e^2}{4\pi \varepsilon_0 m_{ie}}\right)^2 n_e \ln \Lambda_{ie}\left[\frac{2}{3}\frac{1}{v_{th}^3} - \frac{4}{10}\frac{(\Delta v)^2}{v_{th}^5} + \frac{6}{42}\frac{(\Delta v)^4}{v_{th}^7} - \cdots\right] \tag{7.52}$$

3. Electron Fluid Model

The electrons in the plume are simulated using a fluid model, and the electron momentum equation is given as:

$$\frac{\partial \mathbf{u}_e}{\partial t} + \mathbf{u}_e \nabla \cdot \mathbf{u}_e = -\frac{e}{m_e}(\mathbf{E} + \mathbf{u}_e \times \mathbf{B}) - \frac{\nabla p_e}{m_e n_e}$$
$$- \sum_i \bar{v}_{ei}(\mathbf{u}_e - \mathbf{u}_i) - \sum_n \bar{v}_{en}(\mathbf{u}_e - \mathbf{u}_n) \tag{7.53}$$

Gatsonis performed a dimensional analysis based on the characteristics of the PPT plume, ignoring the influences of the unsteady state and magnetic field, to obtain

$$0 = -\frac{e}{m_e}\mathbf{E} - \frac{\nabla p_e}{m_e n_e} - \sum_i \bar{\nu}_{ei}(\mathbf{u}_e - \mathbf{u}_i) - \sum_n \bar{\nu}_{en}(\mathbf{u}_e - \mathbf{u}_n) \qquad (7.54)$$

Therefore, the electron velocity is

$$\mathbf{u}_e = -\frac{e}{m_e \bar{\nu}_e}\mathbf{E} - \frac{\nabla p_e}{m_e n_e \bar{\nu}_e} - \frac{\sum_i \bar{\nu}_{ei}(\mathbf{u}_e - \mathbf{u}_i)}{\bar{\nu}_e} - \frac{\sum_n \bar{\nu}_{en}(\mathbf{u}_e - \mathbf{u}_n)}{\bar{\nu}_e} \qquad (7.55)$$

where

$$\bar{\nu}_e = \sum_i \bar{\nu}_{ei} + \sum_n \bar{\nu}_{en} \qquad (7.56)$$

4. Electrodynamic Model

The electromagnetic phenomena of the plasma are described using Maxwell's system of equations, including the law of electromagnetic induction, Ampere's law, Gauss' law, and the Biot–Savart law.

$$\nabla \times \mathbf{E} = -\frac{\partial \mathbf{B}}{\partial t} \qquad (7.57)$$

$$\nabla \times \mathbf{B} = \mu_0 \left(\mathbf{J} + \varepsilon_0 \frac{\partial \mathbf{E}}{\partial t} \right) \qquad (7.58)$$

$$\nabla \cdot \mathbf{E} = \frac{\rho_e}{\varepsilon_0} \qquad (7.59)$$

$$\nabla \cdot \mathbf{B} = 0 \qquad (7.60)$$

Under quasi-neutrality and neglecting magnetic field assumptions, we obtain

$$\mathbf{E} = -\nabla \phi \qquad (7.61)$$

$$\nabla \cdot \mathbf{J} = 0 \qquad (7.62)$$

The total current density is defined as

$$\mathbf{J} = \sum_s n_s q_s \mathbf{u}_s = \sum_i n_i q_i \mathbf{u}_i - e n_e \mathbf{u}_e \qquad (7.63)$$

Using Eq. (7.59), the above equation becomes

$$\mathbf{J} = \sum_i n_i q_i \mathbf{u}_i - e n_e \left[-\frac{e}{m_e \bar{v}_e} \mathbf{E} - \frac{\nabla p_e}{m_e n_e \bar{v}_e} - \frac{\sum_i \bar{v}_{ei}(\mathbf{u}_e - \mathbf{u}_i)}{\bar{v}_e} - \frac{\sum_n \bar{v}_{en}(\mathbf{u}_e - \mathbf{u}_n)}{\bar{v}_e} \right]$$

$$(7.64)$$

Equation (7.64) is expressed as

$$\mathbf{J} = \mathbf{J}_I + \mathbf{J}_E + \mathbf{J}_D + \mathbf{J}_{IC} + \mathbf{J}_{NC} \qquad (7.65)$$

in which

$$\mathbf{J}_I = \sum_i q_i n_i \mathbf{u}_i \qquad (7.66)$$

$$\mathbf{J}_E = \frac{e^2 n_e}{m_e v_e} \mathbf{E} = \sigma \mathbf{E} \qquad (7.67)$$

$$\mathbf{J}_D = \frac{e}{m_e v_e} \nabla p_e \qquad (7.68)$$

$$\mathbf{J}_{IC} = -\frac{e n_e}{v_e} \sum_i v_{ei} \mathbf{u}_i \qquad (7.69)$$

$$\mathbf{J}_{NC} = -\frac{e n_e}{v_e} \sum_n v_{en} \mathbf{u}_n \qquad (7.70)$$

From Eq. (7.71), we have

$$\nabla \cdot (\sigma \mathbf{E}) = -\nabla \cdot (\mathbf{J}_I + \mathbf{J}_D + \mathbf{J}_{IC} + \mathbf{J}_{NC}) \qquad (7.71)$$

$$\nabla \cdot (\sigma \nabla \cdot \phi) = -\nabla \cdot (\mathbf{J}_I + \mathbf{J}_D + \mathbf{J}_{IC} + \mathbf{J}_{NC}) \qquad (7.72)$$

5. Electron Energy Model

The electron energy equation is introduced to solve for the electron temperature.

$$\frac{\partial \varepsilon_e}{\partial t} + \nabla \cdot [\varepsilon_e u] + p_e \nabla \cdot u = -\Delta \dot{\varepsilon}_{ie} + \nabla \cdot (K_e \cdot \nabla T_e) \qquad (7.73)$$

After simplification of the unsteady state, we have

$$0 = -\nabla \cdot \kappa_e \nabla T_e + \sum_{H=1}^n 3 \frac{m_e}{m_H} v_e n_e k (T_e - T_H) \qquad (7.74)$$

where K_e is the thermal conductivity of electrons

$$\kappa_e = \frac{2.4}{1 + \left(\nu_{ei}/\sqrt{2}\nu_{eH}\right)} \frac{k^2 n_e T_e}{m_e \nu_{eH}} \tag{7.75}$$

$\nu_{eH} = \sum_{H=1}^{n} \nu_{eH}$ is the total collision frequency of electrons and simulated molecules.

6. Key Techniques in Simulation

(1) Meshing the Flow Computational Domain

The DSMC and PIC meshes have distinct spatial scales, approximately corresponding to the molecular mean free path and the Debye length, respectively, necessitating two separate sets of meshes in the simulation. The two sets of meshes are in the same coordinates, and the addresses of the particles in the two sets of meshes are determined based on the spatial positions of the particles.

(2) Selection of the Time Step

Ions in the plume have a higher collision frequency than neutral particles; therefore, different time steps are needed for neutral–neutral collisions and neutral-ion collisions. The time step for ions is smaller than that for neutral particles. As a result, the motion time of ions forms a sub-cycle process within the motion time of neutral particles.

(3) Statistical Solution of Physical Quantities at Cell Points

In the Gatsonis model, the same weight factor expression is used in the axial and radial directions when statistically analyzing the physical quantities at cell points, with some improvements.

The flow field is simulated axisymmetrically. Therefore, the axial and Ruytan-developed radial weight factors are used in the calculation of the charge density at the cell points (Fig. 7.3).

$$S_i = \frac{z_{i+1} - z}{z_{i+1} - z_i} \tag{7.76}$$

$$S_{i+1} = \frac{z - z_i}{z_{i+1} - z_i} \tag{7.77}$$

$$S_j = \frac{(r_{j+1} - r)(2r_{j+1} + 3r_j - r)}{2(r_{j+1}^2 - r_j^2)} \tag{7.78}$$

$$S_{j+1} = \frac{(r - r_j)(3r_{j+1} + 2r_j - r)}{2(r_{j+1}^2 - r_j^2)} \tag{7.79}$$

where Z represents the axial coordinate and r represents the radial coordinate.

$$m_{i,j} = \sum_{p=1}^{s} m_p S_{pi} S_{pj} \tag{7.80}$$

$$\mathbf{M}_{i,j} = \sum_{p=1}^{s} \mathbf{M}_p S_{pi} S_{pj} \tag{7.81}$$

$$\overline{\mathbf{u}}_{i,j} = \frac{\mathbf{M}_{i,j}}{m_{i,j}} \tag{7.82}$$

where M is the momentum of the particle.

(4)   Solving the Electric Potential Equation and Temperature Equation

The solution processes of the electric potential equation and the temperature equation are similar. Here, only the solution process of the electric potential equation is described. From Eq. (7.76), we have

$$\mathbf{J}^* = \mathbf{J}_I + \mathbf{J}_D + \mathbf{J}_{IC} + \mathbf{J}_{NC} \tag{7.83}$$

Discretizing Eq. (7.76) by the finite difference method, the left-hand side of the equation becomes

$$[\nabla \cdot (\sigma \nabla \phi)]_{i,j} = A_W \phi_{i-1,j} + A_E \phi_{i+1,j} + A_S \phi_{i,j-1} + A_N \phi_{i,j+1} + A_C \phi_{i,j} \tag{7.84}$$

where

$$A_W = \frac{\sigma_{i,j} + \sigma_{i-1,j}}{\Delta z_i (\Delta z_i + \Delta z_{i+1})} \tag{7.85}$$

$$A_E = \frac{\sigma_{i+1,j} + \sigma_{i,j}}{\Delta z_{i+1} (\Delta z_i + \Delta z_{i+1})} \tag{7.86}$$

$$A_S = \frac{\sigma_{i,j} + \frac{r_{j-1}}{r_j} \sigma_{i,j-1}}{\Delta r_j (\Delta r_j + \Delta r_{j+1})} \tag{7.87}$$

Fig. 7.3 Assignment of weights in a particle cell

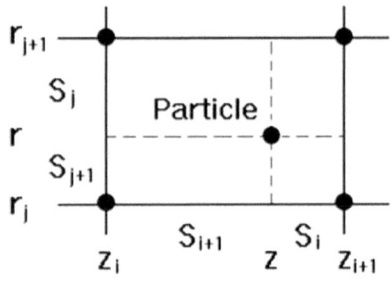

$$A_N = \frac{\sigma_{i,j} + \frac{r_{j+1}}{r_j}\sigma_{i,j+1}}{\Delta r_{j+1}(\Delta r_j + \Delta r_{j+1})} \tag{7.88}$$

$$A_C = -(A_W + A_E + A_S + A_N) \tag{7.89}$$

The right-hand side of the equation becomes

$$\text{RHS}_{i,j} = \frac{1}{r_j}\frac{\frac{1}{2}\left[r_{j+1}(J_r^*)_{i,j+1} - r_{j-1}(J_r^*)_{i,j-1}\right]}{\frac{1}{2}(\Delta r_j + \Delta r_{j+1})} + \frac{\frac{1}{2}\left[(J_z^*)_{i+1,j} - (J_z^*)_{i-1,j}\right]}{\frac{1}{2}(\Delta z_i + \Delta z_{i+1})} \tag{7.90}$$

where

$$A_W\phi_{i-1,j} + A_E\phi_{i+1,j} + A_S\phi_{i,j-1} + A_N\phi_{i,j+1} + A_C\phi_{i,j} = \text{RHS}_{i,j} \tag{7.91}$$

Usually, the generalized minimal residual (GMRES) method is employed to solve the system of equations formed by this pentadiagonal matrix. It is crucial to perform a preconditioning technique. Here, a preconditioning method described in the literature[8, 9] is used to improve the computational efficiency.

7.2 Numerical Simulation of the PPT Plume

The PPT plume is simulated with a two-dimensional (2D) axisymmetric model using the hybrid DSMC/PIC fluid algorithm to investigate the variations of the neutral components, electric field, and temperature field of the plume at initial voltages of 1100, 1500, and 2000 V), respectively, which correspond to initial energies of 7.26 J, 13.5 J and 24 J, respectively. On this basis, the flow field without CEX collisions under an initial voltage of 1500 V (initial energy of 13.5 J) is calculated to study the CEX collisions.

7.2.1 Meshing and Boundary Conditions

Using an axisymmetric configuration, the simulation region is set to $L_Z = 1$ m and $L_R = 1$ m. The laboratory PPT is placed inside an enclosure (Fig. 7.4), which has a length $Z_{S/C}$ of 0.2 m, a radius $R_{S/C}$ of 0.05 m, and an equivalent radius R_T of 0.02 m at the exit.

The computational domain and boundary conditions are shown in Fig. 7.5, where AB is the axisymmetric boundary, DE, CD and BC are vacuum boundaries, and EF is the surface of the enclosure. The DSMC mesh has 320 × 150 cells, and the PIC mesh has 640 × 300 cells.

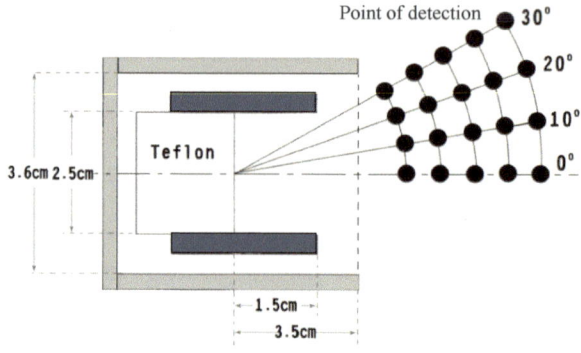

Fig. 7.4 Configuration of the PPT thrust chamber and its enclosure

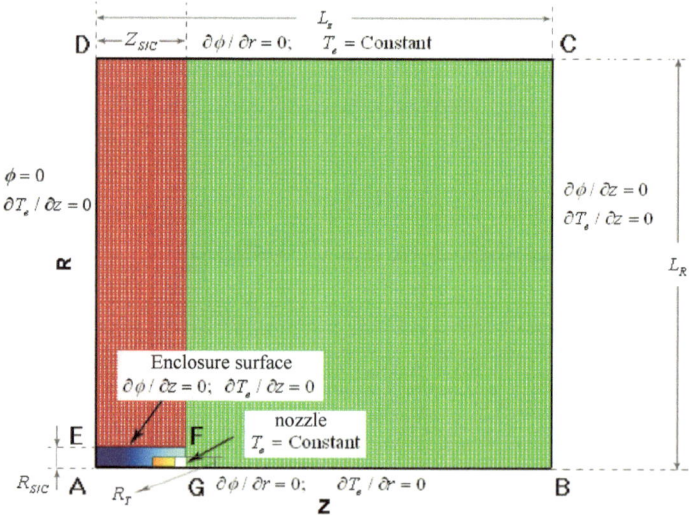

Fig. 7.5 Computational domain and boundary conditions

It is assumed that the plume only contains C, C^+, F and F^+, and the model of the thruster inlet used by Gatsonis is adopted. The electron background temperature is $Tbg\ e = 0.1$ eV, the maximum electron temperature is $Tmax\ e = 5$ eV, the ion background number density is $nbg\ i = 10^{12}$ m^{-3}, and the neutral particle background number density is $nbg\ n = 10^{15}$ m^{-3}.

$$n_s(r, z, t) = n_{s,\max} \sin\left[\frac{\pi}{P}(t - t_1)\right]\left[1 - (1 - C_c)\left(\frac{r}{R_T}\right)^2\right] \tag{7.92}$$

where pulse duration $P = t_1 - t_2$, and C_c is the density coefficient with a value of 0.1. We have

$$n_{s,\max} = \frac{M_s}{\frac{1}{4} W_s \bar{c}_s F(s_s) P(1 + \hat{C}_c) R_T^2} \tag{7.93}$$

The exit electron temperature is

$$T_e = (T_e^{\max} - T_e^{bg}) \sin\left[\frac{\pi}{P}(t - t_1)\right] + T_e^{bg} \tag{7.94}$$

Table 7.4 lists the specific inlet parameters. The mass fluxes of various components are shown in Fig. 7.6.

Table 7.4 Laboratory PPT plume simulation parameters

Known parameters			
$E_D(\mathrm{J})$	7.26	13.5	24
$P(\mu s)$	15	15	15
$I_{bit}(\mu \mathrm{N} \cdot \mathrm{s})$	72	193	343
$M_a(\mu g)$	11	25	40
Assumed parameters			
s	C^+, F^+	C^+, F^+	C^+, F^+
$M_i(\mu g)$	3.3	7.5	12
$u_i(\mathrm{km/s})$	8.8	12	18
$T_i(\mathrm{eV})$	1	1	1
$T_n(\mathrm{eV})$	1	1	1
Derived parameters			
$M_n(\mu g)$	7.7	17.5	28
$u_n(\mathrm{km/s})$	5.58	5.83	4.53
$n_{C\max}(\mathrm{m}^{-3})$	1.04×10^{21}	2.2×10^{21}	3.67×10^{21}
$n_{C^+\max}(\mathrm{m}^{-3})$	3.1×10^{20}	5.41×10^{20}	7.93×10^{20}
$n_{F\max}(\mathrm{m}^{-3})$	2.33×10^{21}	5.33×10^{21}	9×10^{21}
$n_{F^+\max}(\mathrm{m}^{-3})$	5.73×10^{20}	9.39×10^{20}	8.65×10^{20}

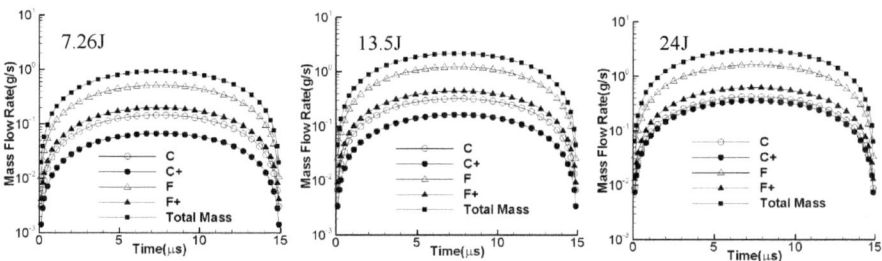

Fig. 7.6 Exit mass fluxes

7.2.2 Flow Field Analysis

At 7.26 J, the weight factor is 4×10^{10}. In the calculation cycle, there are a total of 7,606,894 simulated molecules, including 2,221,083 ions and 5,385,811 neutral particles, and a total of 35,052,749 collisions are calculated, of which there are 13,898,312 CEX collisions, accounting for 39% of the total number of collisions, and there are 4.608 collisions on average per simulated molecule. At 13.5 J, the weight factor is 1.1×10^{11}. In the calculation cycle, there are a total of 6,415,336 simulated molecules, including 1,845,943 ions and 4,569,393 neutral particles, a total of 58,545,918 collisions are calculated, of which there are 24,222,515 CEX collisions, accounting for 41% of the total number of collisions, and there are 9.126 collisions on average per simulated molecule. At 13.5 J, there is no CEX collision that adopts the same calculation parameters as those at 13.5 J, a total of 33,926,152 collisions are calculated, and there are 5.288 collisions on average per simulated molecule. At 24 J, the weight factor is 1.5×10^{11}. In the calculation cycle, there are a total of 6,680,657 simulated molecules, including 2,274,748 ions and 4,405,909 neutral particles, a total of 72,369,304 collisions are calculated, of which there are 32,228,576 CEX collisions, accounting for 44% of the total number of collisions, and there are 10.833 collisions on average per simulated molecule.

1. Plume Field Distribution

(1) Density Distribution

Figures 7.7, 7.8, 7.9, 7.10, 7.11 and 7.12 show the density distributions of ions and neutrals in the plume field at different initial capacitive energies. During the pulse discharge cycle, the plasma plume diffuses rapidly, with faster diffusion occurring at higher energy states.

The case of 13.5 J is analyzed as follows. At 2 μs, the front end of the ion cluster diffuses to a distance of 0.04 m from the exit of the thruster. Due to the acceleration effect of the electric field, this distance is longer than the distance traveled by the ions at the initial inlet velocity of 12 km/s. At the same time, the cluster of neutral particles diffuses to a distance of 0.03 m from the exit. This distance is far greater than the distance traveled at the initial inlet velocity of 5.83 km/s. These high-velocity neutral particles are produced by CEX collisions. At 7.5 μs, the front end of the ion cluster moves to a distance of 0.16 m from the exit, while the neutral particles move to a distance of 0.12 m. The radial diffusion distance of the ion cluster is approximately 0.1 m from the central axis, and the radial diffusion distance of the neutral particle cluster is approximately 0.1 m. At this time, the ion backflow has already started to occur. At the end of the cycle, the front end of the ion cluster has diffused to 0.32 m, a distance twice that traveled in half a cycle, and the neutral particle cluster has diffused to 0.26 m, which is more than twice the distance traveled in half a cycle. Radially, both the ion cluster and the neutral particle cluster diffused to a distance of approximately 0.2 m from the central axis, twice the distance traveled in half a cycle. At this time, both ions and neutral particles exhibit backflow phenomena. The neutral particles generated by CEX collisions have a high velocity. Therefore, the

Fig. 7.7 Ion distribution at
7.26 J (m^{-3})

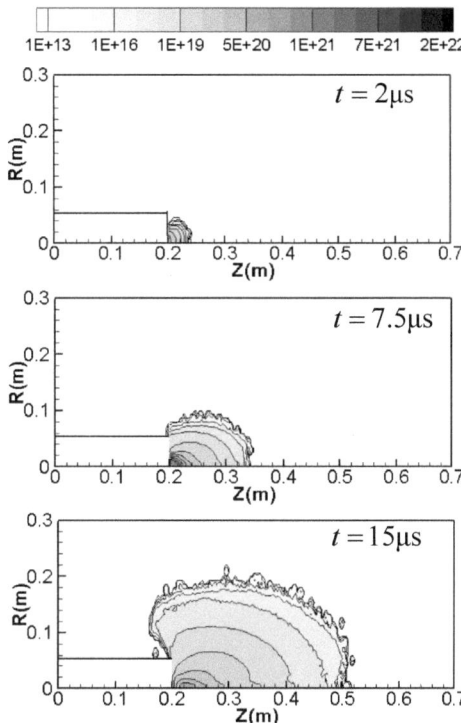

neutral particle cluster has almost the same axial and radial diffusion distances as those of the ion cluster.

The cases of 7.26 and 24 J A are compared as follows. The ion exit velocity at 7.26 J is 49% of that at 24 J. At 2 μs, the diffusion distance of the front end of the ion cluster at 7.26 J (0.04 m) is approximately 67% of that at 24 J (0.06 m); at 7.5 μs, the diffusion distance at 7.26 J (0.14 m) is approximately 70% of that at 24 J (0.2 m); and at 15 μs, the diffusion distance at 7.26 J (0.31 m) is approximately 73% of that at 24 J (0.42 m). As a comparison, for the neutral particle cluster, at 2 μs, the diffusion distance at 7.26 J (0.035 m) is approximately 87% of that at 24 J (0.04 m); at 7.5 μs, the diffusion distance at 7.26 J (0.12 m) is approximately 80% of that at 24 J (0.15 m); and at 15 μs, the diffusion distance at 7.26 J (0.24 m) is approximately 72% of that at 24 J (0.33 m). Regarding the radial diffusion distances, for the ion cluster, at 7.5 μs, the diffusion distance at 7.26 J (0.1 m) is approximately 83% of that at 24 J (0.12 m), and at 15 μs, the diffusion distance at 7.26 J (0.2 m) is approximately 83% of that at 24 J (0.24 m). In comparison, for the neutral particle cluster, at 7.5 μs, the diffusion distance at 7.26 J (0.09 m) is approximately 81% of that at 24 J (0.11 m), and at 15 μs, the diffusion distance at 7.26 J (0.19 m) is approximately 79% of that at 24 J (0.24 m). The ejection velocity of ions in the high-energy state is higher than that in the low-energy state, and the ions diffuse faster along the axial direction. When the initial voltage is high, the content of ionized components in the products is greater,

Fig. 7.8 Ion distribution at 13.5 J (m^{-3})

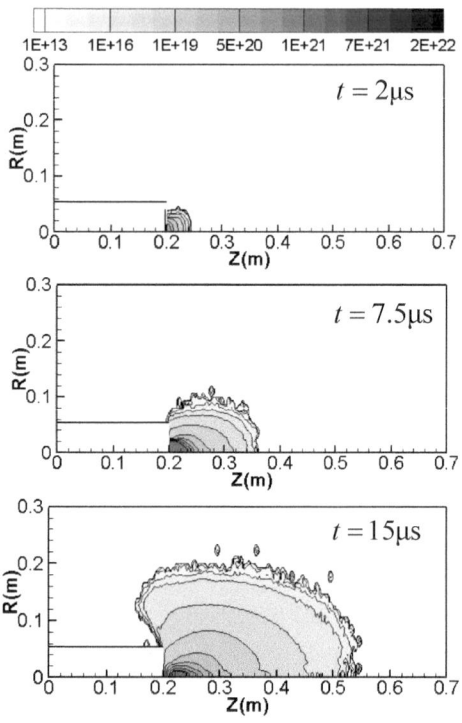

and, at this time, the frequency of CEX collisions is higher, meaning the formation of a larger proportion of low-velocity ions and high-velocity neutrals. Therefore, the axial acceleration of ions in the high-energy state is lower than that in the low-energy state, while the axial acceleration of neutral particles in the high-energy state is higher than that in the low-energy state. In the radial direction, the acceleration of ions is mainly caused by the electric field. These ions have very similar diffusion rates in different energy states, while high-energy neutral particles gain more acceleration due to CEX collisions. Therefore, the radial acceleration of the neutral particles is higher in the high-energy state than in the low-energy state. In different energy states, at the end of the calculation time, both the ion cluster and the neutral particle cluster have different degrees of backflow, and the backflow of the ion cluster occurs earlier than the backflow of the neutral cluster. Ions have higher axial velocities in the high-energy state, and therefore, the ion backflow in the high-energy state is lower than that in the low-energy state. Moreover, the higher CEX collision rate under a high initial voltage results in a higher backflow of neutral particles than that under a low initial voltage.

(2) Electric Potential Field Distribution

Figures 7.13, 7.14 and 7.15 show the electric potential field distributions at different initial capacitive energies. The electric potential field changes rapidly as the ion

Fig. 7.9 Ion distribution at
24 J (m^{-3})

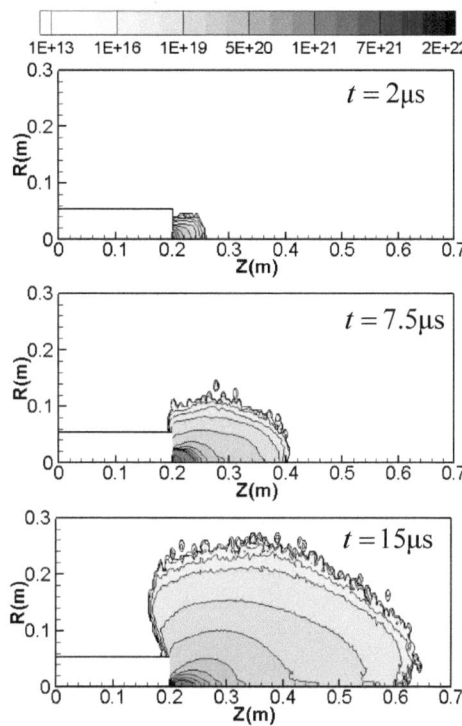

cluster diffuses. A high ion density corresponds to a high potential. There is a high
potential inside the ion cluster. The high potential (~50 V) rapidly decreases in
the neutral high ion density region of the plasmoid to below 10 V outside the ion
cluster. At the beginning of the pulse discharge, the plasmoid has just formed and
is located at the exit of the thruster, forming a high potential region at the exit.
Then, this high potential region diffuses rapidly as the ion cluster diffuses. In the
backflow region of the thruster, the potential gradually decreases to 0.1 V. A drastic
potential change occurs inside the ion cluster, while the variation in the potential
field outside the ion cluster is relatively slow. An analysis of the electric field based
on the electric potential field distribution reveals the following: radially, there is an
upward accelerating electric field; axially, downstream of the thruster exit, there is
an accelerating electric field along the axis; and upstream of the exit in the backflow
region, there is a decelerating electric field along the axis. For the backflow cluster,
there is an electric field that accelerates backflow particles. This region also has a high
electric potential, which is a detrimental factor leading to the backflow contamination
of the plume.

(3) Electron Temperature Distribution

Figures 7.16, 7.17 and 7.18 show the electron temperature distribution at different
initial capacitive energies. The electron temperature exhibits a diffused distribution

Fig. 7.10 Neutral particle
distribution at 7.26 J (m^{-3})

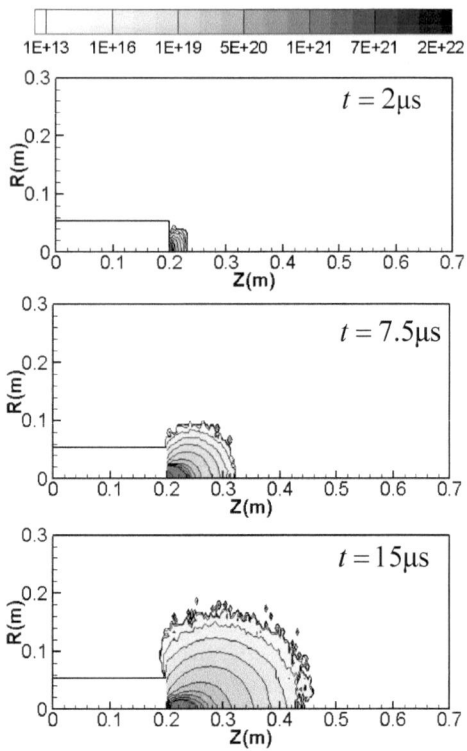

along the exit toward the thruster casing. Unlike the electric potential distribution, the electron temperature distribution at the outer edge of the plume is smoother, with more noticeable variations in the first half of the cycle. According to the inlet conditions, the maximum exit electron temperature occurs at 7.5 μs of the half-cycle, which is also the maximum electron temperature in the flow field during the cycle. At the end of the cycle, the electron temperature at the exit is lower than the ambient temperature. The electron temperature at the exit of the thruster is the main factor affecting the entire electron temperature field. It has similar distribution changes in different energy states. The influence of the temperature at the exit continuously diffuses to the surrounding area. The electron temperature of the backflow region at the exit of the thruster is high. Moreover, the plume diffuses faster in the high-energy state, causing the influence domain of the electron temperature to shift downstream. Therefore, the electron temperature in the backflow region in the high-energy state is lower than that in the high-energy state.

2. Particle Velocity

(1) Velocity Distribution

Figures 7.19, 7.20, 7.21, 7.22, 7.23 and 7.24 show the velocity distributions of ions and neutral particles at different initial capacitive energies. For both ions and neutral

Fig. 7.11 Neutral particle
distribution at 13.5 J (m^{-3})

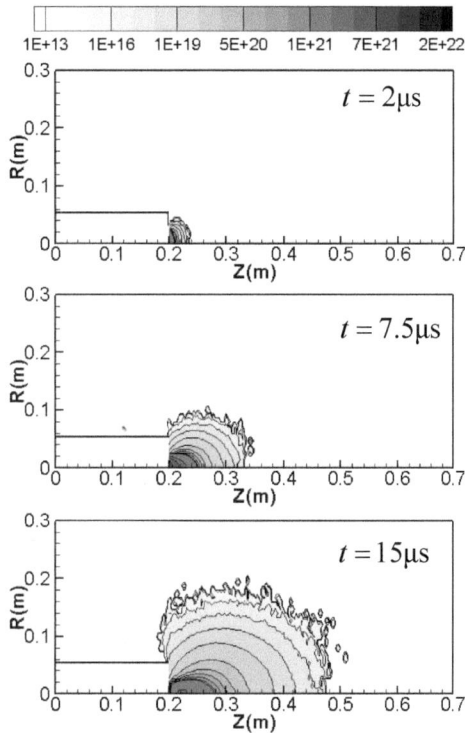

particles, the minimum velocity is located near the exit of the thruster. The ions and neutral particles have approximately equal velocities at the same position inside the plume. Analysis of the case of 13.5 J at 15 μs shows that at the exit, the ion velocity is approximately 9 km/s, which is lower than the exit velocity of 12 km/s, while the neutral particle velocity is approximately 6.5 km/s, which is greater than the exit velocity of 5.58 km/s. This is because the electrons at the exit have high temperature and density, resulting in a high frequency of CEX collisions, generating a large amount of high-velocity neutral particles and low-velocity ions and thus causing the velocities of ions and neutral particles at the exit tend to be equal. These particles in the plasmoid are the main source of particles in the backflow cluster. As the plume diffuses, the electron temperature and density gradually decrease. The frequency of CEX collisions also decreases. Therefore, the frequency of velocity exchange between ions and neutral particles is low. An analysis of backflow streamlines shows that in the backflow region, the backflow plasma has a low density, the probability of CEX collisions is low, and the velocity equilibrium between ions and neutral particles cannot be reached. Therefore, the ion backflow velocity is higher. The influence region of charged particles extends within a range of 130° counterclockwise from the thruster exit plane, while the backflow of neutral particles has a low diffusion rate, and the corresponding influence region of extends within a range of 110° counterclockwise from the thruster exit plane. The impact of high-velocity particles and

Fig. 7.12 Neutral particle
distribution at 24 J (m^{-3})

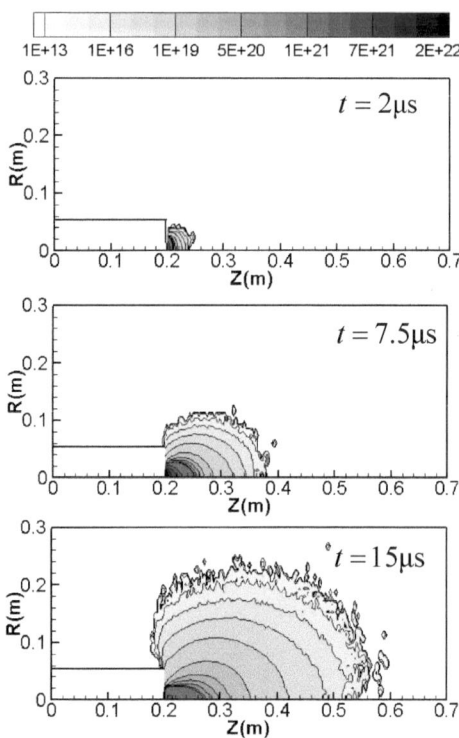

the deposition, adsorption and charging of charged particles may all cause certain damage to a spacecraft. The influence region and degree of the backflow of charged particles is far greater than that of the backflow of neutral particles, making it the main factor of backflow contamination.

(2) Velocity Sampling

Figures 7.25, 7.26, 7.27, 7.28, 7.29 and 7.30 show the velocity samples of ions and neutral particles at different initial capacitive energies. An analysis of the axial velocity shows that at the beginning of the cycle, a large number of low-velocity ions and high-velocity neutral particles are generated. This is due to the high density and high temperature of the plasma at the exit, resulting in a large number of CEX collisions. At the end of the cycle, a certain number of particles with reverse axial velocities are generated, including many ions. These ions have a high forward radial velocity. An analysis of the radial velocity shows that at the beginning of the cycle, the average velocity is 0 km/s. As time progresses, due to the radial acceleration of the electric field, the average radial velocity of the ions increases. Due to the velocity balance caused by CEX collisions, the average radial velocity also increases. At the same time, at the edge of the plume, due to the decrease in the probability of CEX

Fig. 7.13 Potential
distribution (V) at 7.26 J

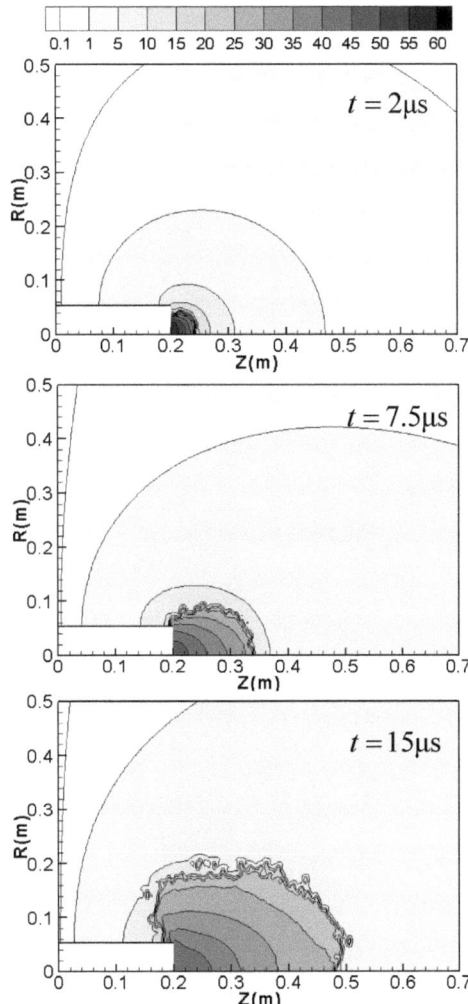

collisions, the CEX collisions between ions and neutral particles decrease. Therefore, the ions have a higher velocity than do the neutral particles. A comparison of the cases of 7.26 and 24 J at 15 μs shows that the maximum sample axial velocity of ions at 7.26 J (24 km/s) is 80% of that at 24 J (30 km/s), the maximum sample radial velocity of ions at 7.26 J (16 km/s) is 80% of that at 24 J (20 km/s), the maximum sample axial velocity of neutral particles at 7.26 J (20 km/s) is 77% of that at 24 J (26 km/s), and the maximum sample radial velocity of neutral particles at 7.26 J (12 km/s) is 71% of that at 24 J (17 km/s). These results correspond to the plots showing the distributions of ions and neutral particles in the plume field.

Fig. 7.14 Potential
distribution (V) at 13.5 J

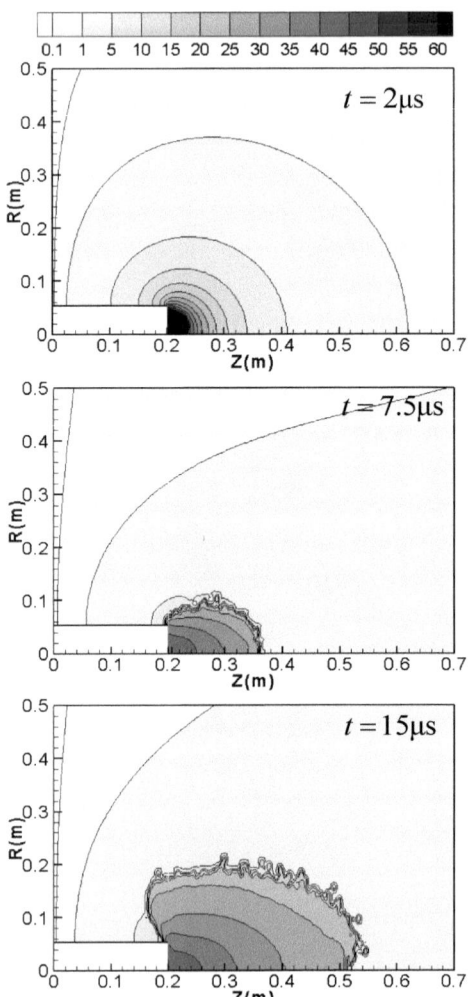

3. Mass Fluxes

(1) Axial Mass Fluxes

Figures 7.31, 7.32 and 7.33 show the axial mass fluxes at different initial capacitive energies. Near the exit, F atoms have the highest mass flux, followed by F^+, C, and C^+. Along the axial direction, the number of CEX collisions gradually decreases, and the velocity of ions continuously increases. At the tail of the plume, the F^+ ions become the dominant particle flow, followed by C^+, F, and C. Before the half cycle, the axial mass flux at the exit is the highest mass flux. After the half cycle, due to the

Fig. 7.15 Potential
distribution (V) at 24 J

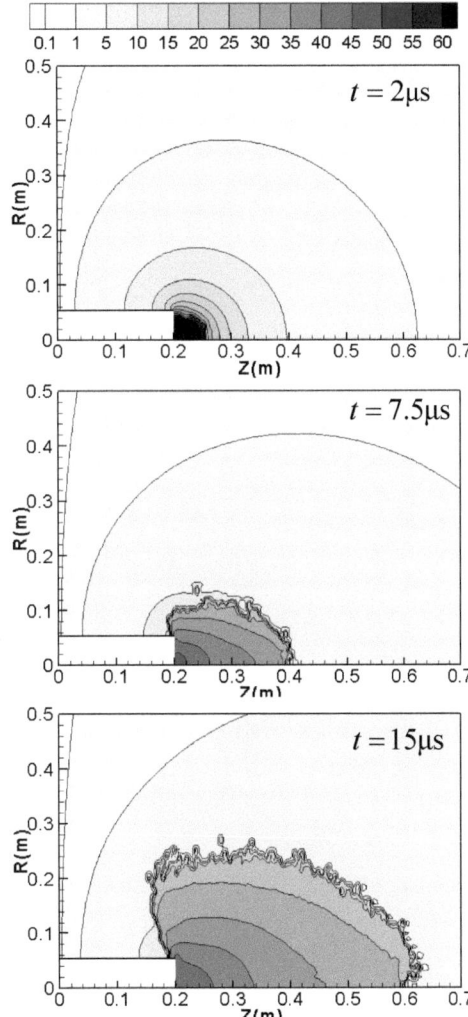

decreasing density at the exit, the mass flux increases sequentially and then gradually decreases.

The difference between the maximum and minimum mass fluxes is nearly six orders of magnitude. A comparison of the cases of 7.26 and 24 J shows that during the half cycle, the ion cluster diffuses to a distance of 0.14 and 0.2 m from the exit at 7.26 and 24 J, respectively, and the atom cluster diffuses to a distance of 0.12 and 0.15 m from the exit at 7.26 J and 24 J, respectively. At the end of the cycle, the ion cluster diffuses to a distance of 0.31 and 0.42 m from the exit at 7.26 and 24 J, respectively, and the atom cluster diffuses to a distance of 0.24 and 0.33 m from the exit at 7.26 J and 24 J, respectively. The value of the minimum mass flux at the

Fig. 7.16 Electron
temperature distribution at
7.26 J (eV)

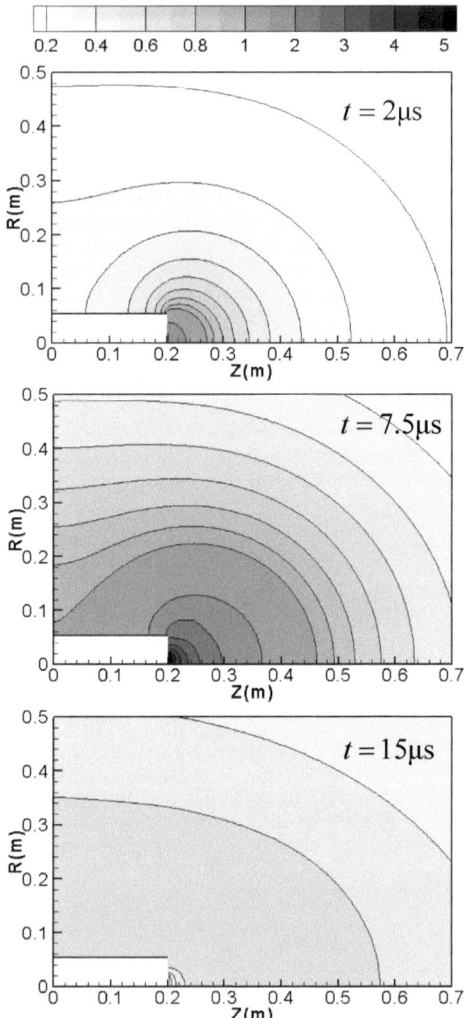

end of the cycle is greater than that in the low-energy state. At the end of the cycle, due to the statistical fluctuations of the particles, the mass flux oscillates within a certain range. The magnitude of the mass flux is determined by both the density and velocity. In the high-energy state, with higher density and velocity, the mass flux is relatively high. At the same time, the high rate of CEX collisions caused by the high degree of ionization makes the mass fluxes of ions and neutral particles on the axis more similar.

(2) Backflow Mass Flux

Figures 7.34, 7.35 and 7.36 show the backflow mass fluxes above the exit plane of the thruster at different initial capacitive energies. The backflow does not appear above

Fig. 7.17 Electron
temperature distribution at
13.5 J (eV)

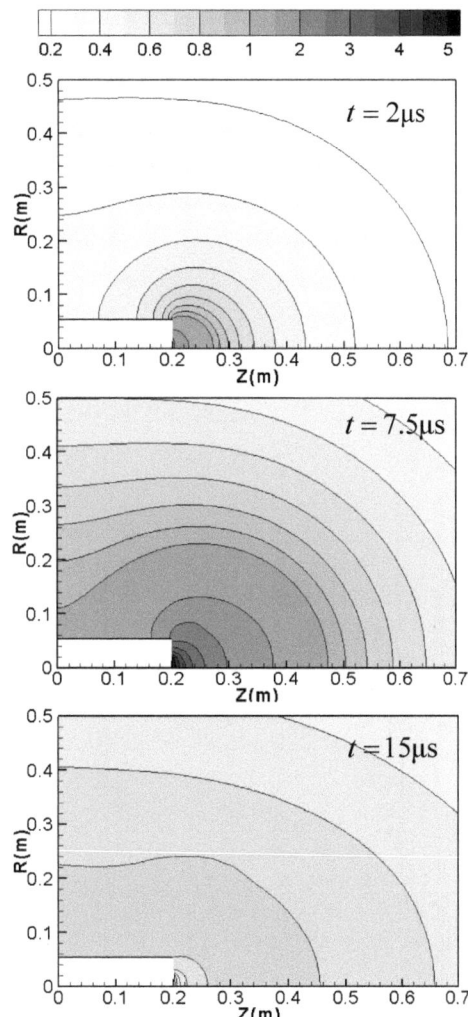

the thruster until after a certain ejection time, i.e., 6.4 μs at 7.26 J, 5.3 μs at 13.5 J, and 3.9 μs at 24 J. The backflow of charged particles begins first. Because the C atom has the largest charge-to-mass ratio, the backflow of C^+ occurs first, followed by the backflows of C, F^+, and F. The backflow continuously diffuses toward the backflow region over time, while the highest mass flux is maintained at a certain level after gradually increasing. The cases of 7.26 and 24 J are compared as follows. At 7.5 μs, 10^{-5} kg/m^3 s backflows of C^+ and C occur at 7.26 J; at this time, the backflows of C^+, F^+, C, and F have already appeared at 24 J, with the maximum backflow of 10^{-4} kg/m^3·s. At 11 μs, the backflow at 7.26 J reaches 6 cm from the thruster exit along the radial direction, with a mass flux of 10^{-4} kg/m^3 s, while the backflow at 24 J reaches 11 cm from the thruster exit in the radial direction, with the mass flux of 10^{-3} kg/m^3

Fig. 7.18 Electron
temperature distribution at
24 J (eV)

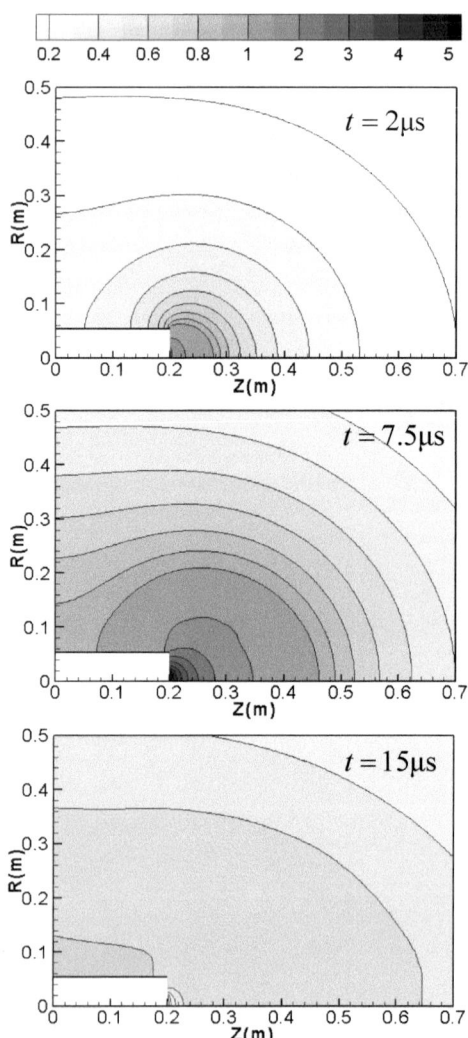

s. At 15 μs, the backflow at 7.26 J reaches 12 cm, with a mass flux of 10^{-4} kg/m^3·s, while the backflow at 24 J reaches 16 cm, with the mass flux maintained at 10^{-3} kg/m^3 s.

At 7.26 J, C$^+$ has the highest mass flux of at 2×10^{-4} kg/m^3 s. After the mass flux of ions is maintained at a certain level, the mass flux of neutral particles continues to increase steadily. C has the highest mass flux of neutral particles of 3×10^{-5} kg/m^3 s. At 24 J, C$^+$ has the highest mass flux of 10^{-3} kg/m^3 s; the trend for neutral particles is similar to that at 7.26 J; and C has the highest mass flux of neutral particles, with a value of 8×10^{-4} kg/m^3 s. It can be observed that the edge of the plume is mainly composed of high-velocity charged particles, while in the central part, ions

Fig. 7.19 Ion velocity
distribution at 7.26 J (km/s)

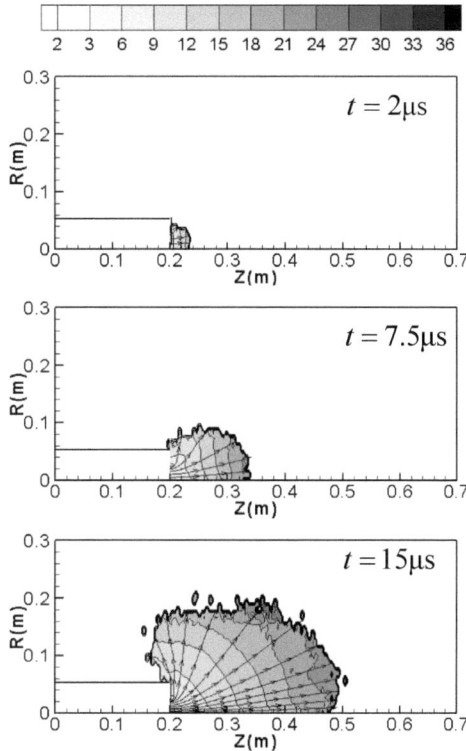

and neutral particles have similar velocities due to CEX collisions. Charged particles
are the factor affecting the backflow region initially. However, as time passes, the
influence of the neutral particles should not be ignored. A comparison of different
energy states shows that the inlet velocity of neutral particles at 24 J is the lowest.
During the early stage of discharge, due to CEX collision, CEX ions with the lowest
velocity are generated at 24 J; thus, the ion backflow occurs first. The occurrence
time of backflow largely depends on the exit ejection velocity of neutral particles.
In addition, the high degree of ionization at 24 J increases the occurrence of CEX
collisions, which makes the increase in the neutral particle backflow relative to the
increase in the ion backflow at the end of the calculation time the highest among all
energy states.

7.2.3 CEX Collision Analysis

CEX collisions result in the exchange of charges in the plume, which directly affects
the composition changes of the plume. Here, we perform a special analysis on CEX
collisions.

Fig. 7.20 Ion velocity
distribution at 13.5 J (km/s)

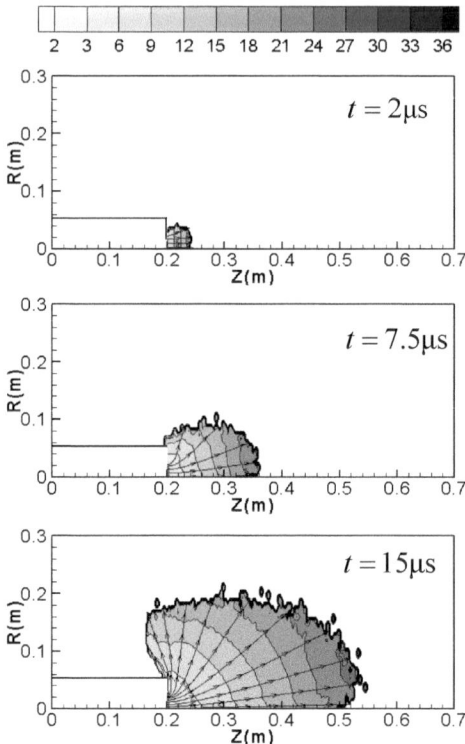

1. CEX Collisions

Figures 7.37, 7.38, 7.39, 7.40, 7.41 and 7.42 show the proportions of the ions and
neutral particles that have undergone CEX collisions in the computational cells in
each energy state given the total numbers of local ions and neutral particles in the
vicinity. As shown in these Figures, the proportion of CEX ions is high near the exit
while the proportion of CEX neutral particles is high within the plume downstream
of the exit. A comparison of the distributions in different energy states reveals that
the proportion of CEX particles in the flow field increases in the high-energy state,
regardless of whether they are ions or neutral particles. Previous statistical calculations on the number of CEX collisions have shown that in the high-energy state,
there is a higher probability of CEX collisions due to the highest degree of ionization,
resulting in an increased proportion of particles undergoing CEX collisions.

Fig. 7.21 Ion velocity distribution at 24 J (km/s)

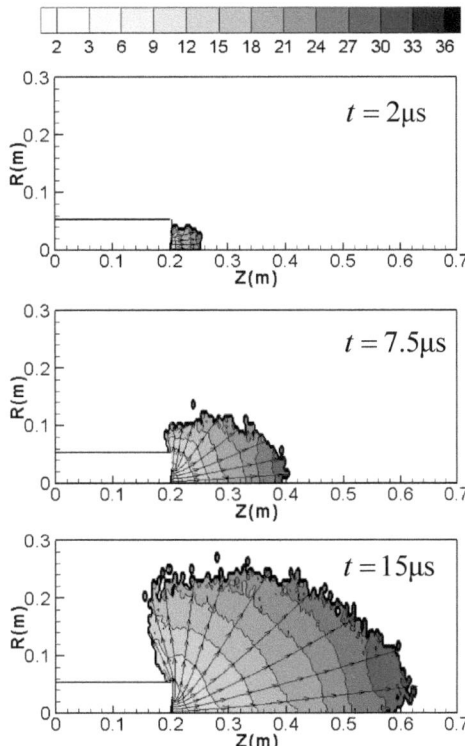

Figures 7.43, 7.44, 7.45, 7.46, 7.47 and 7.48 show the velocity distributions of CEX ions and CEX neutral particles in different energy states. The velocity distributions of ions and neutral particles are similar to the velocity distribution of total particles, with the minimum velocities located near the exit of the thruster. A comparison of cases of different energies reveals that the exit velocity at high energy is greater, and therefore, the velocity is higher. An analysis of the case of 13.5 J at 15 μs shows that at the exit, the CEX ions have a velocity of 8 km/s, which is lower than the average ion velocity of 9 km/s, while the CEX neutral particle have a velocity of 7 km/s, which is higher than the average neutral particle velocity at the exit. CEX collisions lead to a decrease in the ion velocity but an increase in the neutral particle velocity.

Figures 7.49 and 7.50 show the proportions of CEX collisions and non-CEX collisions among the total collisions in the axial cells in the energy state of 13.5 J, respectively. A comparison with Fig. 7.38 reveals that particle collisions mainly occur in regions with high density near the plume exit. From Eqs. (7.41) and (7.44), it is found that the cross section of elastic collision momentum exchange between ions and neutral particles decreases as the velocity difference between the two increases, while the CEX collision cross section increases with increasing velocity difference between the two. Near the exit, due to the high density and high collision frequency,

Fig. 7.22 Neutral particle velocity distribution at 7.26 J (km/s)

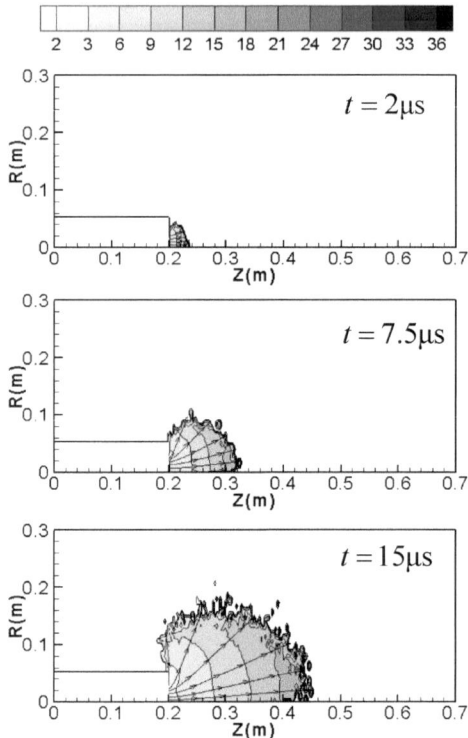

ions and neutral particles undergo frequent velocity exchanges. At this time, ions are just beginning to accelerate in the electric field. Therefore, the proportion of non-CEX collisions among all collisions is greater. As the plume continuously diffuses, ions are gradually accelerated, and at the same time, the decreases in density and temperature lead to a decrease in the collision frequency, therefore CEX collisions finally occur.

There are two types of CEX collisions: those between high-velocity ions and low-velocity neutral particles, referred to as CEX1 collisions, and those between low-velocity ions and high-velocity neutral particles, referred to as CEX2 collisions. Figures 7.51 and 7.52 show the proportions of these two types of collisions on the axis in the CEX collisions in the energy state of 13.5 J. It is observed that one type of collision primarily occurs in the first 2 μs of the pulse, which corresponds to the fact that the ion concentration is higher than the neutral particle concentration at the beginning of the pulse. After half a cycle, the densities of both components reach a certain level. Therefore, the occurrences of two types of collisions are essentially comparable. In addition, according to Figs. 7.49 and 7.50, a large number of non-CEX collisions also occur at the early stage of the pulse. Under the concentrated action of various collisions, the velocity distribution of CEX particles is similar to that of the total particles (Figs. 7.53 and 7.54).

Fig. 7.23 Neutral particle velocity distribution at 13.5 J (km/s)

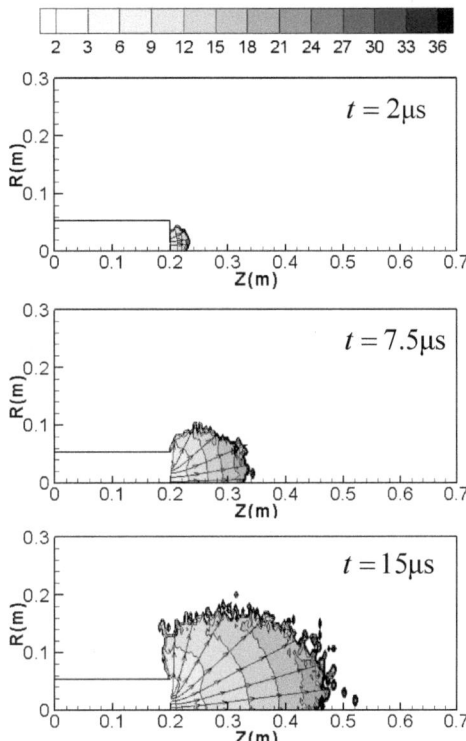

2. Calculation Without CEX Collisions

To further investigate the influence of CEX collisions on the plume, a calculation without CEX collisions is performed for the case of 13.5 J.

Figure 7.55 shows the axial mass flux without CEX collisions. Comparing with Fig. 7.32, each component exhibits a similar variation pattern, and the plume has basically the same diffusion rate along the axis. However, the mass flux of each component is greater when there is no CEX collision. The distributions of ions and neutral particles at 15 μs are compared under the two conditions (Figs. 7.56, 7.48, and 7.11). In the absence of CEX collisions, ions have a longer radial diffusion distance, while the distance for neutral particles is shorter. CEX collisions enable neutral particles to reach a high radial velocity, which promotes the radial diffusion of the plume, thereby leading to a decrease in the density of each component in the axial direction and thus affecting the axial mass flux.

Fig. 7.24 Neutral particle velocity distribution at 24 J (km/s)

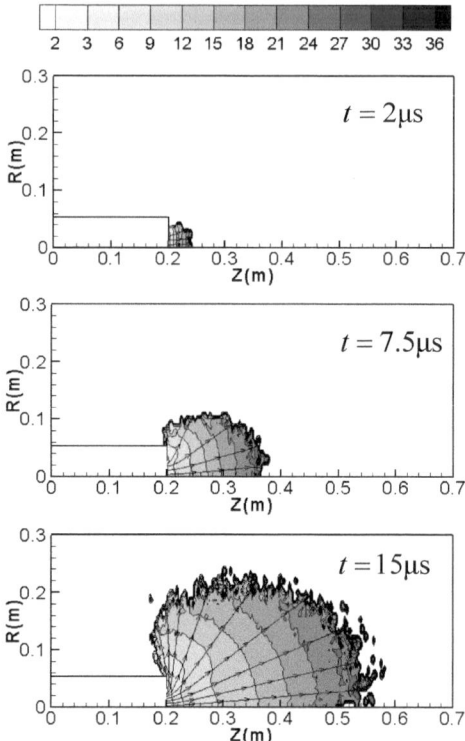

Figure 7.57 shows the backflow mass flux above the exit plane of the thruster without CEX collisions. Even in the absence of CEX collisions, the backflow occurs still at 5.3 μs. A comparison with Fig. 7.35 reveals that each component of the backflow particles has roughly similar behavior, and the part of particles introduced by radial diffusion increases the density of the backflow particles. As a result, the plume without CEX collisions has a higher mass flux and faster diffusion.

Fig. 7.25 Ion velocity
sampling at 7.26 J (km/s)

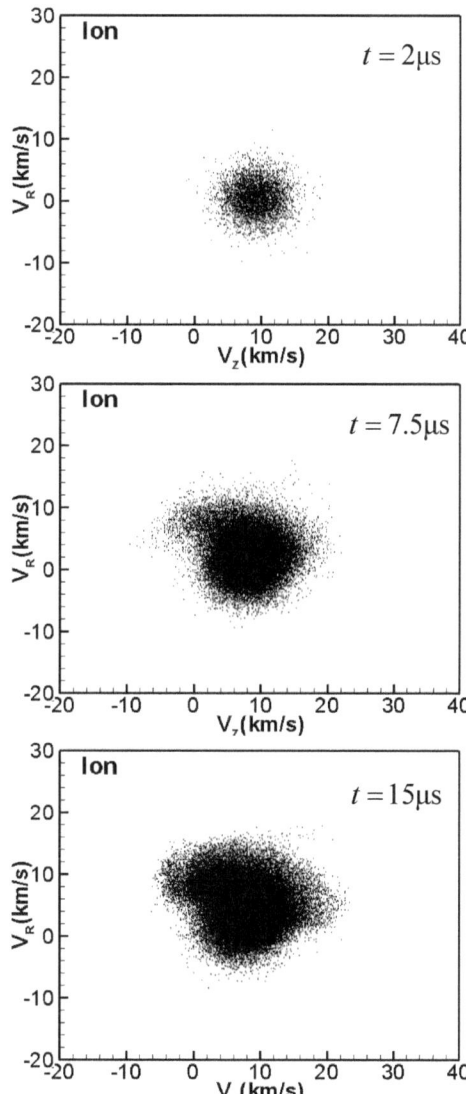

Fig. 7.26 Ion velocity
sampling at 13.5 J (km/s)

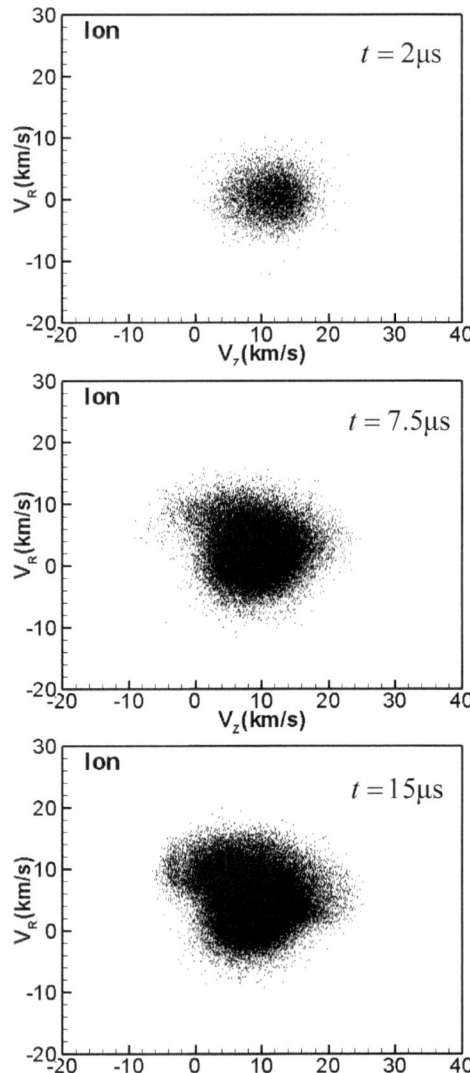

Fig. 7.27 Ion velocity
sampling at 24 J (km/s)

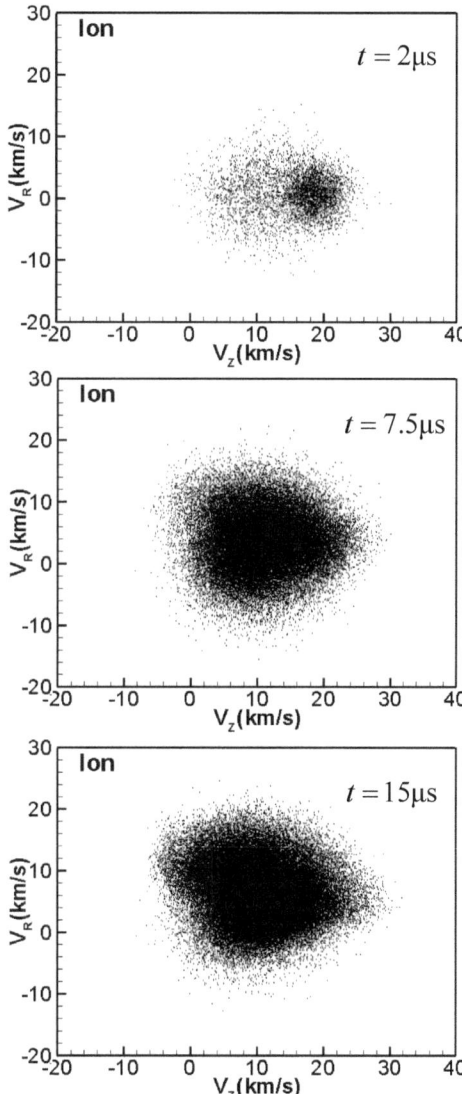

Fig. 7.28 Neutral particle
velocity sampling at 7.26 J
(km/s)

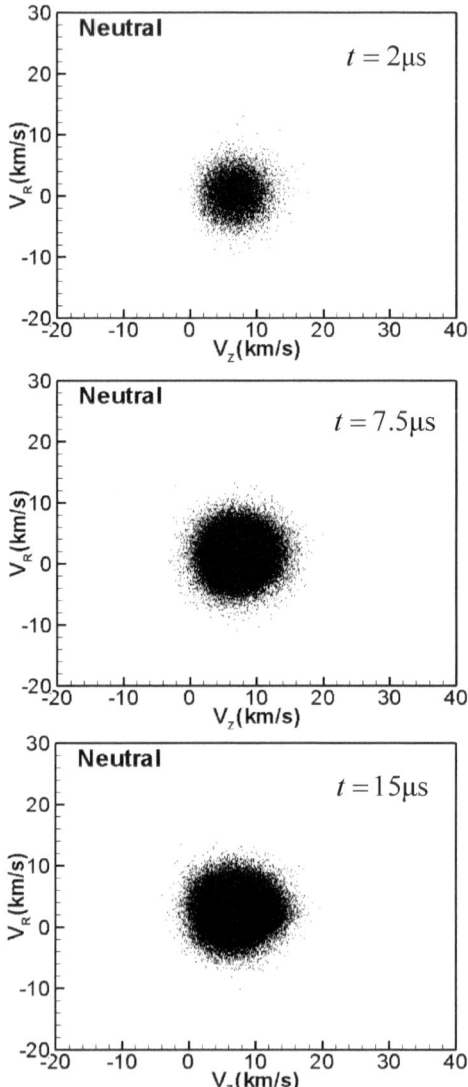

Fig. 7.29 Neutral particle
velocity sampling at 13.5 J
(km/s)

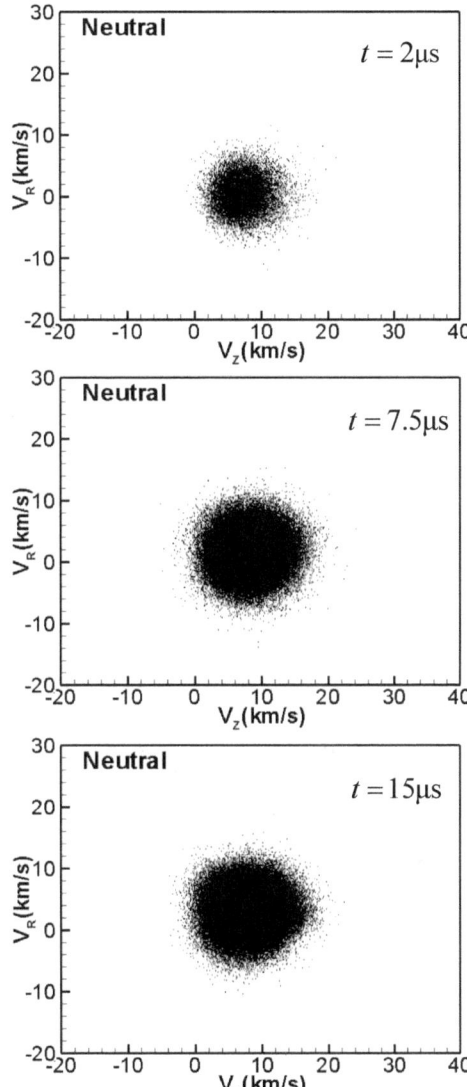

Fig. 7.30 Neutral particle velocity sampling at 24 J (km/s)

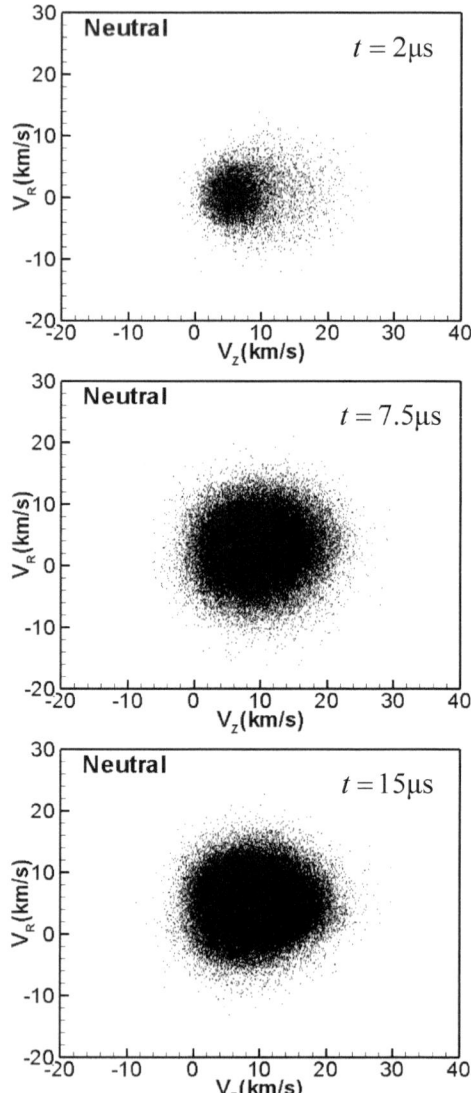

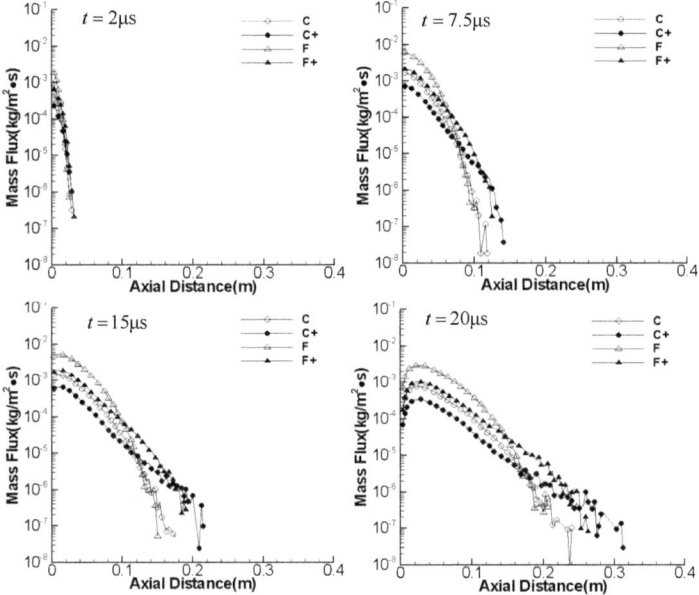

Fig. 7.31 Axial mass fluxes at 7.26 J

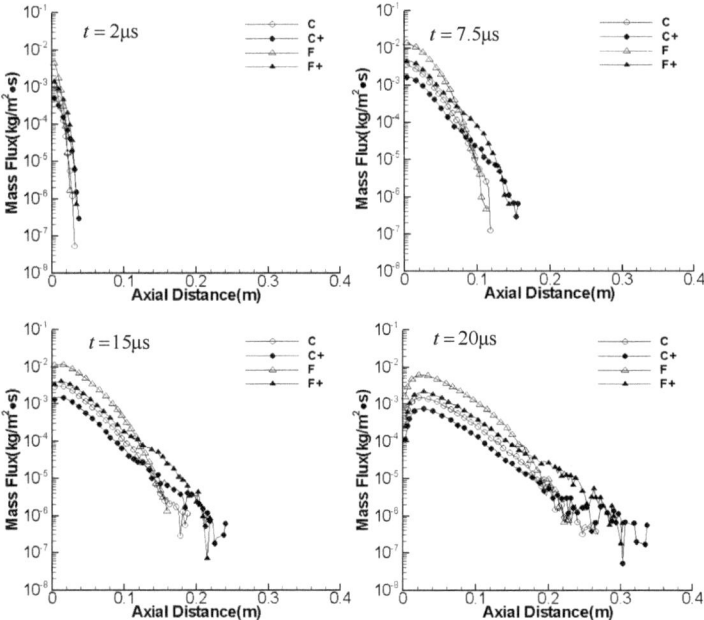

Fig. 7.32 Axial mass fluxes at 13.5 J

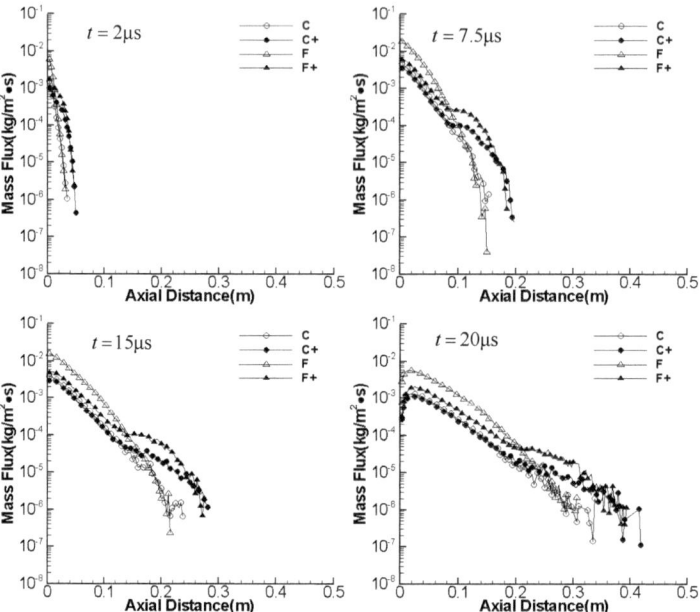

Fig. 7.33 Axial mass fluxes at 24 J

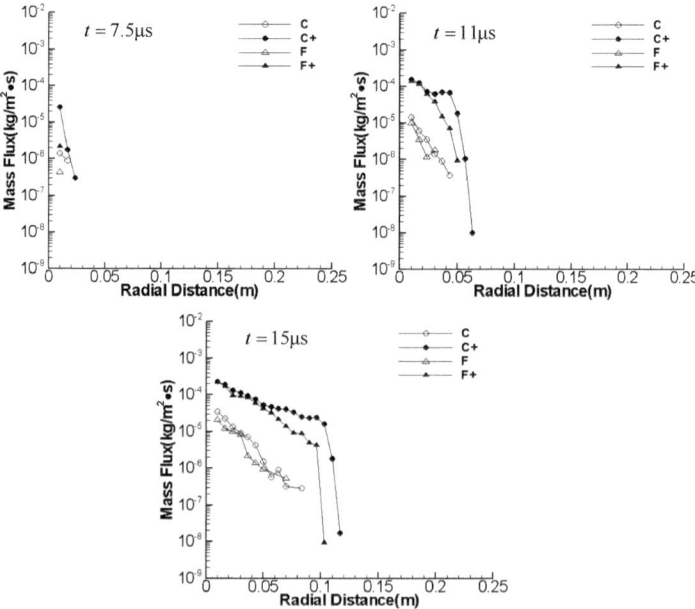

Fig. 7.34 Backflow mass fluxes at the thruster exit at 7.26 J

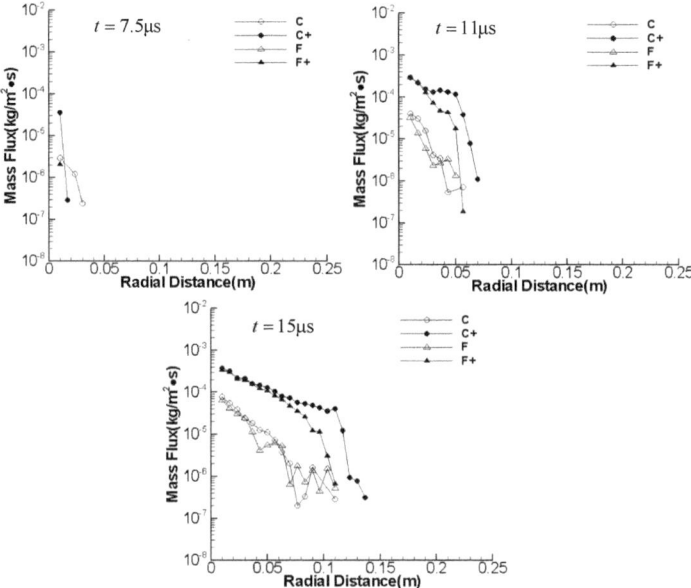

Fig. 7.35 Backflow mass fluxes at the thruster exit at 13.5 J

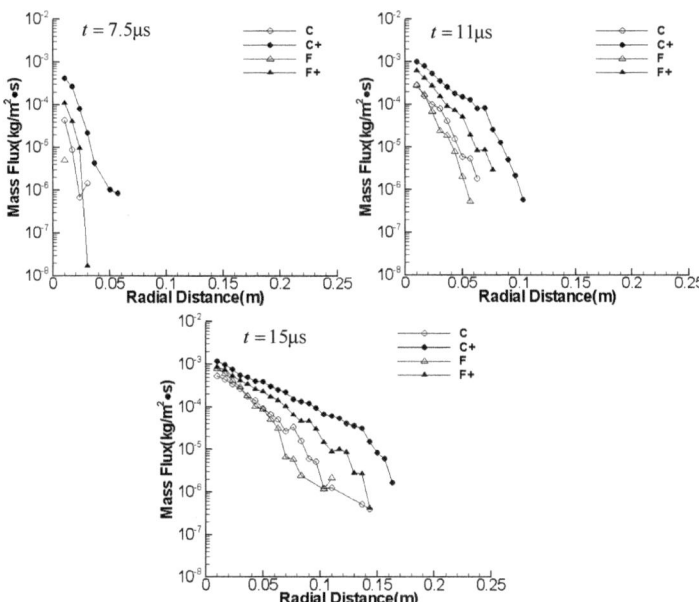

Fig. 7.36 Backflow mass fluxes at the thruster exit at 24 J

Fig. 7.37 Proportion of
CEX ions at 7.26 J

Fig. 7.38 Proportion of
CEX ions at 13.5 J

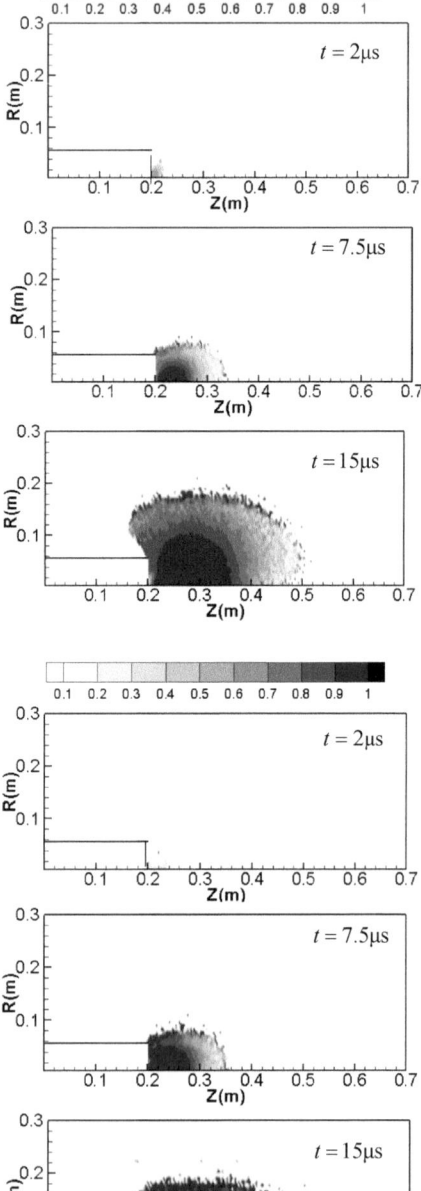

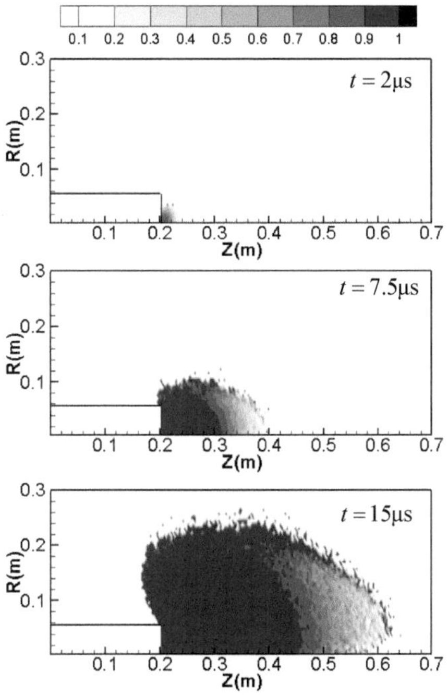

Fig. 7.39 Proportion of CEX ions at 24 J

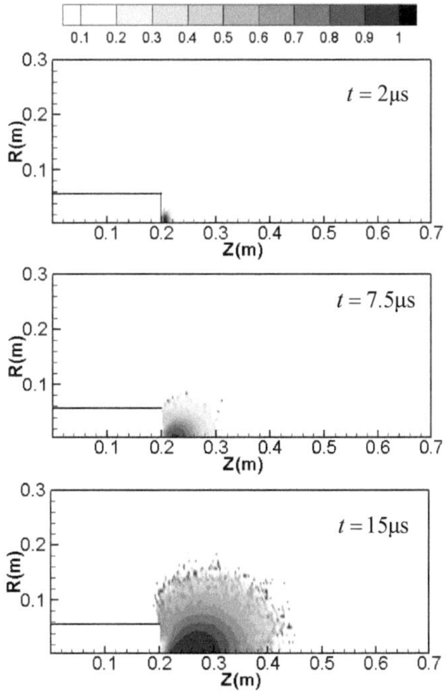

Fig. 7.40 Proportion of CEX neutral particles at 7.26 J

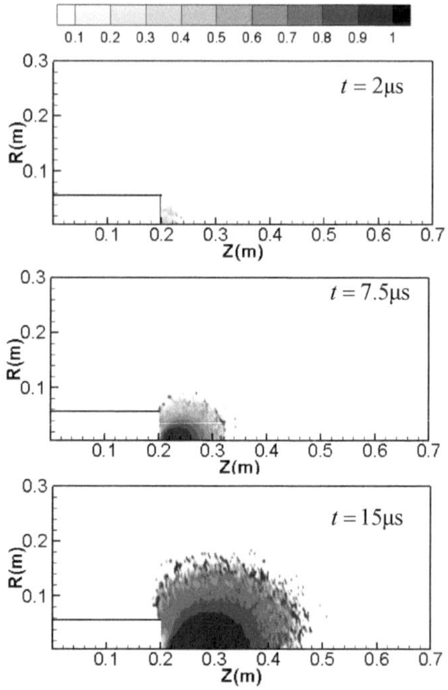

Fig. 7.41 Proportion of
CEX neutral particles at 13.5
J

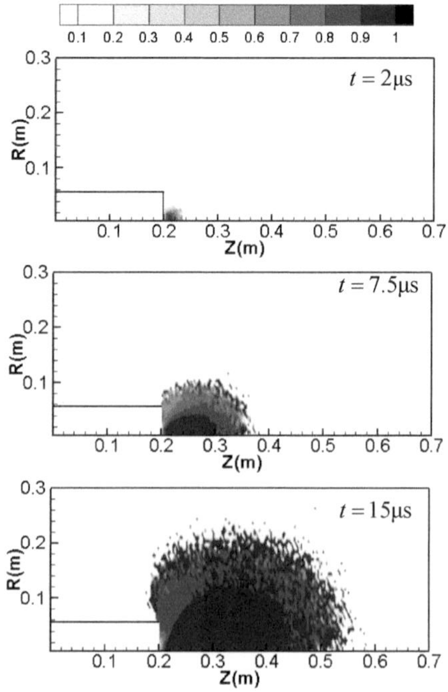

Fig. 7.42 Proportion of
CEX neutral particles at 24 J

Fig. 7.43 CEX ion velocity
at 7.26 J (km/s)

Fig. 7.44 CEX ion velocity
at 13.5 J (km/s)

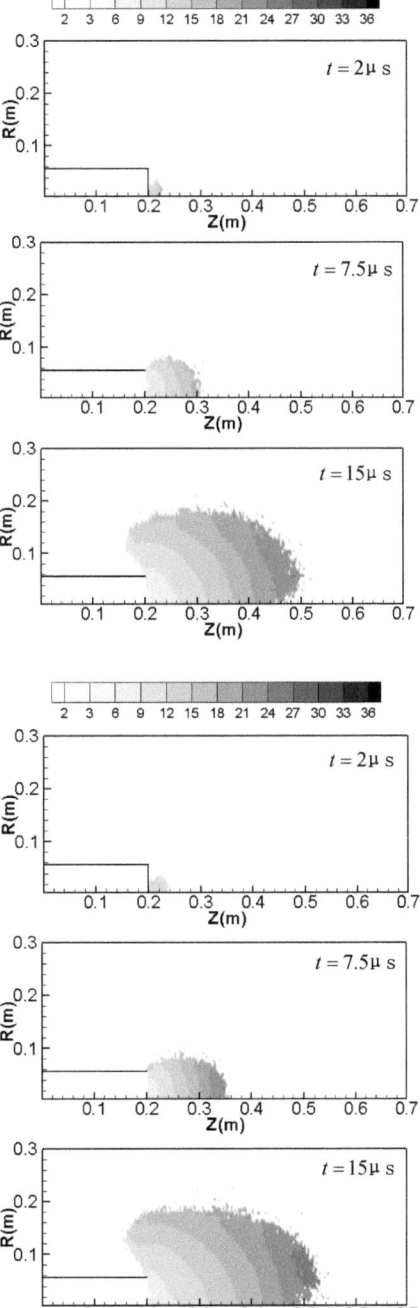

Fig. 7.45 CEX ion velocity
at 24 J (km/s)

Fig. 7.46 CEX neutral
particle velocity at 7.26 J

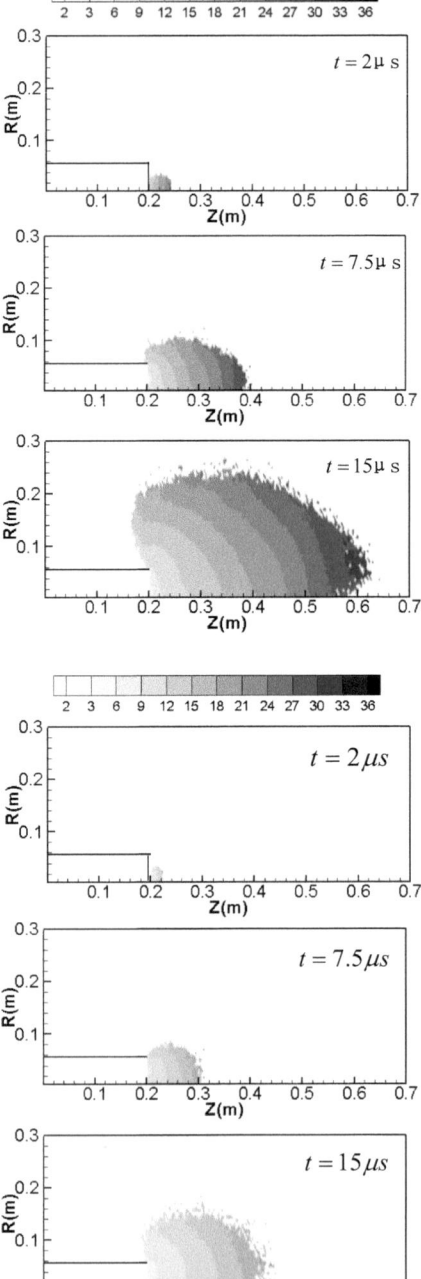

Fig. 7.47 CEX neutral
particle velocity at 13.5 J

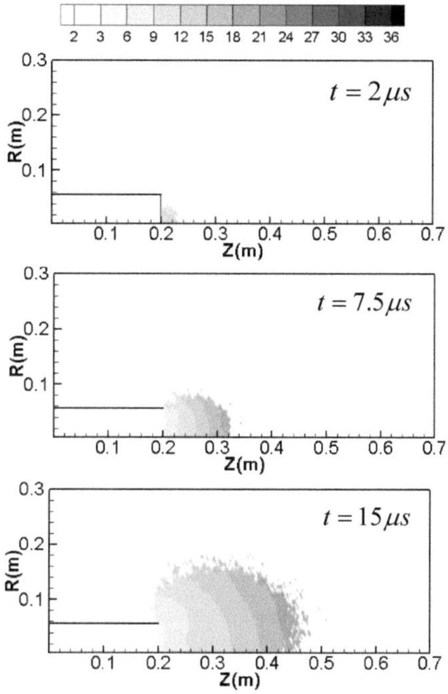

Fig. 7.48 CEX neutral
particle velocity at 24 J

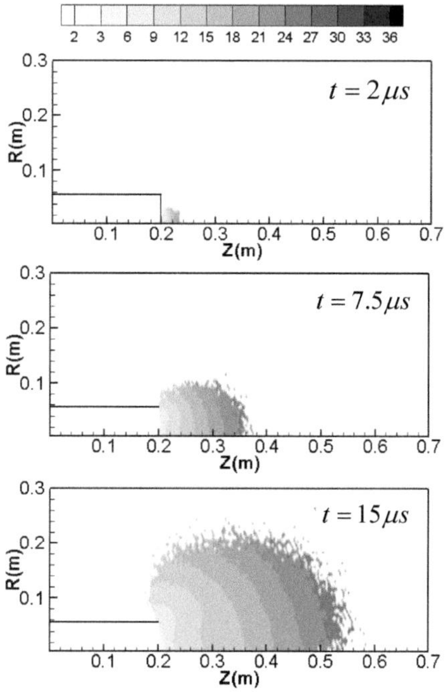

Fig. 7.49 Proportion of CEX collisions on the axis among all collisions

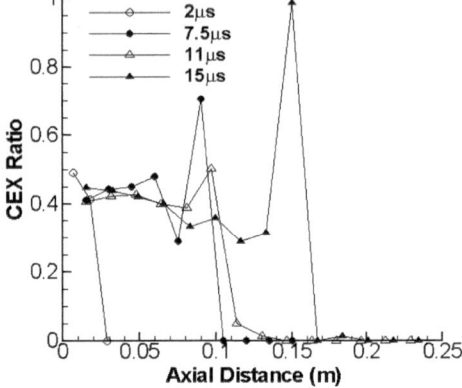

Fig. 7.50 Proportion of non-CEX collisions on the axis among all collisions

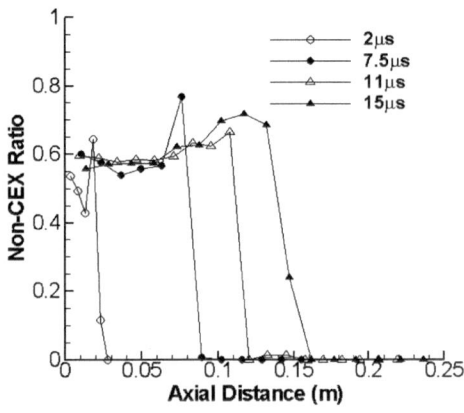

Fig. 7.51 Proportion of CEX1 collisions on the axis among the total CEX collisions

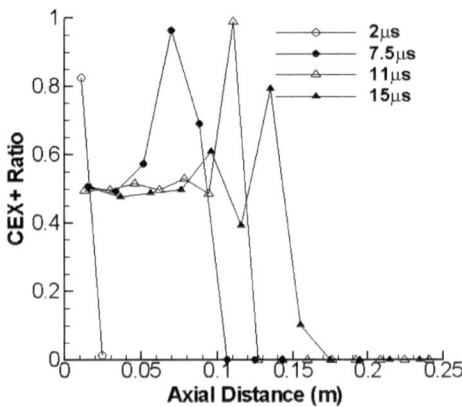

Fig. 7.52 Proportion of
CEX2 collisions on the axis
among the total CEX
collisions

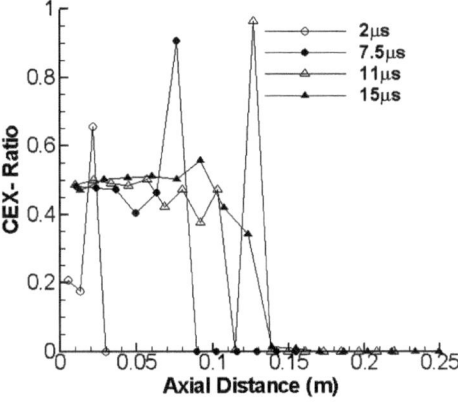

Fig. 7.53 Velocity
distribution of CEX particles
on the axis at 2 μs

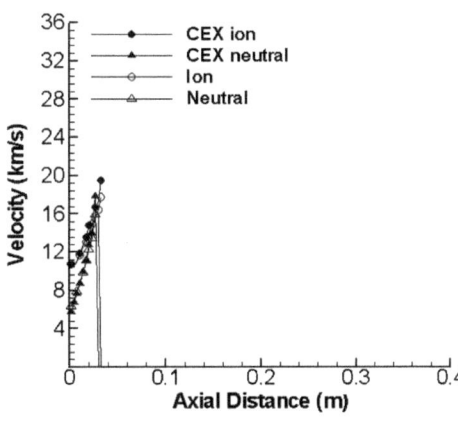

Fig. 7.54 Velocity
distribution of CEX particles
on the axis at 11 μs

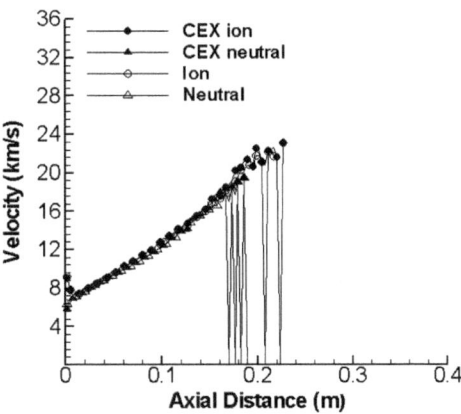

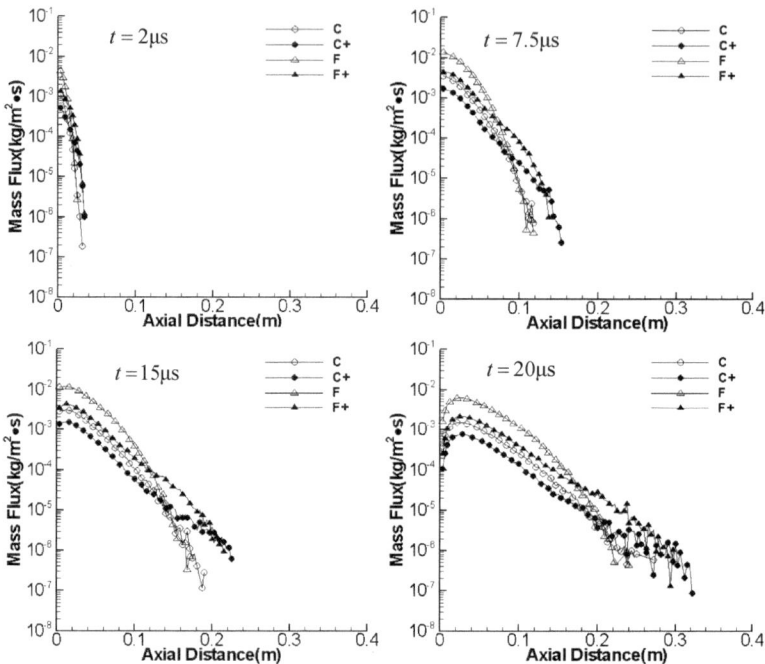

Fig. 7.55 Axial mass flux at 13.5 J without CEX collisions

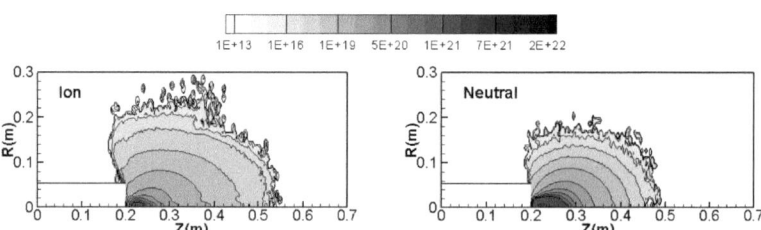

Fig. 7.56 Distributions of ions and neutral particles at 15 μs at 13.5 J without CEX collisions

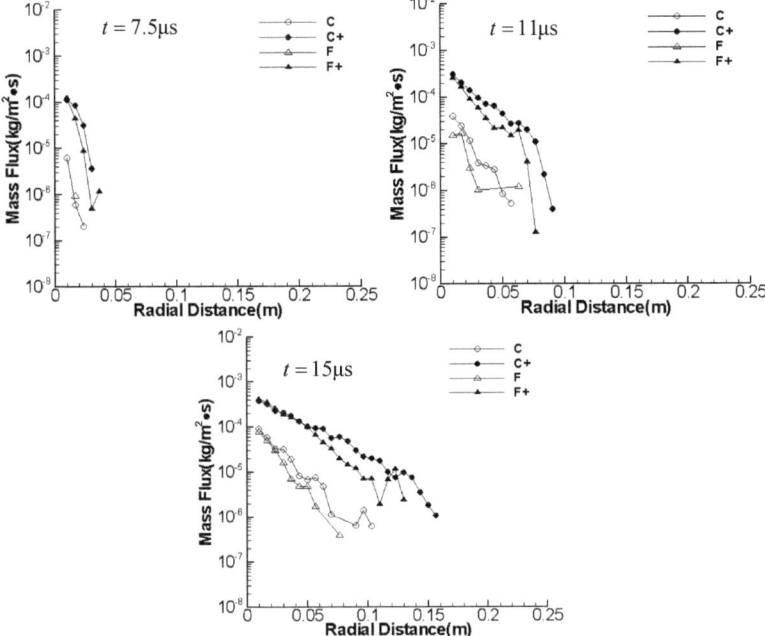

Fig. 7.57 Distributions of CEX particles in the backflow without CEX collisions at the thruster exit at 13.5 J

References

1. Bird GA. Molecular gas dynamics and the direct simulation of gas flows. Oxford: Clarendon Press;1994.
2. Birdsall CK, Langdon AB. Plasma physics via computer simlation. New York: McGraw-Hill; 1985.
3. Wu Q, Chen W. DSMC method for the thermochemical nonequilibrium flow of high-temperature rarefied gas. Changsha, China: National University of Defense Technology Press; 1999.
4. Gatsonis NA, Yin X. Particle/fluid modelling of pulsed plasma thruster plumes. In: AIAA 99–2299, 37th AIAA Aerospace Science Meeting and Exhibit. Reno, NV; 1999.
5. Lieberman MA, Lichtenberg AJ. Principles of plasma discharges and material processing. John Wiley & Sons; 1994.
6. Sakabe S, Izawa Y. Cross sections for resonant charge transfer between atoms and their positive ions: collision velocity ≤ 1 a.u. Atomic Nuclear Data Table.1991;49:257–314.
7. Mitchner M, Charles H. Kruger J. Partially Ionized Gases. John Wiley; 1994.
8. Yu C, Mu Y. Preconditioning generalized minimal residual algorithm. Math Theory Appl. 2003;25(2):38–42.
9. Jian Li, et al. Study of scaling law for particle-in-cell/Monte Carlo simulation of low temperature magnetized plasma for electric propulsion. J Phys D: Appl Phys. 2019; 52:455203.

Chapter 8
Integrated Numerical Simulation of PPTs

Unsteady and strong transient behavior are important characteristics of the pulsed plasma thruster (PPT) pulse discharge process. During the discharge process, rapidly heating the propellant surface leads to the ablation and ionization of the wall material. The plasmoid is rapidly formed in the acceleration channel and then accelerated and ejected under the Lorentz force. During the pulse, the chemical composition and kinetic energy of the plasmoid in the acceleration channel of the thruster undergo drastic changes, causing the thruster plume field to also exhibit transient characteristics. To obtain more accurate calculation results, it is necessary to perform integrated numerical simulations of the discharge process and plume motion [1–3].

In this chapter, the one-dimensional (1D) two-temperature magnetohydrodynamic (MHD) discharge model and the three-dimensional (3D) two-temperature MHD model are used as the inlet model, together with the hybrid plume particle–fluid model established in the previous chapter, to conduct integrated numerical simulation of PPTs. Using the unsteady exit, results calculated by the discharge process model as the inlet boundary conditions for the PPT plume, the characteristics of the plume field, such as the mass flux and temperature field, of the PPT under different initial voltages and different capacitances are investigated.

8.1 Two-Temperature Model and Integrated Boundary Conditions

8.1.1 Two-Temperature Model

The thermochemical model for plasma used in the MHD model, as described in Chap. 6 of this book, is based on the assumption of local thermodynamic equilibrium of the plasma. Due to the strong transient characteristics of the PPT operational

© The Author(s) 2025
J. Wu et al., *Numerical Simulation of Pulsed Plasma Thruster*,
https://doi.org/10.1007/978-981-97-7958-1_8

process, the temperature of both electrons and heavy particles changes rapidly. If the plasma is regarded as existing in a thermodynamic equilibrium state, accurate numerical simulation results cannot be obtained. Therefore, it becomes important to establish a two-temperature model for electrons and heavy particles.

The internal energy of a plasma fluid consists of particle and electron energies.

$$\varepsilon_{\text{int}} = \varepsilon_i + \varepsilon_e \tag{8.1}$$

For a 1D two-temperature model, the conservation of electron energy equation is

$$\frac{\partial \varepsilon_e}{\partial t} + \frac{\partial(\varepsilon_e u)}{\partial x} + \frac{\partial(p_e u)}{\partial x} = Q_j - Q_{\text{rad}}(T_e) - Q_a(T_e) - \Delta \dot{\varepsilon}_{ie} \tag{8.2}$$

$$\varepsilon_e = \frac{p_e}{(\gamma - 1)} \qquad \Delta \dot{\varepsilon}_{ie} = \frac{3 \rho_e \nu_{ei}}{m_i} k(T_e - T_h) \tag{8.3}$$

where ε_e is the electron internal energy, p_e is the electron pressure, $\Delta \dot{\varepsilon}_{ie}$ is the rate of energy exchange between electrons and ions, k is the Boltzmann constant, and T_h is the temperature of heavy particles.

For a 3D two-temperature model, the heat conduction term is divided into the sum of the electronic term and the particle term as follows:

$$\kappa \nabla T = \kappa_e \nabla T_e + \kappa_i \nabla T_h \tag{8.4}$$

Assuming that the ohmic heat mainly affects the electron energy, the electron energy equation is

$$\frac{\partial(\varepsilon_e)}{\partial t} + \nabla \cdot (\varepsilon_e U) + \nabla \cdot (p_e U) = \frac{J^2}{\sigma_e} + \nabla \cdot q_e - \Delta \dot{\varepsilon}_{ie} \tag{8.5}$$

Similarly, the particle energy equation is

$$\frac{\partial(\varepsilon_i)}{\partial t} + \nabla \cdot (\varepsilon_i U) + \nabla \cdot (p_i U) = \nabla \cdot q_i + \Delta \dot{\varepsilon}_{ie} \tag{8.6}$$

The energy exchange between electrons and particles is manifested through collisions.

$$\Delta \dot{\varepsilon}_{ie} = \frac{3 \rho_e \nu_{ei}}{m_i} k(T_e - T_h) \tag{8.7}$$

Replacing the original thermochemical model with the above 1D and 3D two-temperature models can yield more accurate simulation results for the PPT discharge process, which can serve as the inlet conditions for the plume simulation in this chapter.

8.1.2 Integrated Boundary Condition Processing

During the PPT operation, the thruster is in a vacuum state, and the high-density plasmoid generated in the thrust chamber is ejected outwards to form a plasma plume. The charged components in the plume field form a changing electric field environment outside the thruster. This electric field environment has little impact on the intense discharge process of the thruster and thus is ignored in the calculation process.

The exit parameters calculated for the PPT at initial capacitive energies of 7.26, 13.5, and 24 J and the exit parameters calculated for the PPT at capacitances of 2, 12, and 20 c are used as the inlet parameters of the plume. The hybrid DSMC/PIC algorithm is employed for the plume field calculations.

Figure 8.1 shows the variations in the exit flux over time provided by the 1D model at different initial capacitive energies. Figure 8.2 presents the variations in the exit flux over time under different capacitances using the 1D model. The thruster inlet model used by Gatsonis is still adopted here.

$$n_s(r, z, t) = n_{s,\text{exit}} \left[1 - (1 - C_c)\left(\frac{r}{R_T}\right)^2 \right] \tag{8.8}$$

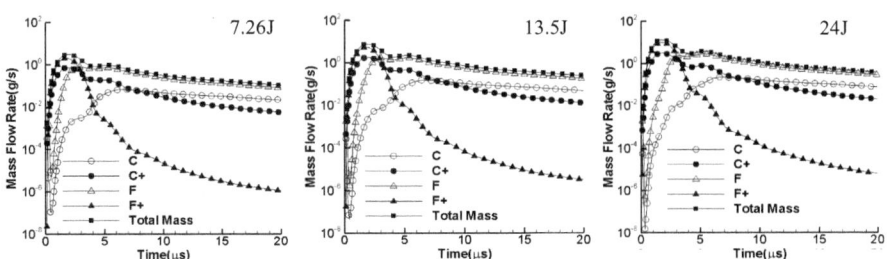

Fig. 8.1 Exit mass fluxes at different initial capacitive energies

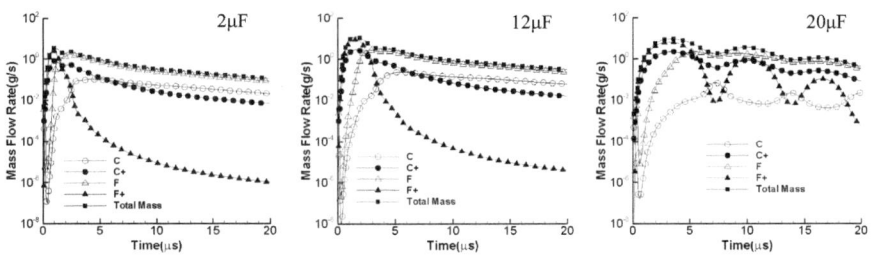

Fig. 8.2 Exit mass fluxes under different capacitances

where $n_{s,\,exit}$ is the density of each component at the exit at the current moment provided by the 1D model, and C_c is the density coefficient with a value of 0.1. The exit electron temperature and particle temperature are obtained using the calculated values at each time step.

For the exit parameters provided by the 3D model, due to the differences between the thruster exit and the thruster casing, the following geometric processing is needed:

$$r_x = R'_{h_{PPT}} r_{exit}/R_T \tag{8.9}$$

where r_{exit} is the inlet radius of the plume, and $R'_{hppt} = (h_{PPT} + w_{PPT})/2$ is the effective exit radius of the thruster. The inlet parameters at r_{exit} are provided by the parameters at r_x, which are obtained from the average values at the exit center of the thruster in the vertical and horizontal directions.

Figure 8.3 shows the variations in the exit flux over time provided by the 1D model at different initial capacitive energies. Figure 8.4 presents the variations in the exit flux over time under different capacitances using the 1D model.

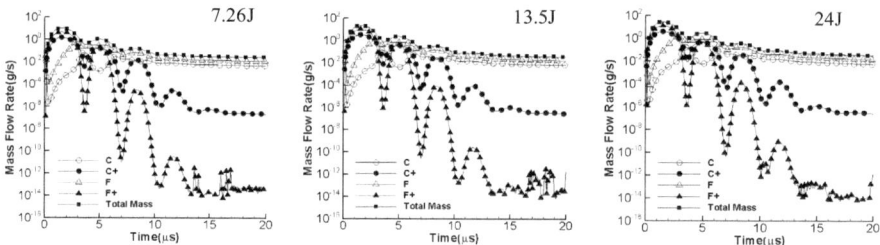

Fig. 8.3 Exit mass fluxes at different initial capacitive energies for geometric treatment

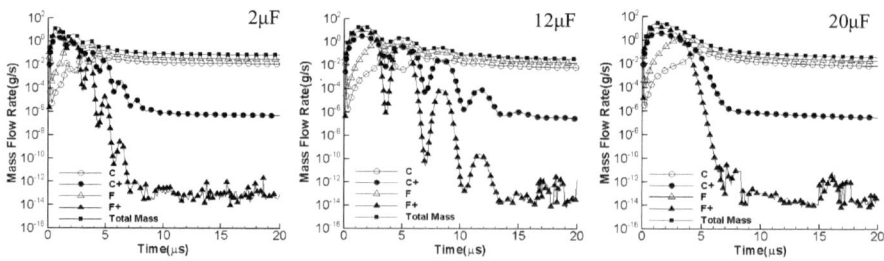

Fig. 8.4 Exit mass fluxes under different capacitances for geometric treatment

8.2 Plume Simulation Using the 1D Boundary Model

8.2.1 Model Validation

To verify the reliability of the model, the PPT prototype in the laboratory of the Glenn Research Center [1] is calculated at a discharge energy of 20 J using the meshing method described in Sect. 7.2.1.

Figures 8.5 and 8.6 show the electron density at distances of 6 and 16 cm from the propellant, respectively, with a probe angle of 10°. The time axis of the calculation results is shifted backwards accordingly due to the time delay in the experimental data. Comparison with the experimental results reveals that the electron density near the surface matches better and that the model generally achieves a good fit.

Figures 8.7 and 8.8 present the maximum electron temperature and electron density, respectively, at different distances with probe angles of 10° and 30°. A comparison with the experimental results shows that the electron density results are in better agreement with the electron temperature results, and the results are closer to the experimental data at large angles and near the surface. The calculation results are in general agreement within the error range of the experimental results.

Fig. 8.5 Electron density at $r = 6$ cm and $\theta=10°$(1D simulation)

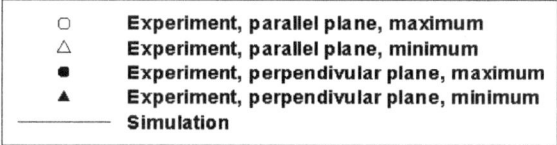

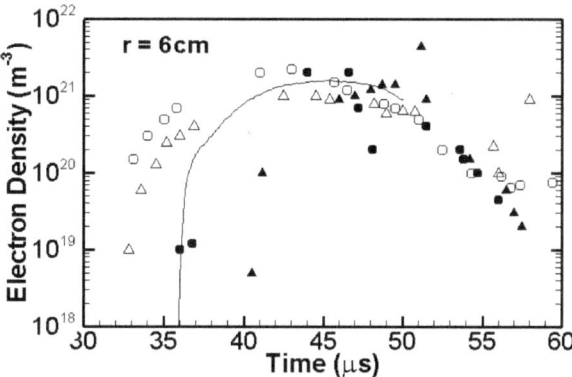

Fig. 8.6 Electron density at $r = 16$ cm and $\theta=10°$(1D simulation)

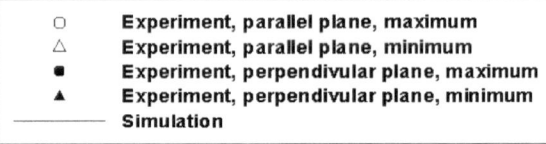

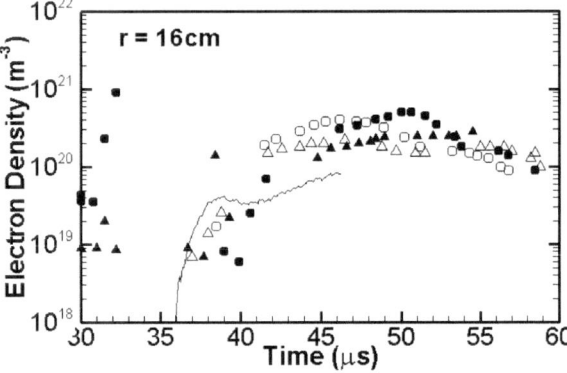

Fig. 8.7 Electron temperature at axis angles of $10°$ and $30°$(1D simulation)

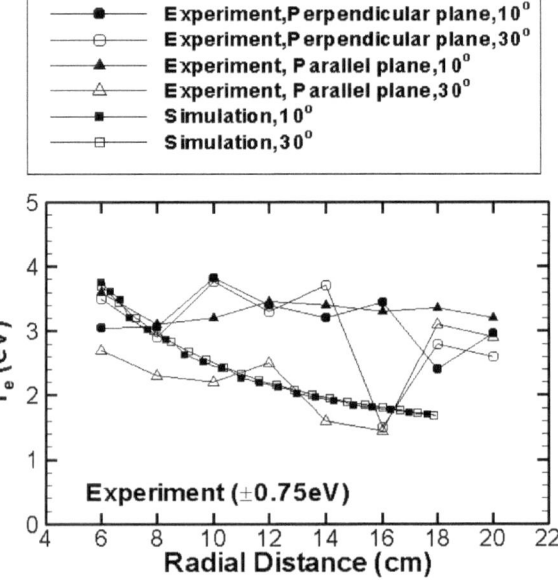

Fig. 8.8 Electron density at axis angles of 10° and 30°(1D simulation)

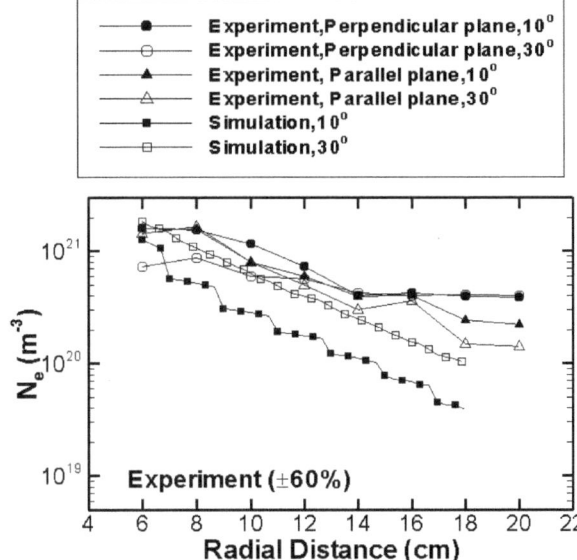

8.2.2 Calculation Result Analysis

1. Different Initial Voltages

Different initial voltages are considered. At 7.26 J, the weight factor is 8×10^{10}; in the calculation cycle, there are a total of 5,293,917 simulated molecules, including 2,693,210 ions and 2,600,707 neutral particles, and a total of 28,527,691 collisions are calculated, of which there are 5,271,241 CEX collisions, accounting for 19% of the total number of collisions. At 13.5 J, the weight factor is 1.5×10^{11}; in the calculation cycle, there are a total of 6,581,338 simulated molecules, including 3,566,604 ions and 3,014,734 neutral particles, and a total of 69,310,753 collisions are calculated, of which there are 10,470,284 CEX collisions, accounting for 15% of the total number of collisions. At 24 J, the weight factor is 2×10^{11}; there are a total of 8,356,471 simulated molecules, including 4,877,955 ions and 3,478,516 neutral particles, and a total of 117,828,834 collisions are calculated, of which there are 13,122,889 CEX collisions, accounting for 11% of the total number of collisions.

Figures 8.9, 8.10 and 8.11 show axial mass fluxes at different initial capacitive energies. The charged components mainly appear at the beginning of the pulse. At 2 μs, F⁺ ions have the highest mass flux, followed by C⁺, F, and C. As the pulse time increases, the charged components decrease continuously, the neutral particles continuously increase. However, the gas ejection velocity continuously decreases. At the exit, F atoms have the highest mass flux, followed by C, F⁺, and C⁺. Along the axial direction, the number of CEX collisions gradually decreases, the velocity of ions continuously increases, and the tail of the plume consists of charged particle flow. At the beginning of the pulse, C, due to its small atomic weight, is easier to accelerate under the action of an electric field. C^+ has the maximum particle flow at the tail of the plume. In the later stage of the pulse, the generation of a large amount of F^+ ions leads to these ions achieving the maximum particle flow, followed by F and C. A comparison of the cases of 7.26 and 24 J shows that at 7.5 μs, the ion cluster diffuses to distances of 0.16 and 0.18 m from the exit at 7.26 and 24 J, respectively, and the atom cluster diffuses to the same distances of 0.12 m from the exit at 7.26 and 24 J. At 20 μs, the ion cluster diffuses to distances of 0.51 and 0.58 m from the exit at 7.26 and 24 J, respectively, and the atom cluster diffuses to distances of 0.32 and 0.46 m from the exit at 7.26 J and 24 J, respectively. Comparison with the results obtained for the Gatsonis inlet model in Sect. 7.2 shows that the diffusion rate of the plume obtained by the 1D inlet model is higher than that of the Gatsonis model. In the high-energy state, both the lowest and highest mass fluxes at the end of the cycle are greater than those in the low-energy state.

Figures 8.12, 8.13 and 8.14 show the backflow mass fluxes above the exit plane of the thruster under different initial capacitive energy states. The backflow occurs above the thruster after a certain ejection time. Specifically, backflow occurs at 3.1, 3.4, and 2.8 μs at 7.26 J, 13.5 J, and 24 J, respectively. The charged components remain the first to backflow. Among them, because the C atom has the largest charge-to-mass ratio, the backflow of C^+ occurs first, followed by the backflows of F^+, F, and C. The backflow continuously diffuses toward the backflow region over time.

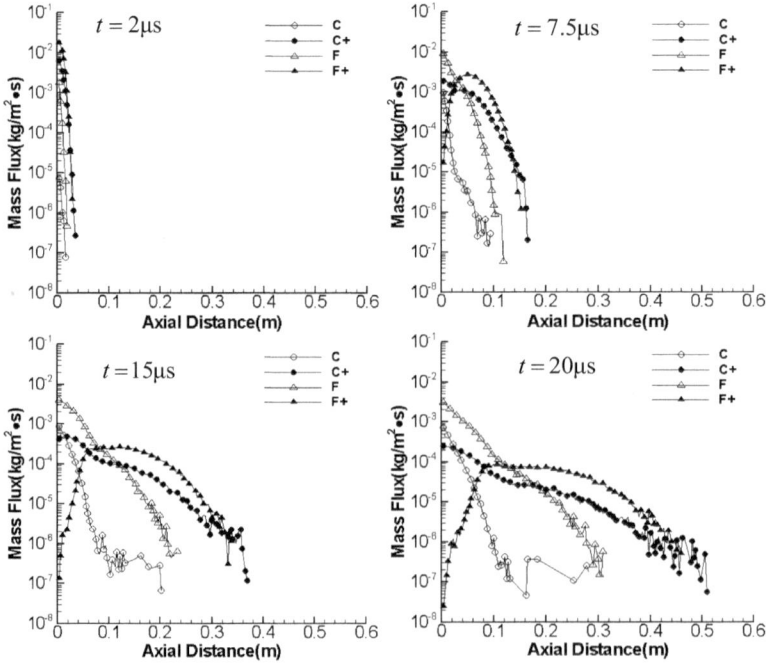

Fig. 8.9   Axial mass fluxes at 7.26 J (1D simulation)

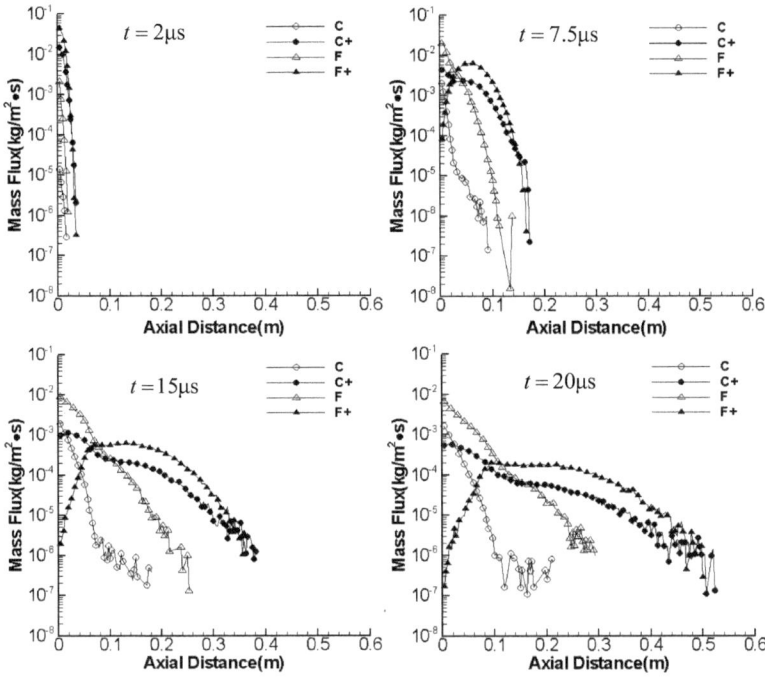

Fig. 8.10   Axial mass fluxes at 13.5 J (1D simulation)

The highest mass flux is always located near the ejection exit, and a certain level is maintained. In the early stage, the charged components are dominant in the backflow, while in the later stage, the neutral components near the exit become non-negligible. A comparison of the cases of 7.26 and 24 J shows that at 7.5 μs, the backflows of C^+, F^+ and F occur at 7.26 J, with a maximum backflow of 4×10^{-4} kg/m^3·s; at this time, the backflows of C^+, F^+ and F appear at 24 J, with a maximum backflow of 10^{-3} kg/m^3·s. At 11 μs, the backflow at 7.26 J reaches 8 cm from the thruster exit in the radial direction, with the highest mass flux of 10^{-3} kg/m^3·s, while the backflow at 24 J reaches 12 cm from the thruster exit in the radial direction, with the highest mass flux of 3×10^{-3} kg/m^3 s. At 20 μs, the backflow at 7.26 J reaches 18 cm, with the highest mass flux of 10^{-3} kg/m^3·s, while the backflow at 24 J reaches 28 cm, with the mass flux of 2×10^{-3} kg/m^3·s. At 7.26 J, C^+ has the highest mass flux of 10^{-3} kg/ m^3 s, and when the mass flux of ions is maintained at a certain level, the mass flux of neutral particles still continuously increases. F has the highest mass flux among the neutral particles, with a value of 10^{-4} kg/m^3·s. At 24 J, C^+ has the highest mass flux of 2×10^{-3} kg/m^3·s, and the trend of the neutral particles is similar to that at 7.26 J. F achieves the maximum neutral particle mass flux of 10^{-4} kg/m^3·s. It is worth mentioning that, unlike the Gatsonis model, among the three energy states, although the diffusion rate is the fastest and the backflow is the earliest to occur at 24 J, the time points at which backflows appear at 7.26 J and at 13.5 J are very close, with

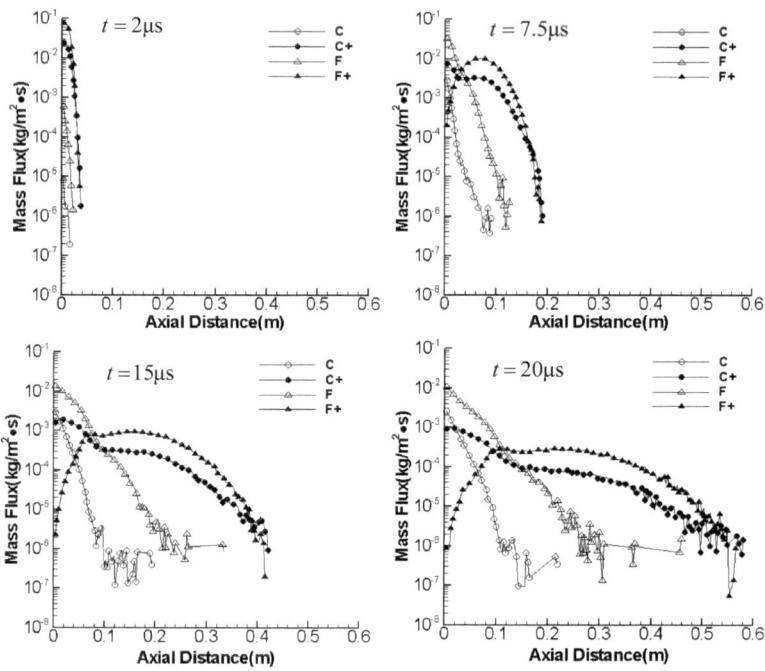

Fig. 8.11 Axial mass fluxes at 24 J (1D simulation)

the occurrence at 7.26 J is even earlier than that at 13.5 J. Therefore, a higher energy leads to a greater magnitude of the particle backflow.

Figure 8.15 shows the ion distribution at 20 μs under different initial voltages. At 24 J, the influence angle of the backflow reaches 130°, and the influence region of the plume is wider at high voltages. Figure 8.16 shows the proportion distribution of CEX ions at different initial voltages. The CEX ions are concentrated inside the plume near the exit. At high voltages, CEX ions account for a greater proportion at the exit, and their content is also higher in the backflow.

2. Different Capacitances

Different capacitances are considered. At 2 μF, the weight factor is 8×10^{10}; in the calculation cycle, there are a total of 5,233,613 simulated molecules, including 2,013,535 ions and 3,220,078 neutral particles, and a total of 42,410,628 collisions are calculated, of which there are 4,952,563 CEX collisions, accounting for 12% of the total number of collisions. At 12 μF, the weight factor is 2×10^{11}; in the calculation cycle, there are a total of 4,624,662 simulated molecules, including 2,842,097 ions and 2,863,281 neutral particles, and a total of 53,216,161 collisions are calculated, of which there are 7,454,207 CEX collisions, accounting for 14% of the total number of collisions. At 20 μF, the weight factor is 3×10^{11}; in the calculation cycle, there are a total of 5,106,873 simulated molecules, including 3,462,801 ions and 1,644,072

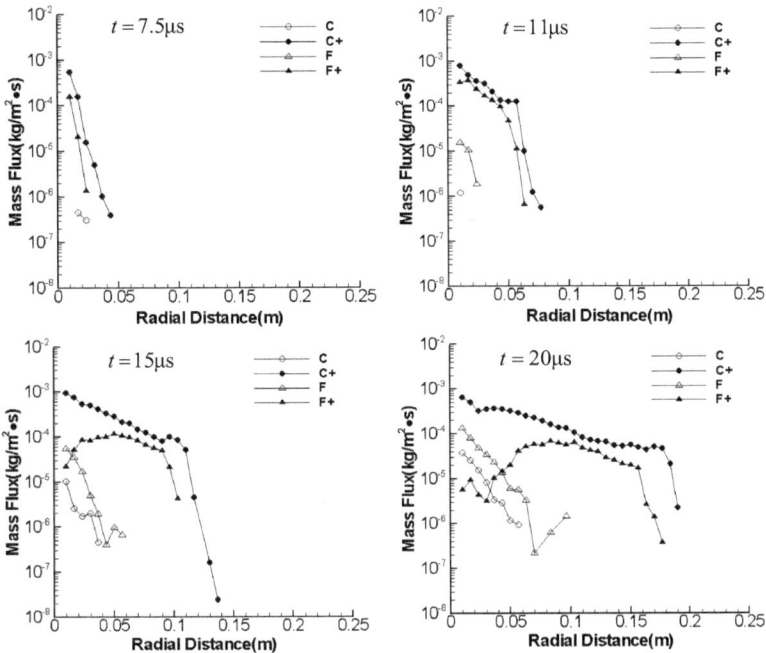

Fig. 8.12 Backflow mass fluxes at the thruster exit at 7.26 J (1D simulation)

neutral particles, and a total of 47,657,140 collisions are calculated, of which there are 8,165,476 CEX collisions, accounting for 17% of the total number of collisions.

Figures 8.17, 8.18 and 8.19 show axial mass fluxes at different capacitances. The formation pattern of charged components is similar to that under different initial voltages. The earliest F^+ ions appear at the beginning of the pulse and have the highest mass flux, followed by C^+, F, and C. As the pulse time increases, at the exit, F atoms become the species with the highest mass flux, followed by C, F^+, and C^+. Along the axial direction, the tail of the plume is composed of charged particle flow, with C^+ having the maximum particle flow at the tail of the plume. In the later stage of the pulse, the generation of a large number of F^+ ions causes them to have the maximum particle flow, followed by F and C. A comparison of the cases of 2 and 20 μF shows that at 7.5 μs, the ion cluster diffuses to distances of 0.18 and 0.19 m from the exit at 2 μF and 20 μF, respectively, and the atom cluster diffuses to distances of 0.11 and 0.14 m from the exit at 2 μF and 20 μF, respectively; at 20 μs, the ion cluster diffuses to distances of 0.53 and 0.56 m from the exit at 2 μF and 20 μF, respectively, and the atom cluster diffuses to distances of 0.28 and 0.36 m from the exit at 2 μF and 20 μF, respectively. The calculation results reveal that high capacitance leads to the fastest diffusion rate, but the advantage is not pronounced. Therefore, the diffusion rates of the plume are similar under different capacitances, and the mass flux is higher at high capacitance.

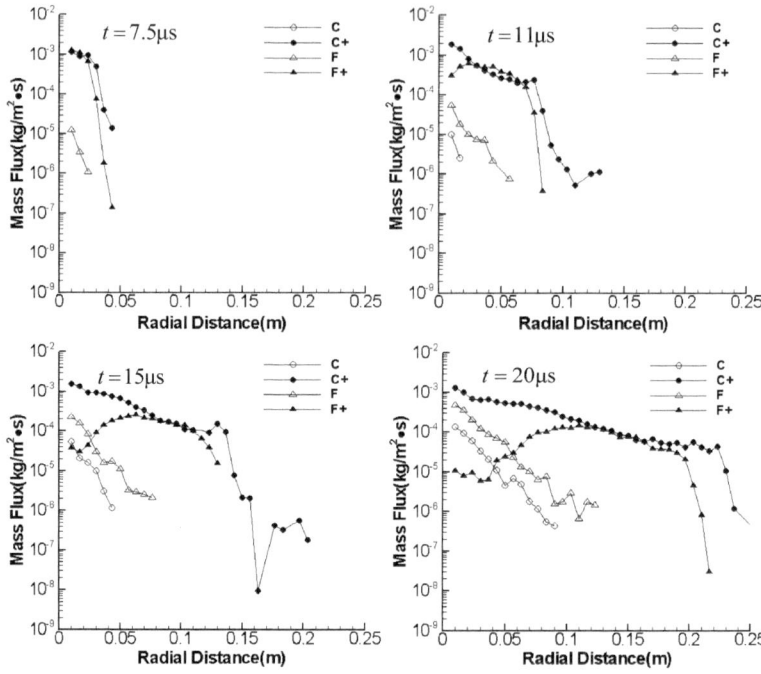

Fig. 8.13 Backflow mass fluxes at the thruster exit at 13.5 J (1D simulation)

Figures 8.20, 8.21 and 8.22 show the backflow mass fluxes above the exit plane of the thruster under different capacitances. Backflow occurs above the thruster after a certain ejection time. Backflow occurs at 2.8, 3.4, and 3.4 μs at capacitances of 2 μF, 12 μF, and 20 μF, respectively. The charged components remain the first to backflow; in particular, the backflow of C^+ appears first, followed by F^+, F, and C. The backflow continuously diffuses toward the backflow region over time, while the highest mass flux is always located near the ejection exit, and a certain level is maintained. In the early stage, the backflow is dominated by charged components, while in the later stage, the neutral components near the exit become non-negligible. A comparison of the cases of 2 μF and 20 μF shows that at 7.5 μs, the backflows of C^+, F^+, F, and C occur at 2 μF, with a maximum backflow of 10^{-3} kg/m³·s. At this time, the backflow of C does occurs at 20 μF, with a maximum backflow of 10^{-3} kg/m³·s. At 11 μs, the backflow at 2 μF reaches a radial distance of 9 cm from the thruster exit, with the highest mass flux maintained at 10^{-3} kg/m³·s. The backflow at 20 μF reaches a radial direction of 10 cm from the thruster exit, with the highest mass flux maintained at 2×10^{-3} kg/m³·s. At 20 μs, the backflow at 2 μF reaches 22 cm, with the highest mass flux maintained at 10^{-3} kg/m³·s, while the backflow at 20 μF reaches 22 cm, with the mass flux maintained at 2×10^{-3} kg/m³·s. For 2 μF, C^+ has the highest mass flux of 10^{-3} kg/m³ s. When the mass flux of ions is maintained at a certain level, the mass flux of neutral particles still shows a rising

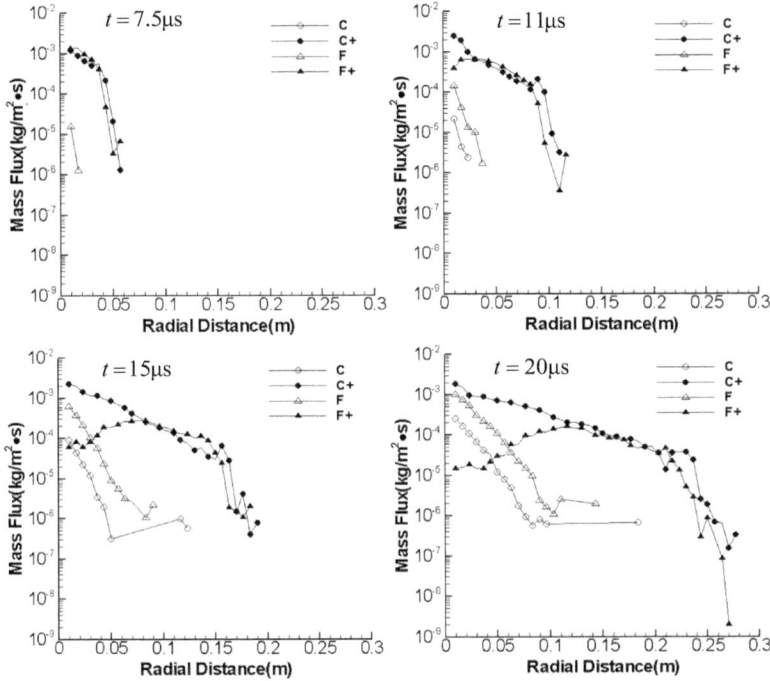

Fig. 8.14 Backflow mass fluxes at the thruster exit at 24 J (1D simulation)

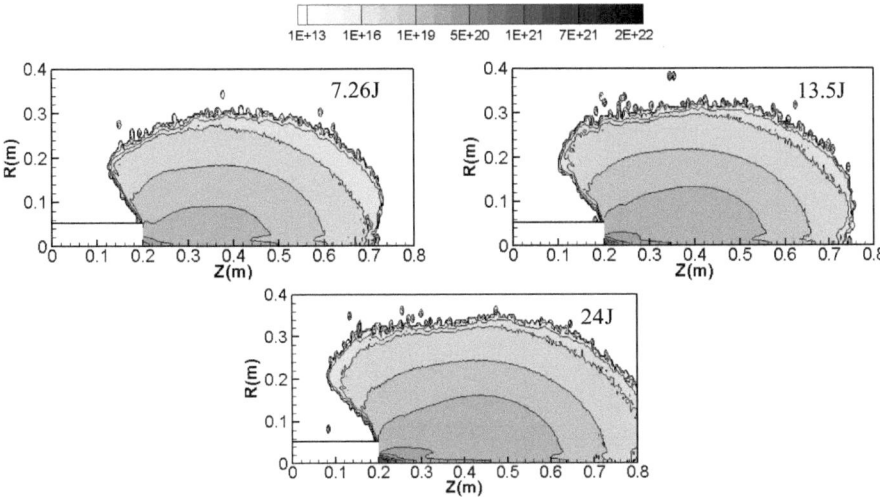

Fig. 8.15 Distribution of ions at 20 μs under different initial voltages (1D simulation)

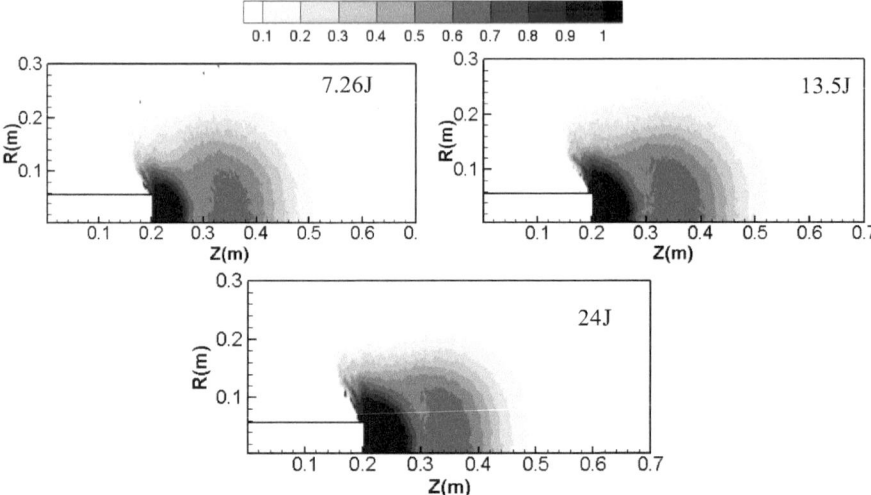

Fig. 8.16 Distribution of CEX ions at 20 μs under different initial voltages (1D simulation)

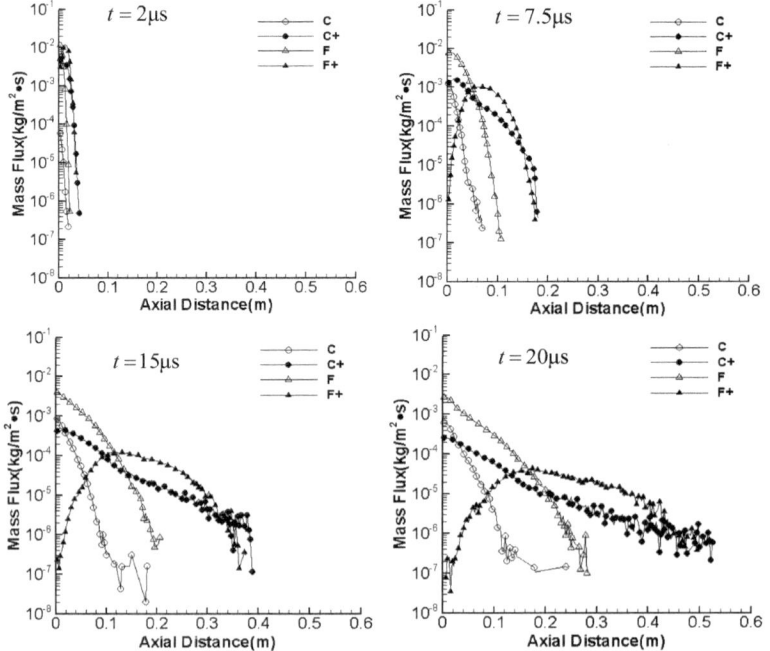

Fig. 8.17 Axial mass flux at 2 μF (1D simulation)

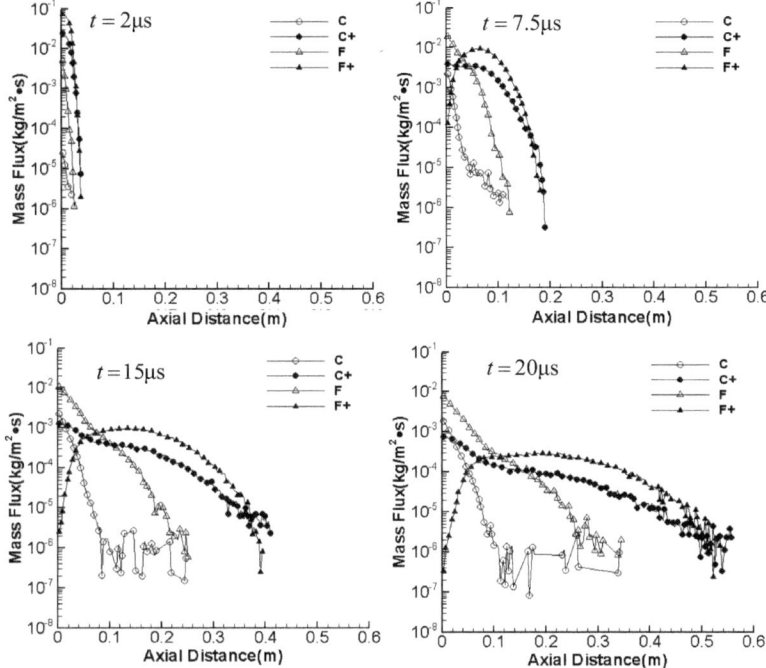

Fig. 8.18 Axial mass flux at 12 μF (1D simulation)

trend. F has the highest mass flux among the neutral particles of 3×10^{-4} kg/m³·s. At 20 μF, C^+ has the highest mass flux of 2×10^{-3} kg/m³ s. The trend of neutral particles is similar to that at 2 μF, and F had the highest mass flux among neutral particles of 8×10^{-4} kg/m³ s. At the same time, similar results are obtained under different capacitances. Under higher capacitance, the overall mass flux and the mass flux of neutral particles are higher, causing more destruction.

Figure 8.23 shows the ion distribution at 20 μs under different capacitances. Under different capacitances, the axial influence distance is longer under higher capacitance, but the influence is not pronounced in the radial direction. The backflow angle reaches 150° under 20 μF, and the influence region of the backflow is wider under higher capacitance.

The distribution of CEX ions under different capacitances is analyzed in Fig. 8.24. It is observed that the CEX ions are concentrated near the exit, and the content of CEX ions near the exit is very high at low capacitance. A comparison with Fig. 8.23 shows that the content of CEX ions in the earliest backflow is not high, but their later influence should not be ignored.

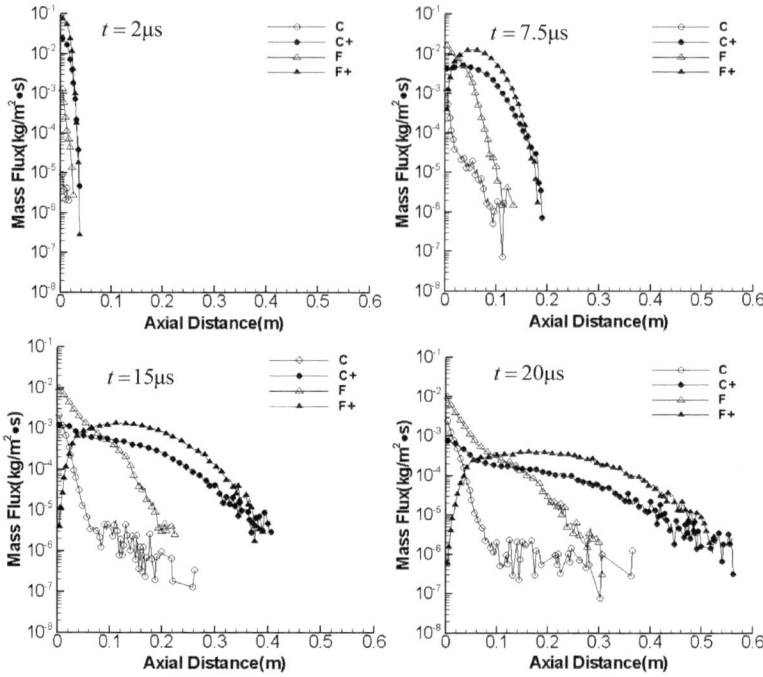

Fig. 8.19 Axial mass flux at 20 μF (1D simulation)

8.3 Plume Simulation Using the 3D Boundary Inlet Model

8.3.1 Model Validation

To verify the reliability of the model, calculations are performed for the PPT prototype in the laboratory of the Glenn Research Center at a discharge energy of 20 J using the meshing method described in Sect. 7.2.1.

Figures 8.25 and 8.26 show the electron density at distances of 6 and 16 cm from the propellant, respectively, with a probe angle of 10°. The time axis of the calculation results is shifted backwards due to the time delay of the experimental data. Comparison with the experimental results reveals that the model generally follows the variation trend. Compared with results for the 1D inlet, the results for the 3D inlet match better in the far field, but the results fail to reflect the plume ejection due to ablation lag.

Figures 8.27 and 8.28 show the maximum electron temperature and electron density at different distances with probe angles of 10° and 30°, respectively. Similar to the 1D results, a comparison with the experimental results finds that the electron density results are in better agreement than the electron temperature results, and the results are closer to the experimental data at large angles and near the surface. The

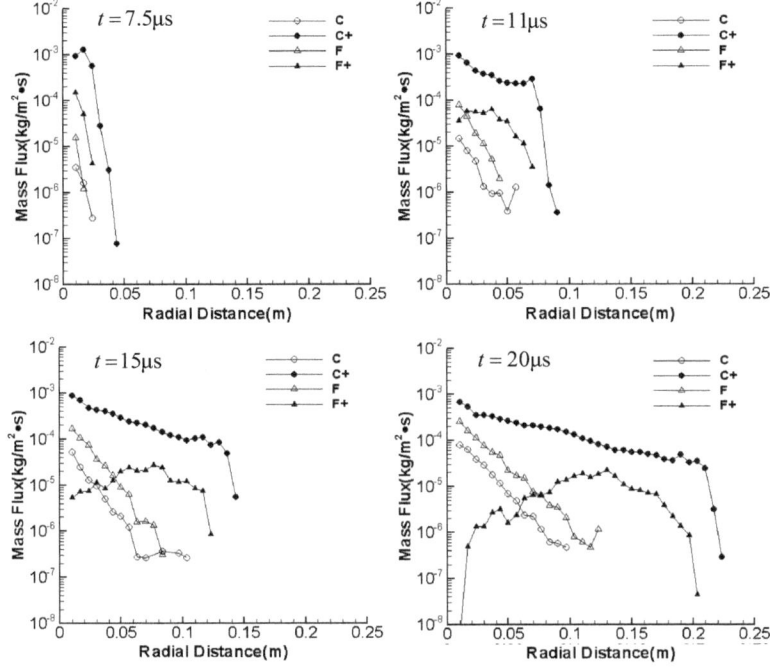

Fig. 8.20   Backflow mass flux at the thruster exit at 2 μF (1D simulation)

calculation results are in general agreement within the error range of the experimental results.

8.3.2   Calculation Result Analysis

1.   Different Initial Voltages

Different initial voltages are considered. At 7.26 J, the weight factor is 8×10^{10}; in the calculation cycle, there are a total of 6,131,759 simulated molecules, including 4,795,615 ions and 1,336,144 neutral particles, and a total of 11,048,206 collisions are calculated, of which there are 4,546,155 CEX collisions, accounting for 41% of the total number of collisions. At 13.5 J, the weight factor is 1.5×10^{11}; in the calculation cycle, there are a total of 6,999,327 simulated molecules, including 5,490,438 ions and 1,508,889 neutral particles, and a total of 22,317,990 collisions are calculated, of which there are 7,457,077 CEX collisions, accounting for 33% of the total number of collisions. At 24 J, the weight factor is 2×10^{11}; in the calculation cycle, there are a total of 8,143,460 simulated molecules, including 6,743,050 ions and 1,400,410 neutral particles, and a total of 22,205,115 collisions are calculated,

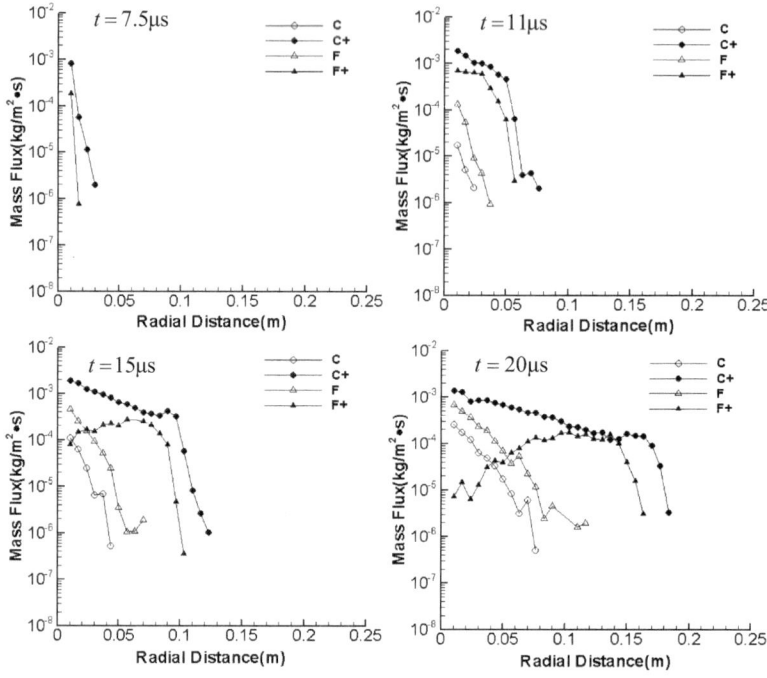

Fig. 8.21 Backflow mass flux at the thruster exit at 12 μF (1D simulation)

of which there are 6,068,390 CEX collisions, accounting for 27% of the total number of collisions.

Figures 8.29, 8.30 and 8.31 show axial mass fluxes at different initial capacitive energies. Compared to the results for the 1D inlet, the change characteristics of various components are similar. At the beginning of discharge, F^+ ions have the highest mass flux, followed by C^+, F, and C. As the pulse time increases, at the exit, F atoms become the species with the highest mass flux, followed by C, F^+, and C^+. At the beginning of the pulse, C^+ has the maximum particle flow at the tail of the plume. At the later stage of the pulse, the generation of a large number of F^+ ions makes it become the maximum particle flow, followed by F and C. A comparison of the cases of 7.26 and 24 J reveals that at 7.5 μs, the ion cluster diffuses to distances of 0.23 and 0.23 m from the exit at 7.26 J and 24 J, respectively, and the atom cluster diffuses to distances of 0.15 and 0.16 m from the exit at 7.26 J and 24 J, respectively. At 20 μs, the ion cluster diffuses to distances of 0.60 and 0.63 m from the exit at 7.26 J and 24 J, respectively, and the atom cluster diffuses to distances of 0.46 and 0.32 m from the exit at 7.26 J and 24 J, respectively. In the high-energy state, the lowest mass flux and the highest mass flux at the end of the cycle are both greater than those in the low-energy state. A comparison with the results obtained by the 1D inlet model shows that because the exit peak density calculated by the 3D model is greater than that obtained by the 1D model, the exit velocity is very low at the end of

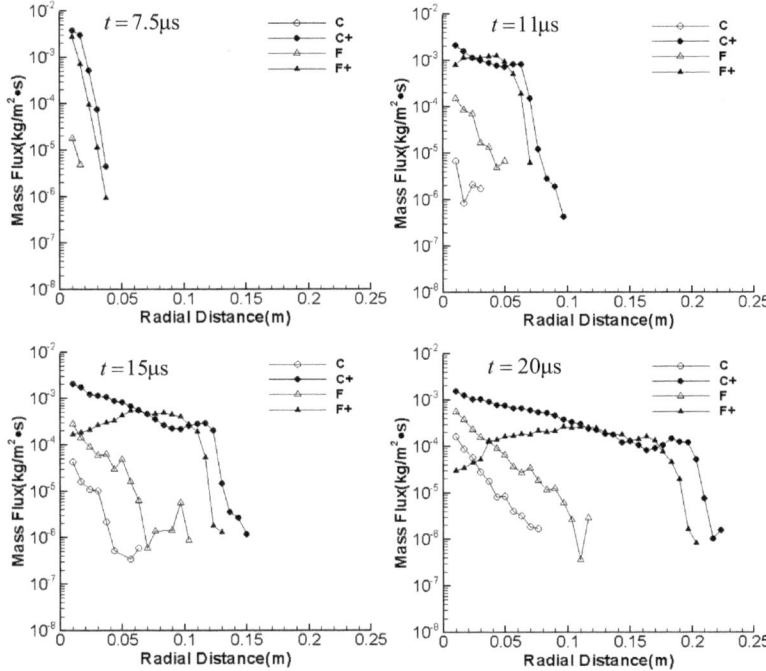

Fig. 8.22   Backflow mass flux at the thruster exit at 20 μF (1D simulation)

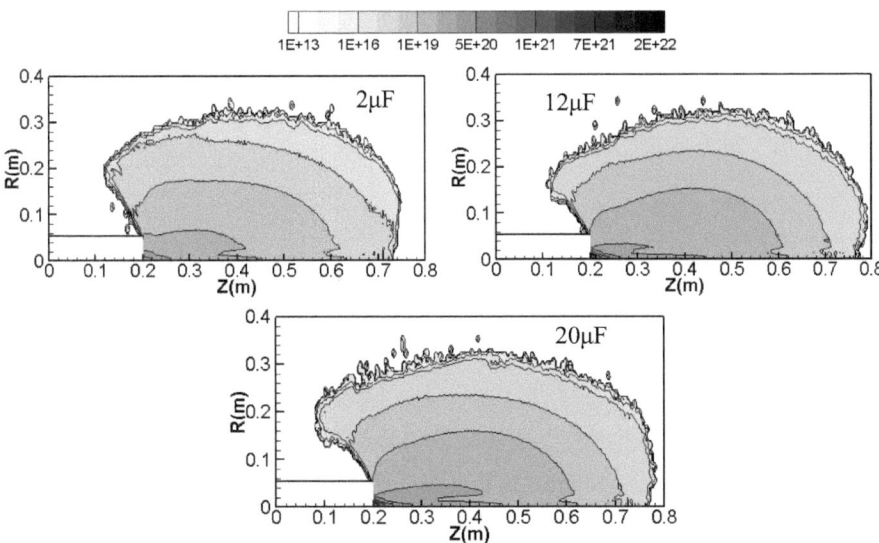

Fig. 8.23   Ion distribution at 20 μs under different capacitances (1D simulation)

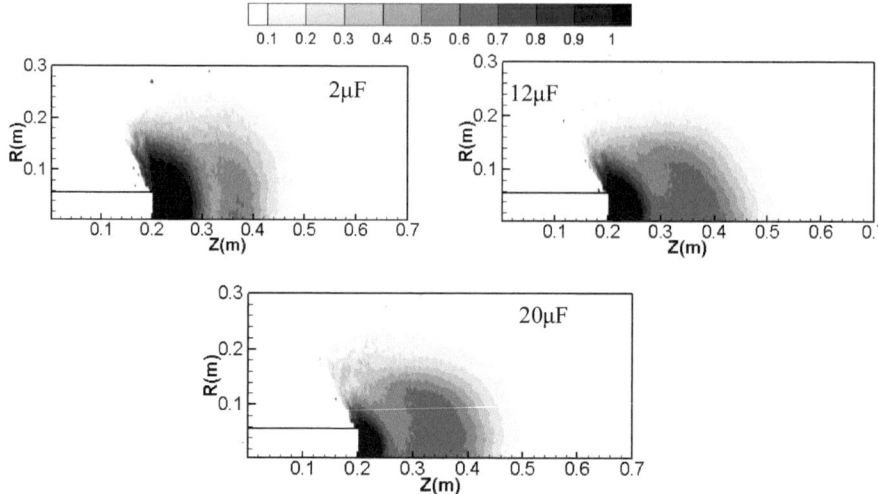

Fig. 8.24 Distribution of CEX ions at 20 μs under different capacitances (1D simulation)

Fig. 8.25 Electron density
at r = 6 cm and θ=10° (3D
simulation)

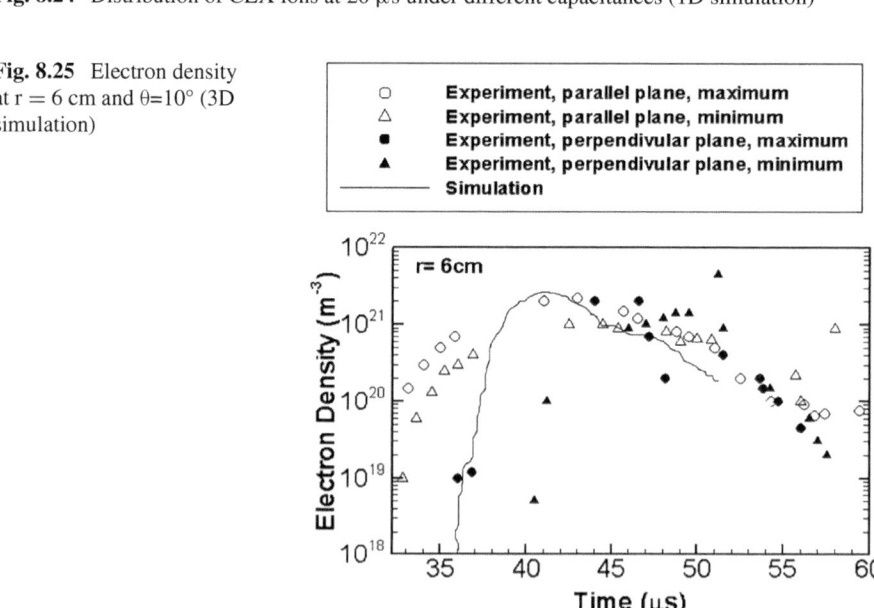

the pulse, and the ion diffusion rate in the plume obtained by the 3D model is greater than that obtained by the 1D model. However, the atomic diffusion rate is lower than that of the 1D inlet model.

Figures 8.32, 8.33 and 8.34 present the backflow mass fluxes above the exit plane of the thruster at different initial capacitive energies. According to the calculation results using the 3D inlet model, the backflows at 7.26, 13.5, and 24 J occur at

Fig. 8.26 Electron density at r = 16 cm and θ=10° (3D simulation)

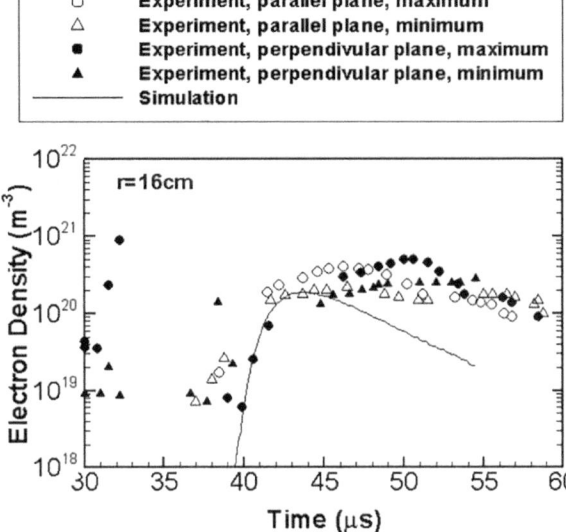

Fig. 8.27 Electron temperature at axis angles of 10° and 30° (3D simulation)

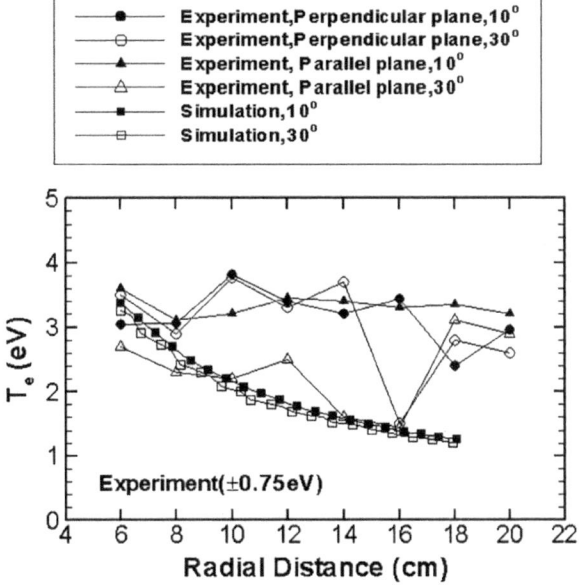

4.4 μs, 4.7 μs, and 3.9 μs, respectively. Similar to the 1D results, the backflow of C^+ appears first, followed by F^+, F, and C. In the early stage, the backflow is dominated by charged components, while in the later stage, the neutral components near the exit became non-negligible. A comparison of the cases of 7.26 and 24 J shows that at

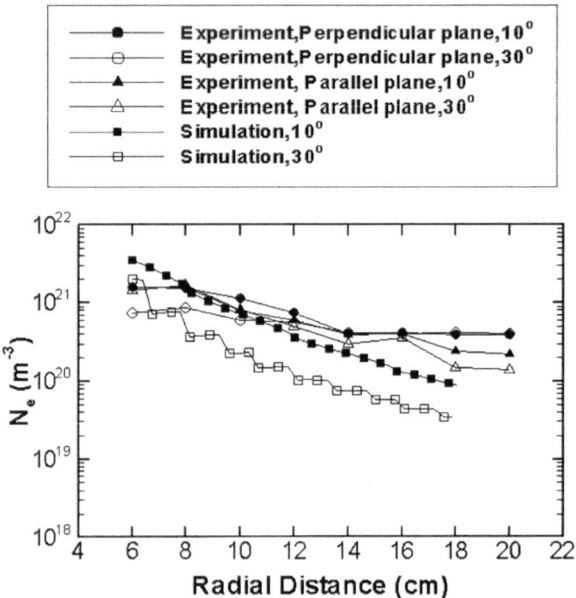

Fig. 8.28 Electron density at axis angles of 10° and 30° (3D simulation)

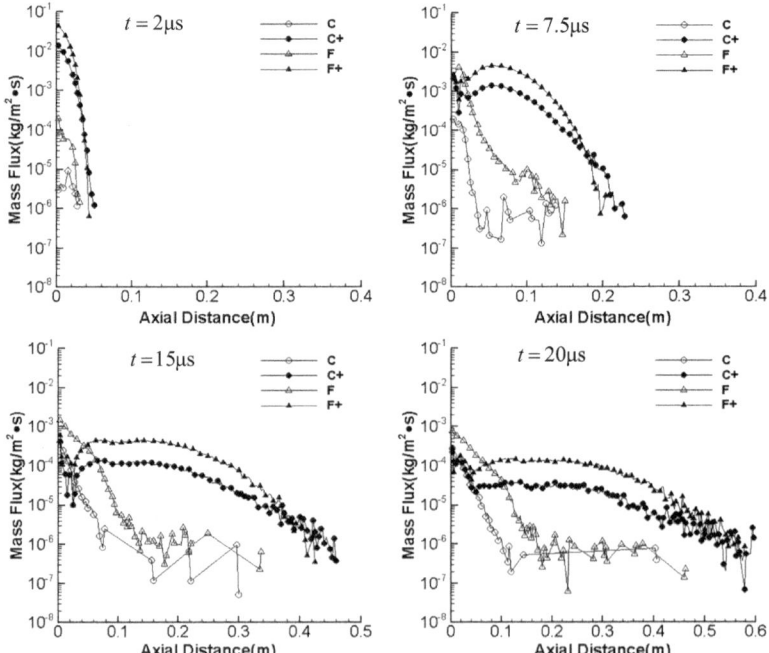

Fig. 8.29 Axial mass fluxes at 7.26 J

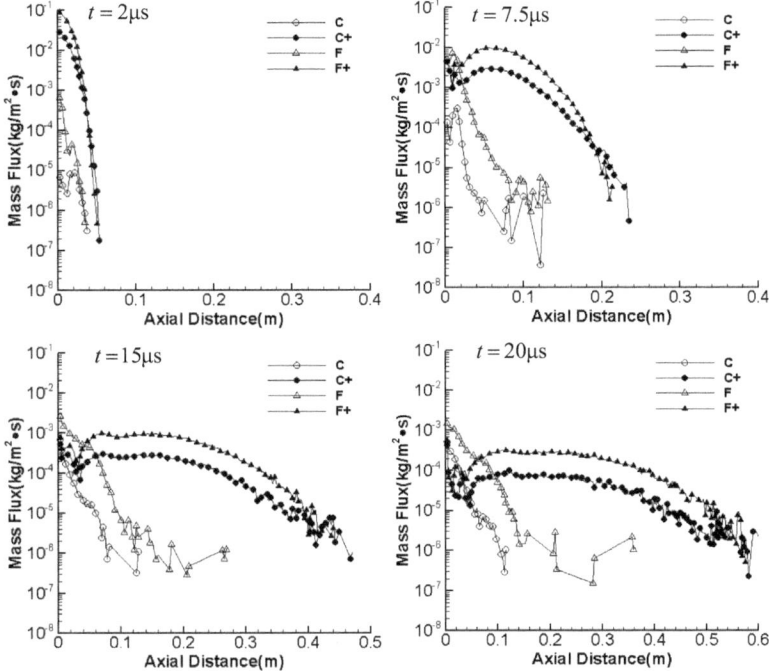

Fig. 8.30 Axial mass fluxes at 13.5 J

7.5 μs, the backflows of C⁺, F⁺, F, and C occur at 7.26 J, with a maximum backflow of 4×10^{-3} kg/m³·s; at this time, backflows of C⁺, F⁺, F, and C occur at 24 J, with a maximum backflow of 9×10^{-3} kg/m³·s. At 11 μs, the backflow at 7.26 J reaches a radial distance of 19 cm from the thruster exit, with a highest mass flux of 2×10^{-3} kg/m³·s, while the backflow at 24 J reaches a radial distance of 16 cm from the thruster exit, with a highest mass flux of 4×10^{-3} kg/m³ s. At 20 μs, the backflow at 7.26 J is maintained at 19 cm, with a highest mass flux of 2×10^{-3} kg/m³·s, while the backflow at 24 J reaches 33 cm, with a mass flux maintained at 4×10^{-3} kg/m³·s. In different energy states, C⁺ has the highest mass flux, F has the highest mass flux among the neutral particles, and the mass flux is higher in the high-energy state. Similar to the 1D results, the backflow first occurs at 24 J, and the greater the energy is, the greater the magnitude of the particle backflow. The difference from the 1D results is that the diffusion in the backflow region is faster at 13.5 J. The inlet velocity in the 3D inlet model has a certain ejection angle, which has some influence on the radial ejection velocity. The calculation results show that the ejection angle of the inlet velocity has a great impact on the backflow. In practice, this angle is a factor worthy of consideration.

The ion distributions at 20 μs under different initial voltages are presented in Fig. 8.35. Compared with the 1D results, the 3D results show that the axial diffusion distance at 24 J is longer, and the influence angle of backflow reaches a maximum

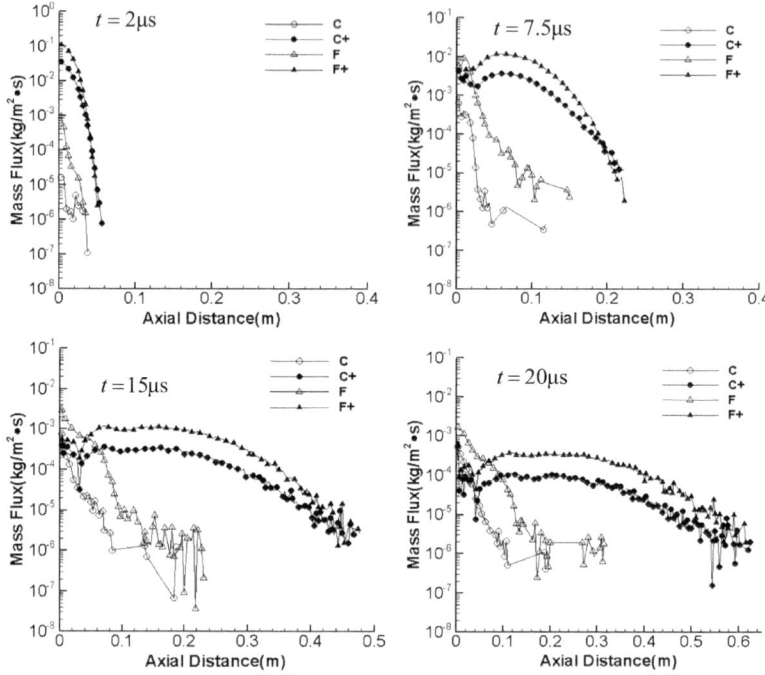

Fig. 8.31 Axial mass fluxes at 24 J

of 160° at 13.5 J. Figure 8.36 shows the proportion distribution of CEX ions under different initial voltages. Similar to the 1D result, the CEX ions are concentrated inside the plume near the exit. At higher voltages, CEX ions account for a greater proportion at the exit, and their content in the backflow is also greater.

2. Different Capacitances

Different capacitances are considered. At 2 μF, the weight factor is 8×10^{10}; in the calculation cycle, there are a total of 5,880,639 simulated molecules, including 4,218,643 ions and 1,661,996 neutral particles, and a total of 18,926,712 collisions is calculated, of which there are 7,100,766 CEX collisions, accounting for 38% of the total number of collisions. At 12 μF, the weight factor is 1.5×10^{11}; in the calculation cycle, there are a total of 6,999,327 simulated molecules, including 5,490,438 ions and 1,508,889 neutral particles, and a total of 22,317,990 collisions are calculated, of which there are 7,457,077 CEX collisions, accounting for 33% of the total number of collisions. At 20 μF, the weight factor is 2×10^{11}; in the calculation cycle, there are a total of 7,375,617 simulated molecules, including 5,986,294 ions and 1,389,323 neutral particles, and a total of 28,007,446 collisions are calculated, of which there are 8,895,029 CEX collisions, accounting for 31% of the total number of collisions.

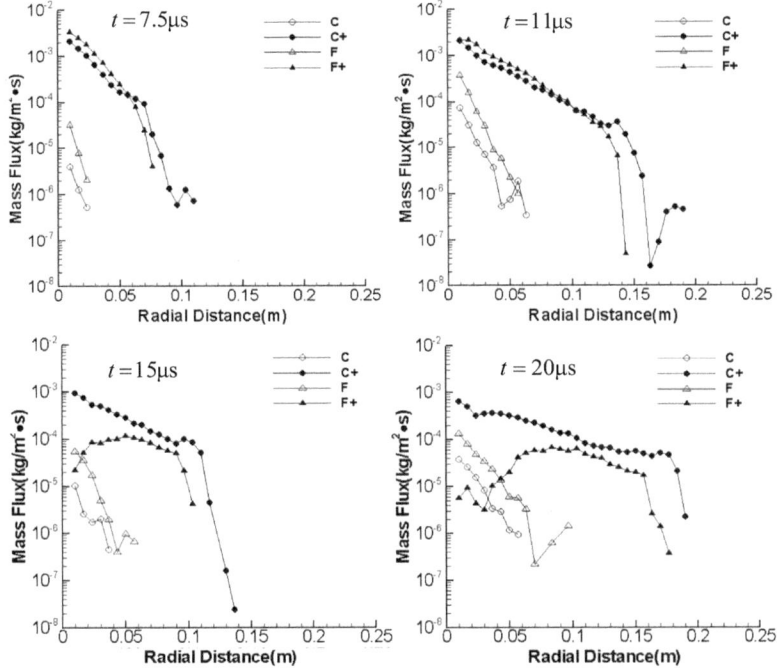

Fig. 8.32 Backflow mass fluxes at the thruster exit at 7.26 J

Figures 8.37, 8.38 and 8.39 show axial mass fluxes under different capacitances. Compared to the results for the 1D inlet, the change characteristics of various components are similar. F^+ ions first appear at the beginning of the pulse, followed by C^+, F, and C. As the pulse time increases, at the exit, F atoms become the species with the highest mass flux, followed by C, F^+, and C^+. Along the axial direction, the tail of the plume is composed of charged particle flow, and C^+ has the maximum particle flow at the tail of the plume. In the later stage of the pulse, the generation of a large number of F^+ ions makes it become the maximum particle flow, followed by F and C. A comparison of the cases of 2 μF and 20 μF reveals that at 7.5 μs, the ion cluster diffuses to distances of 0.23 and 0.24 m from the exit at 2 μF and 20 μF, respectively, and the atom cluster diffuses to distances of 0.18 and 0.16 m from the exit at 2 μF and 20 μF, respectively. At 20 μs, the ion cluster diffuses to distances of 0.63 and 0.66 m from the exit at 2 μF and 20 μF, respectively, and the atom cluster diffuses to distances of 0.18 and 0.22 m from the exit at 2 μF and 20 μF, respectively. Similar to the 1D results, in the high-capacitance state, the mass flux is higher, and the ions have the fastest diffusion rate at high capacitance, but the advantage is not significant. Unlike the 1D results, the atomic diffusion rate in the 3D results is lower at high capacitance, which is caused by the low velocity in the later stage of pulse in the 3D calculations.

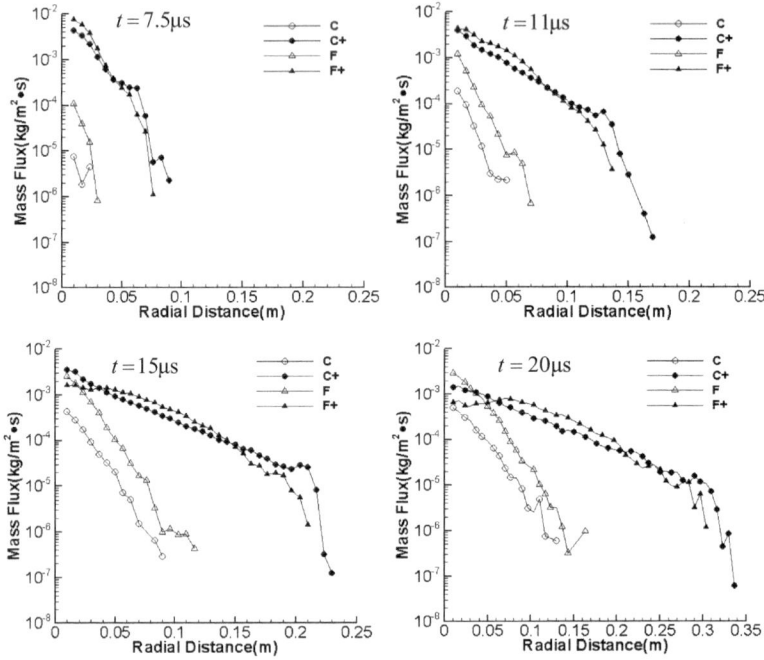

Fig. 8.33   Backflow mass fluxes at the thruster exit at 13.5 J

Figures 8.40, 8.41 and 8.42 show the backflow mass fluxes above the exit plane of the thruster under different capacitances. According to the 3D inlet calculation results, backflow occurs at 2.8, 3.4, and 3.4 μs under 2 μF, 12 μF, and 20 μF, respectively. Similar to the 1D results, the backflow of C^+ occurs first, followed by F^+, F, and C. In the early stage, the backflow is dominated by the charged components, while in the later stage, the neutral components near the exit become non-negligible. A comparison of the cases of 2 μF and 20 μF shows that at 7.5 μs, the backflows of C^+, F^+, F, and C occur at 2 μF, with the maximum backflow of 3×10^{-3} kg/m³·s. At this time, four types of backflow also occur at 20 μF, with a maximum of 10^{-2} kg/ m³·s. At 11 μs, the backflow at 2 μF reaches a radial distance of 18 cm from the thruster exit, with a highest mass flux maintained at 2×10^{-3} kg/m³·s, while the backflow at 20 μF reaches a radial distance of 16 cm from the thruster exit, with a highest mass flux of 7×10^{-3} kg/m³·s. At 20 μs, the backflow at 2 μF reaches 38 cm, with a highest mass flux maintained at 2×10^{-3} kg/m³·s, while the backflow at 20 μF reaches 32 cm, with a mass flux maintained at 4×10^{-3} kg/m³·s. Regardless of the capacitance, C^+ has the highest mass flux, and F has the maximum neutral particle mass flux. Compared with the 1D results, due to the existence of the radial velocity, the diffusion rate, range, and intensity of the backflow in the 3D results are significantly greater than those in the 1D results. However, regardless of the inlet model, the backflow flow under high capacitance leads to higher mass fluxes.

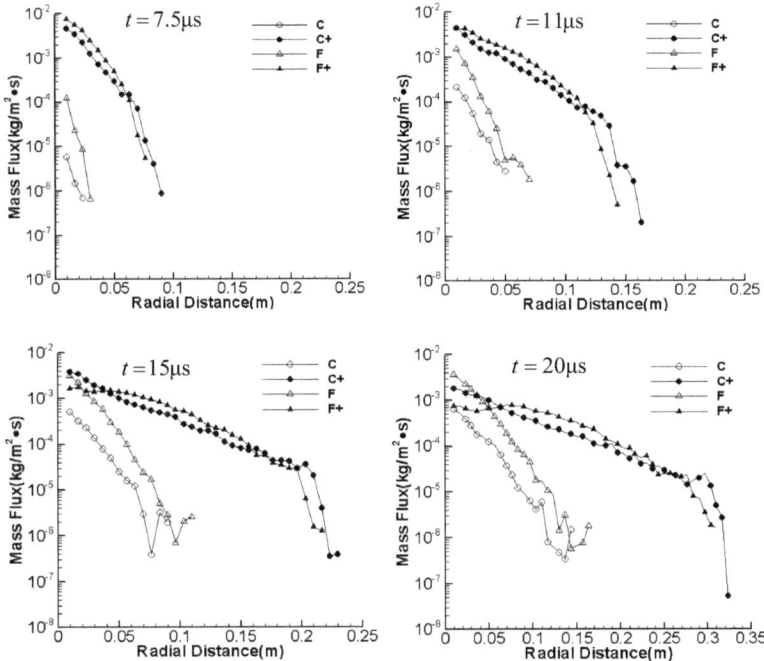

Fig. 8.34 Backflow mass fluxes at the thruster exit at 24 J

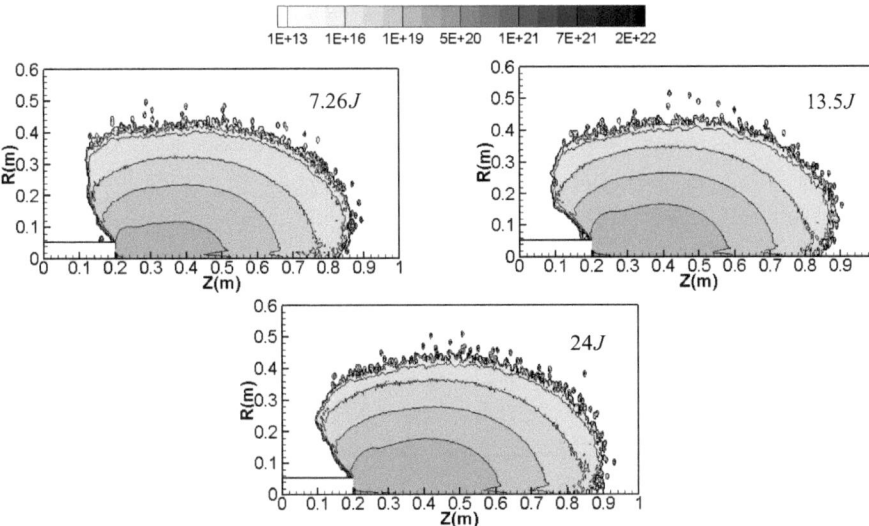

Fig. 8.35 Ion distribution at 20 μs under different initial voltages

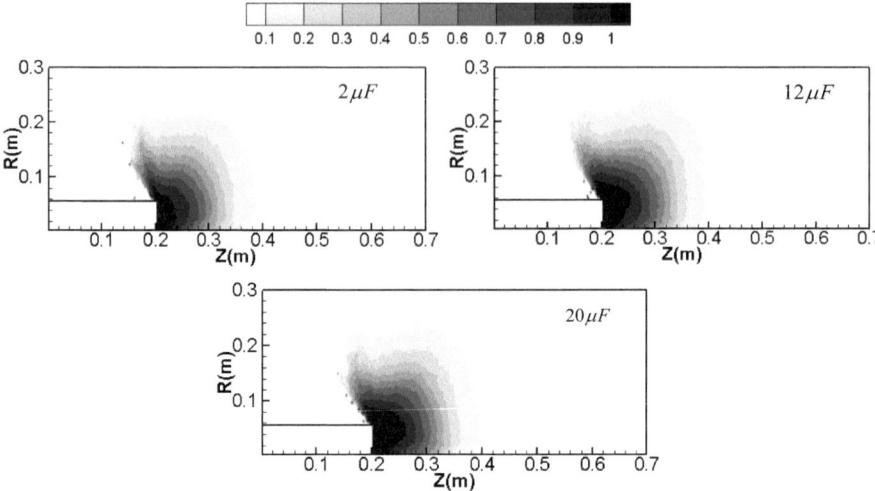

Fig. 8.36   CEX Ion distribution at 20 μs under different initial voltages

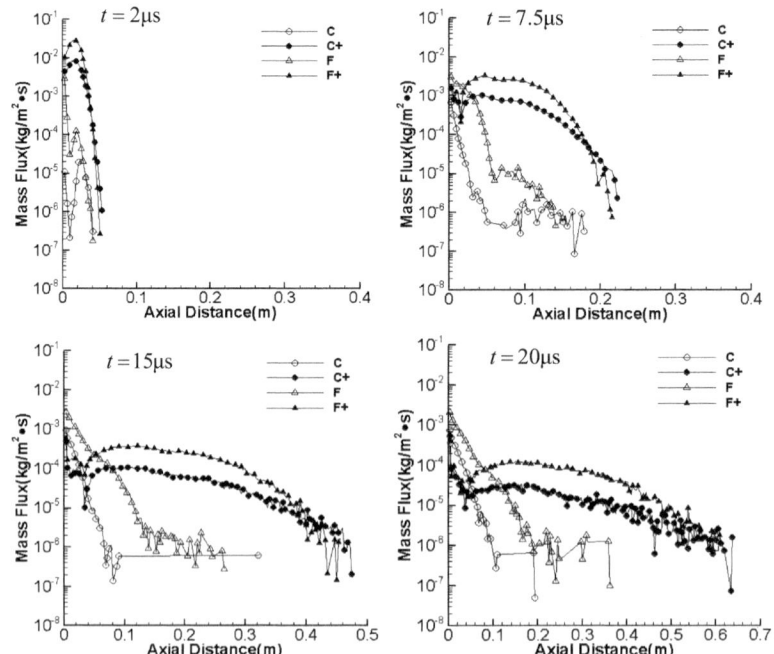

Fig. 8.37   Axial mass fluxes at 2 μF

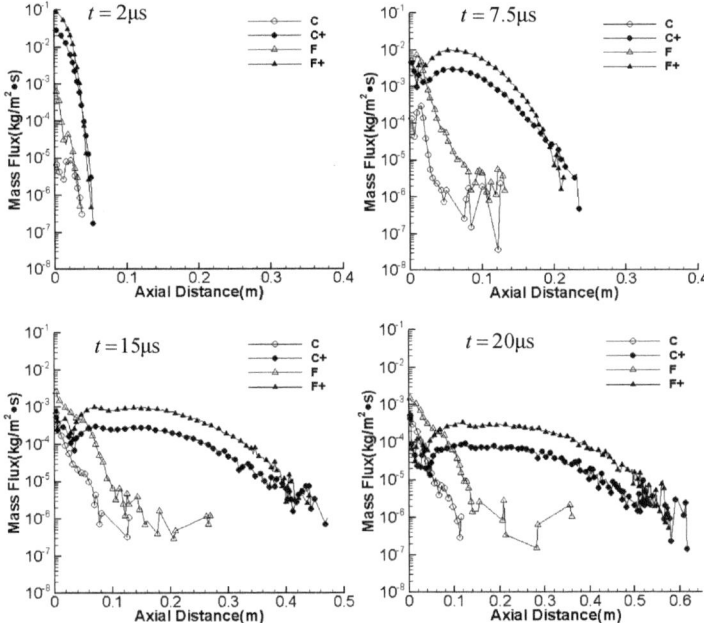

Fig. 8.38   Axial mass fluxes at 12 μF

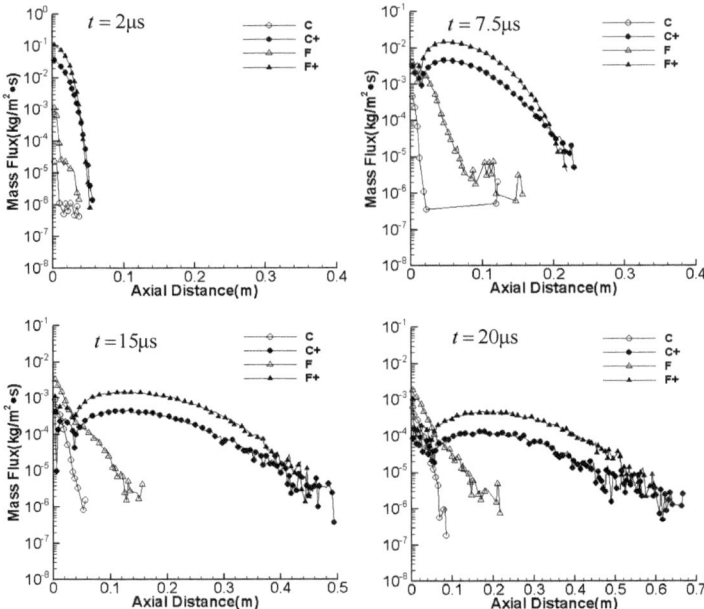

Fig. 8.39   Axial mass fluxes at 20 μF

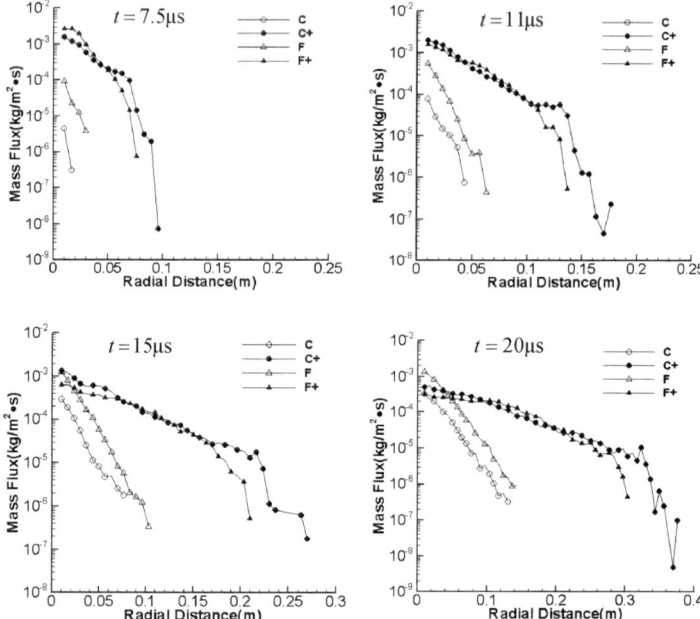

Fig. 8.40 Backflow mass fluxes at the thruster exit at 2 μF

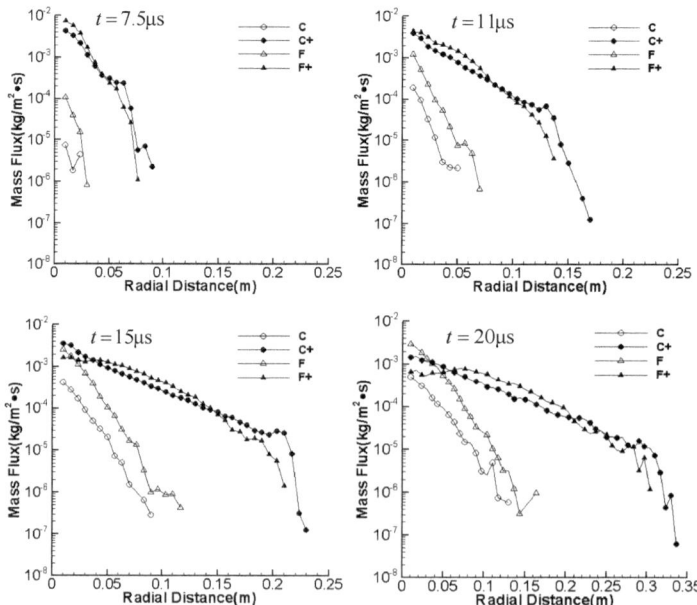

Fig. 8.41 Backflow mass fluxes at the thruster exit at 12 μF

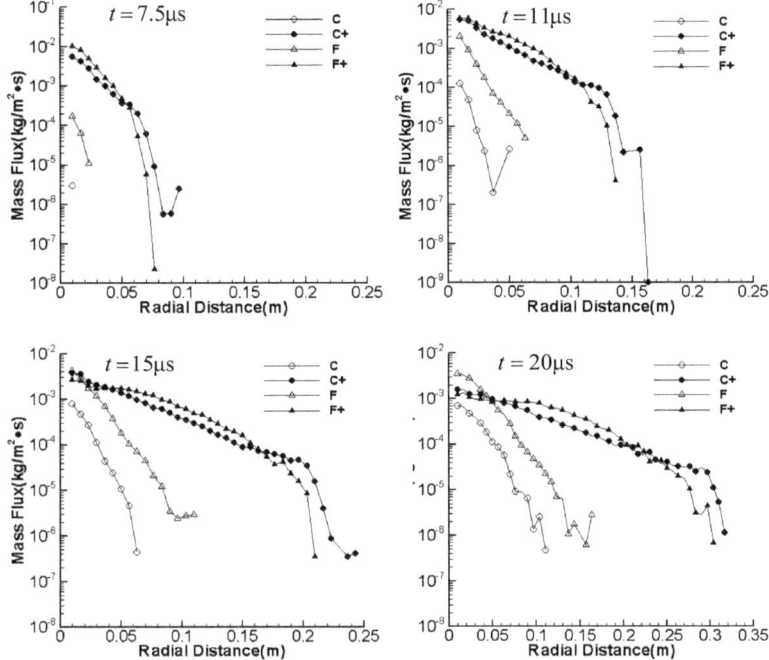

Fig. 8.42 Backflow mass fluxes at the thruster exit at 20 μF

Figure 8.43 presents the distribution of ions under different capacitances at 20 μs. Under different capacitances, the axial influence distance of the high capacitance is longer, and the backflow angle reaches a maximum of 150° under 12 μF.

The distribution of CEX ions under different capacitances is analyzed in Fig. 8.44. The CEX ions are concentrated near the exit, and the content of CEX ions near the exit is very high under low capacitance. The content of CEX ions in the region near the exit is very high. The CEX results indicate that the degree of ionization of the system under low capacitance is greater than that under high capacitance.

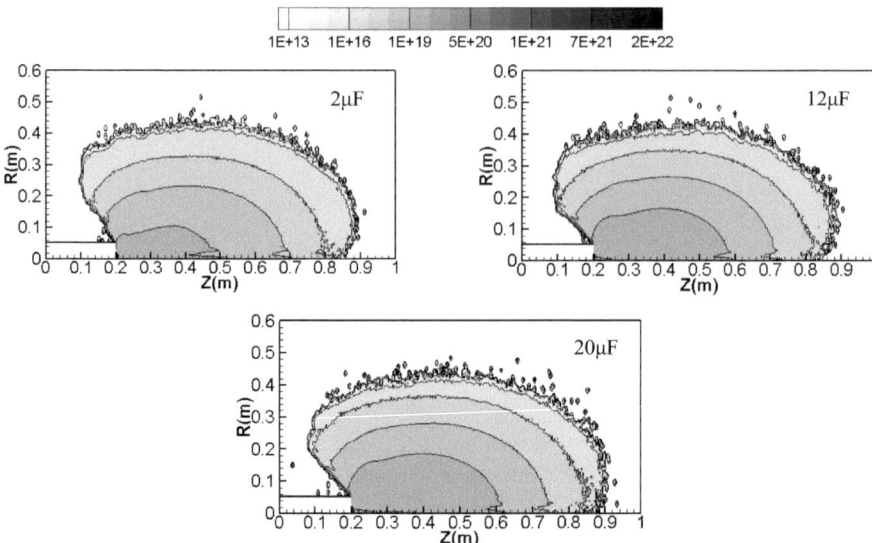

Fig. 8.43 Ion distribution at 20 μs under different capacitances

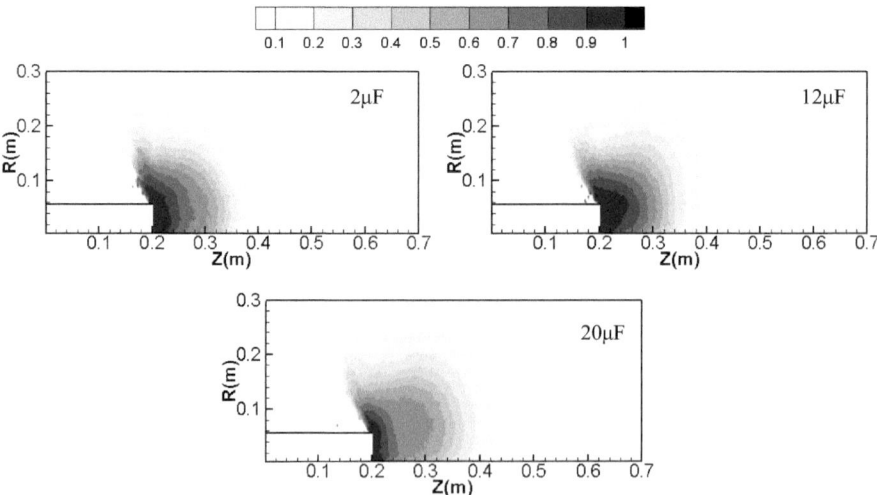

Fig. 8.44 CEX ion distribution at 20 μs under different capacitances

References

1. Keidar M, Boyd ID. Ablation study in the capillary discharge of an electrothermal gun. J Appl Phys. 2006;99(5):053301–8.
2. Keidar M, Boyd I D. Device and plume model of an electrothermal pulsed plasam thruster. In: AIAA 2000–3430, 36th AIAA/ASME/SAE/ASEE joint propulsion conference and exhibit. Huntsville, AL; 2000.
3. Keidar M, Boyed D, Beilis II. Model of particulate interaction with plasma in a teflon pulsed plasma thruster. J Propul Power. 2001;17(1):125–31.
4. Eckman R, Byrne L, Gatsonis NA, et al. Triple langmuir probe measurements in the plume of a pulsed plasma thruster. J Propul Power. 2001;17(4):762–71.